Smart Environmental Monitoring

Sneha Gautam · Robert Blaga ·
Prasanalakshmi Balaji
Editors

Smart Environmental Monitoring

IoT-Based Systems and Applications

Springer

Editors
Sneha Gautam
Division of Civil Engineering
Karunya Institute of Technology
and Sciences
Coimbatore, Tamil Nadu, India

Robert Blaga
Faculty of Physics and Mathematics
West University of Timisoara
Timisoara, Romania

Prasanalakshmi Balaji
Department of Computer Science
King Khalid University
Abha, Saudi Arabia

ISBN 978-981-92-0293-5 ISBN 978-981-92-0294-2 (eBook)
https://doi.org/10.1007/978-981-92-0294-2

This Springer imprint is published by the registered company Springer Nature Singapore Pte Ltd.
The registered company address is: 152 Beach Road, #21-01/04 Gateway East, Singapore 189721, Singapore

Preface

Currently, digital technologies are evolving at a pace never witnessed in history. Consequently, there has been a significant change in the fields of engineering and environmental science. Despite the limitations and challenges associated with emerging technologies, their advantages are driving innovation across sectors. In particular, the integration of the Internet of things (IoT) into environmental monitoring is proving to be a potent solution for achieving sustainability, helping maximise the use of available resources and minimise environmental degradation.

The nature and extent of environmental concerns worldwide, including climate change, pollution, depletion of natural resources, and environmental uncertainty, have increased. Traditional environmental monitoring techniques are evolving to intelligent monitoring techniques. Internet of things technologies, such as distributed sensor networks, cloud-edge computing, artificial intelligence, and real-time analytics, are increasingly being employed for environmental monitoring, including air quality, water management, waste management, biodiversity, energy management, and climate change monitoring. In this edited book, several researchers, academicians, and industry professionals have come together to present a collection of articles on the integration of IoT technologies into environmental monitoring from a multidisciplinary perspective. The book is divided into parts that provide a comprehensive understanding of IoT technologies in environmental monitoring, including foundational concepts, intelligent technologies, smart cities, decision-making frameworks, and domain-specific applications.

The chapters cover a wide range of topics, from the basics and development of IoT technology for environmental monitoring to smart sensor technology, communication, and the integration of artificial intelligence. It also focuses on the application of technology, including air and water quality management, smart agriculture, renewable energy management, and IoT for smart cities. All the chapters provide clear insight into the application of technology to environmental management.

The primary objective of the book is to bridge the gap between theory and practice. It is a book for a wide range of readers, including graduate students, researchers, environmental engineers, computer scientists, policymakers, and industry experts who are interested in smart and sustainable technology. It provides a platform for interdisciplinary collaboration between engineers, environmentalists, data scientists, and sustainability experts.

The editors would like to extend their sincere appreciation to all the contributing authors for their valuable insights and dedication towards academic excellence. We would also like to extend our appreciation to the reviewers for their valuable suggestions, which have greatly improved the quality of the chapters. We would like to extend our special appreciation to the publishing team for their support throughout the book's development. It is believed that the book would serve as a valuable reference towards academic excellence and inspire more research, innovation, and development in the field of smart environmental monitoring. It is believed that the increasing complexity of environmental concerns would be addressed through the integration of IoT/digital technologies, thereby playing a pivotal role in the development of intelligent societies.

Coimbatore, India Sneha Gautam
Timisoara, Romania Robert Blaga
Abha, Saudi Arabia Prasanalakshmi Balaji

Contents

Part I
Foundations

Chapter 1
An Overview of IoT Technology and Applications

Harshit Mishra

Abstract Internet of things (IoT) technology has emerged as a transformative force in modern industries, enabling seamless connectivity between devices, sensors, and cloud systems. The evolution of IoT has been marked by significant milestones, from early RFID applications in the 1990s to the present-day integration of AI, edge computing, and 5G. By 2023, the global IoT market was valued at approximately $662 billion, with projections indicating it would reach $3.3 trillion by 2030, driven by advancements in smart cities, health care, and industrial automation. The core components of IoT include sensors, connectivity modules (Wi-Fi, Bluetooth, LoRa, and 5G), and cloud-based data processing, collectively enhancing real-time decision-making. IoT architecture follows a layered approach, encompassing the perception, network, and application layers, thereby facilitating secure and efficient data exchange. The deployment of IoT systems relies on robust communication protocols, such as MQTT, CoAP, and AMQP, ensuring seamless device-to-cloud and device-to-device interactions. However, challenges persist in data security, as IoT networks are vulnerable to cyberthreats, unauthorized access, and privacy concerns. Regulations such as the GDPR and HIPAA aim to mitigate these risks and promote ethical IoT practices. In environmental applications, IoT enables smart air and water quality monitoring, precision agriculture, and disaster management, significantly improving sustainability efforts. Looking ahead, the integration of AI, blockchain, and green IoT solutions will drive further innovations, addressing scalability, energy efficiency, and regulatory challenges. With increasing adoption across diverse sectors, IoT is set to redefine connectivity, automation, and data-driven decision-making, shaping a more efficient and intelligent future.

Keywords Applications · Connectivity · Innovation · IoT · Security

---

H. Mishra (✉)
Department of Agricultural Economics, College of Agriculture, Acharya Narendra Deva University of Agriculture and Technology, Ayodhya, U.P., India
e-mail: wehars@gmail.com

S. Gautam et al. (eds.), *Smart Environmental Monitoring*,
https://doi.org/10.1007/978-981-92-0294-2_1

1.1 Introduction

IoT, or the Internet of things, refers to a network of interconnected devices that communicate and exchange data without human intervention. The concept of IoT has evolved significantly over the past few decades, with its origins rooted in early sensor networks and machine-to-machine communication. Today, IoT is a key enabler of digital transformation across industries, with an estimated 15.14 billion connected devices worldwide as of 2023, projected to reach 29.42 billion by 2030 [2]. The rapid growth of IoT is driven by advances in wireless communication technologies, including 5G, LoRaWAN, and NB-IoT, which enable seamless data transmission across devices. The integration of edge computing and cloud infrastructure further enhances IoT capabilities, allowing for real-time data processing and improved decision-making.

At the core of IoT technology are sensors, actuators, and connectivity modules that facilitate data collection, transmission, and processing. IoT architecture typically follows a layered model, comprising the perception layer (sensors and actuators), the network layer (connectivity protocols such as MQTT and CoAP), and the application layer (user interfaces and analytics platforms). The effectiveness of IoT systems depends on robust communication frameworks, including device-to-device, device-to-cloud, and backend data-sharing models. Despite the benefits, interoperability remains a challenge due to the diversity of IoT platforms and protocols. Standardization efforts, such as the IEEE P2413 IoT framework and ISO/IEC 30141, aim to address these challenges by providing guidelines for system integration [6].

One of the most impactful applications of IoT is in environmental monitoring, where sensor networks track air and water quality, detect pollution levels, and provide real-time data for disaster management. Smart agriculture leverages IoT to enable automated irrigation, soil health monitoring, and livestock tracking. In urban settings, smart cities leverage IoT for traffic management, energy optimization, and waste management, thereby increasing efficiency and sustainability. The economic impact of IoT is significant, with the global IoT market valued at approximately $662 billion in 2023 and expected to exceed $1.5 trillion by 2030 [21]. However, IoT adoption also raises concerns about cybersecurity, data privacy, and ethical considerations. Vulnerabilities in IoT devices, unauthorized access, and data ownership disputes necessitate stringent security measures and regulatory frameworks such as GDPR and HIPAA.

As IoT continues to evolve, emerging trends such as AI-driven analytics, blockchain for secure transactions, and green IoT technologies are shaping its future. The convergence of IoT with next-generation connectivity solutions, such as 6G, will further enhance its capabilities, enabling autonomous systems and self-healing networks. Despite challenges related to scalability, cost, and regulatory compliance, IoT remains a transformative technology with the potential to revolutionize industries, improve quality of life, and drive sustainable development in the years to come.

1.2 Fundamentals of IoT Technology

The IoT has emerged as a transformative technology that enables seamless connectivity and interaction between physical devices, sensors, and digital systems. By integrating data collection, processing, and automation, IoT enhances efficiency and decision-making across various industries. This section discusses the fundamental aspects of IoT, including its definition, evolution, core components, and technical enablers.

The IoT refers to a network of interconnected devices that communicate and exchange data over the Internet without requiring direct human intervention. These devices, embedded with sensors, actuators, and communication technologies, collect and transmit data to centralized or distributed systems for processing and analysis. IoT applications span across multiple domains, including smart cities, health care, agriculture, industrial automation, and environmental monitoring.

At its core, IoT enables:

- *Interconnectivity*: Devices communicate using wired or wireless networks.
- *Automation*: Systems operate autonomously with minimal human input.
- *Data Analytics*: Large-scale data processing enhances decision-making.
- *Remote Monitoring and Control*: Users can access and manage devices remotely.

The fundamental concept behind IoT is to create an intelligent ecosystem where objects interact and respond to real-world scenarios in real time. The growing adoption of AI and ML further strengthens IoT capabilities by enabling predictive analytics and automated decision-making.

1.3 Evolution and Growth of IoT

The development of IoT technology has been driven by advancements in computing, networking, and sensor technologies. The following subsections outline the historical evolution of IoT, its present-day applications, and future trends.

Early Developments and Milestones

The origins of IoT can be traced back to the 1980s and 1990s, when early concepts of machine-to-machine (M2M) communication and wireless sensor networks were introduced. Some key milestones in IoT history include:

- *1991*: Mark Weiser introduced the concept of ubiquitous computing, envisioning a world where digital interactions are seamlessly integrated into everyday life.
- *1999*: Kevin Ashton coined the term Internet of things, emphasizing the role of radio frequency identification (RFID) in connecting physical objects to digital networks.
- *2003–2005*: The adoption of IPv6 enabled an expanded address space, crucial for IoT scalability.

- *2010s*: The rise of cloud computing, low-power sensors, and wireless networks accelerated IoT growth, enabling large-scale deployments in industries and smart cities.

IoT in the Present Era

Today, IoT is a vital component of digital transformation across various sectors. With the advent of 5G networks, edge computing, and AI-driven analytics, IoT applications are more powerful and efficient than ever. Some areas of IoT implementation include:

- *Smart Homes*: Intelligent devices such as smart thermostats, security cameras, and voice assistants (e.g., Amazon Alexa and Google Nest) enhance home automation.
- *Health Care*: Remote patient monitoring, wearable fitness trackers, and AI-assisted diagnostics improve medical care.
- *Agriculture*: IoT-based precision farming optimizes irrigation, pest control, and crop management.
- *Industrial IoT (IIoT)*: Factories deploy IoT-enabled sensors for real-time equipment monitoring and predictive maintenance.
- *Smart Cities*: IoT solutions optimize traffic management, waste collection, and energy distribution.

Prospects of IoT Technology

The future of IoT is poised for exponential growth, driven by innovations in connectivity, AI integration, and automation. Emerging trends include:

- *6G and Advanced Networking*: Ultra-fast, low-latency networks will further enhance real-time IoT applications.
- *Blockchain for IoT Security*: Decentralized ledgers can improve data integrity and device authentication.
- *AI-Powered Automation*: Machine learning algorithms will enable self-optimizing IoT systems.
- *Edge Computing Dominance*: Reduced reliance on centralized cloud processing will improve latency and efficiency.
- *Quantum IoT*: Quantum computing may enable ultra-secured data transmission for critical applications.

A projected 75 billion IoT-connected devices are expected worldwide by 2025, indicating the vast potential and continued expansion of the IoT ecosystem [19].

1.4 Core Components of IoT

IoT systems rely on a combination of hardware and software components to function efficiently. The essential building blocks of IoT include.

Sensors and Actuators

Sensors and actuators serve as the primary interface between the physical and digital worlds in IoT ecosystems.

- *Sensors*: Devices that capture environmental data, such as temperature, humidity, motion, pressure, and light. Examples include:
  - Temperature sensors (e.g., LM35 and DHT11)
  - Motion sensors (e.g., PIR sensors and accelerometers)
  - Gas sensors (e.g., MQ-135 for air quality monitoring).
- *Actuators*: Components that convert digital commands into physical actions, such as motors, relays, and solenoids. Examples include:
  - Servomotors for robotic arms
  - Electromagnetic relays for smart switches
  - Piezoelectric actuators for vibration control.

These components enable IoT devices to interact with the environment and perform automated functions based on sensor feedback.

Connectivity Modules (Wi-Fi, Bluetooth, LoRa, 5G, etc.)

IoT devices require robust connectivity to exchange data with centralized or distributed networks. As shown in Table 1.1, various communication technologies offer different ranges, data rates, and applications.

Each connectivity option has specific advantages with respect to power consumption, data throughput, and operational range. The rapid adoption of 5G, as accentuated in Table 1.1, is expected to revolutionize IoT with ultra-reliable low-latency communication (URLLC) and massive machine-type communication (mMTC).

Edge and Cloud Computing

IoT data processing is facilitated through *edge computing and cloud computing*:

- *Edge Computing*: Data is processed at the device or gateway level, reducing latency and bandwidth usage.
- *Cloud Computing*: Centralized cloud platforms store and analyze large-scale IoT data.

Table 1.1 Some common communication technologies

Technology	Range	Data rate	Application
Wi-Fi	Short (100 m)	High (Mbps)	Smart homes, offices
Bluetooth	Short (10 m)	Medium (Mbps)	Wearable devices, IoT gadgets
Long range (LoRa)	Long (10–15 km)	Low (kbps)	Industrial, rural IoT
5G	Medium (1–10 km)	Ultra-high (Gbps)	Smart cities, autonomous vehicles

A hybrid approach, **fog computing**, enables multi-layered data processing to balance efficiency and scalability in IoT networks.

IoT Protocols and Standards

IoT communication relies on specialized protocols and standards for efficient data transmission and interoperability. Some protocols include:

- *Message queuing telemetry transport (MQTT)*: Lightweight and widely used in IoT messaging.
- *Constrained application protocol (CoAP)*: Optimized for low-power devices.
- *Zigbee and Z-Wave*: Low-power mesh networking standards for smart home automation.
- *HTTP/HTTPS*: Web-based communication for cloud-integrated IoT devices.

Ensuring standardization and security in IoT protocols is crucial for the development of scalable and interoperable IoT ecosystems.

1.5 IoT Architectures and Frameworks

The IoT has revolutionized modern technology by enabling interconnected devices to communicate and share data seamlessly. To facilitate this interconnectivity, structured architectures and frameworks are essential. IoT architecture defines the layers and mechanisms through which devices interact, ensuring efficient communication, processing, and application implementation. Various IoT frameworks exist to standardize the deployment of IoT solutions, promoting interoperability across diverse industries such as health care, agriculture, smart cities, and industrial automation. The following sections provide a detailed discussion on IoT architecture and its layered structure, communication models, and the challenges of system integration and interoperability.

Layered Architecture of IoT

The IoT ecosystem comprises multiple layers, each responsible for specific functions in data collection, transmission, and application. A well-defined architecture helps streamline data flow and ensures security, efficiency, and scalability. The three primary layers of IoT architecture are the perception layer, network layer, and application layer. Each of these layers is crucial in maintaining the overall functionality of IoT systems.

Perception Layer

The perception layer, also known as the sensing layer, is the foundation of IoT architecture. It is responsible for interacting with the physical environment by collecting real-world data using sensors, actuators, and other edge devices. The primary components of this layer include:

- *Sensors*: Measure environmental parameters such as temperature, humidity, motion, pressure, and light intensity.
- *Actuators*: Convert digital signals into physical actions, such as turning on/off a device, adjusting temperature, or opening/closing a valve.
- *RFID and NFC*: Technologies such as radio frequency identification (RFID) and near-field communication (NFC) enable real-time tracking and authentication of objects and individuals.
- *Cameras and Image Sensors*: Used for surveillance, facial recognition, and smart security systems.

Data collected at this layer is typically raw and unprocessed, requiring further transmission to higher layers for analysis and decision-making.

Network Layer

The network layer serves as the communication backbone of IoT systems, ensuring seamless data transfer between the perception and application layers. This layer is responsible for:

- *Data Transmission*: Utilizing protocols such as Wi-Fi, Bluetooth, Zigbee, LoRaWAN, and 5G to transmit data to cloud servers or edge computing devices.
- *Data Routing*: Employing networking models, such as TCP/IP and MQTT, to efficiently direct data between IoT devices and central hubs.
- *Security Mechanisms*: Implementing encryption, authentication, and secure tunneling protocols to prevent cyberthreats and unauthorized access.

The effectiveness of the network layer depends on factors such as latency, bandwidth, and power consumption, which vary across application domains (e.g., industrial IoT vs. consumer IoT).

Application Layer

The application layer is the interface between IoT systems and end-users, providing functionalities such as data analytics, visualization, and remote control. This layer supports a wide range of applications, including:

- *Smart Homes*: Automated lighting, security systems, and connected appliances.
- *Healthcare Monitoring*: Wearable devices that track vital signs and alert medical professionals in case of emergencies.
- *IIoT*: Predictive maintenance and process automation in manufacturing environments.
- *Smart Agriculture*: Precision farming and remote soil monitoring for optimal crop yield.

By integrating AI and ML, the application layer can extract meaningful insights from collected data, enabling real-time decision-making and automation.

1.6 IoT Communication Models

IoT communication models define how devices interact with each other and with centralized systems. Depending on the application, IoT devices may communicate in various ways, including direct communication between devices or through intermediary platforms such as cloud services and gateways [22]. The primary communication models in IoT include:

- ***Device-to-Device (D2D)***

The device-to-device (D2D) model allows IoT devices to communicate directly without an intermediary server. This model is commonly used in:

- Smart home systems where lights, thermostats, and security cameras exchange data for automation.
- Healthcare monitoring where wearable sensors transmit data directly to other medical devices for real-time response.
- Industrial automation where factory sensors and actuators coordinate production processes.

D2D communication reduces latency and bandwidth consumption, making it suitable for real-time applications. However, it requires standardized protocols for device compatibility and efficient data exchange.

Device-to-Cloud (D2C)

The device-to-cloud (D2C) model enables IoT devices to connect directly to cloud platforms, such as AWS IoT, Microsoft Azure IoT Hub, and Google Cloud IoT. This model supports:

- Remote monitoring of industrial machines, vehicles, and home appliances.
- Cloud-based analytics for big data processing and predictive maintenance.
- Centralized storage of IoT data for long-term analysis and decision-making.

D2C communication provides scalability and powerful computing resources, but it requires reliable Internet connectivity and robust cybersecurity measures to prevent data breaches.

- ***Device-to-Gateway (D2G)***

In the device-to-gateway (D2G) model, IoT devices connect to a local gateway before transmitting data to cloud services. The gateway acts as an intermediary that:

- Aggregates and filters sensor data to reduce network congestion.
- Enhances security by encrypting data before transmission.
- Supports edge computing for local data processing, reducing dependency on cloud resources.

This model is commonly used in smart agriculture, industrial automation, and health care, where low-latency and secure data transmission are critical.

- ***Backend Data-Sharing Model***

The backend data-sharing model involves IoT devices transmitting data to centralized repositories, where multiple applications can access and analyze it. This model is widely used in:

- Smart city applications where traffic, pollution, and energy data are shared across municipal services.
- Healthcare ecosystems where patient data is stored in cloud databases for doctors, researchers, and insurance providers.
- Supply chain management where logistics data is shared among manufacturers, suppliers, and distributors.

This model ensures scalability and cross-platform integration, but it raises concerns about data privacy, ownership, and compliance with regulations such as GDPR and HIPAA.

1.6.1 IoT System Integration and Interoperability

The integration of diverse IoT devices into a unified system presents significant challenges due to variations in communication protocols, hardware specifications, and software platforms. Interoperability refers to the ability of different IoT systems to seamlessly communicate and exchange data without compatibility issues. Some challenges in IoT integration include:

1. *Heterogeneous Communication Standards*: Different IoT protocols (e.g., MQTT, CoAP, Zigbee, and LoRaWAN) make it difficult for devices from different vendors to interact.
2. *Data Format and Semantics*: IoT devices generate data in various formats, requiring standardized data models for effective interpretation.
3. *Security and Privacy Concerns*: Ensuring secure data exchange between IoT devices while preventing cyberthreats remains a major concern.
4. *Scalability and Network Congestion*: As the number of IoT devices increases, networks must handle large-scale deployments without performance degradation.

To enhance interoperability, organizations and standardization bodies (e.g., IEEE, IETF, and W3C) are working on unified IoT frameworks such as:

- Open IoT standards (OneM2M, AllJoyn)
- Middleware platforms (IoTivity, EdgeX Foundry)
- Semantic web technologies for data sharing.

By addressing these challenges, IoT can achieve seamless connectivity across industries, enabling smart cities, industrial automation, and healthcare innovations on a large scale.

1.7 IoT Connectivity and Communication Protocols

IoT is a complex ecosystem of interconnected devices that communicate seamlessly to enable intelligent decision-making and automation. One of the fundamental components of IoT is its connectivity and communication infrastructure, which ensures the efficient transmission of data across networks. The communication architecture in IoT is broadly classified into wireless communication technologies and network protocols, each addressing different requirements, such as data rate, latency, power consumption, and range [12]. This section examines the key wireless communication technologies and network protocols used in IoT systems, as well as the challenges associated with IoT networking and connectivity.

1.7.1 Wireless Communication Technologies in IoT

Wireless communication technologies play a vital role in enabling IoT applications by providing seamless connections between devices, gateways, and cloud services. These technologies differ in terms of their range, bandwidth, energy efficiency, and deployment costs. The choice of wireless technology for IoT applications depends on specific use cases, including smart homes, industrial automation, smart agriculture, healthcare monitoring, and urban infrastructure management.

- **Wi-Fi and Bluetooth**

Wi-Fi and Bluetooth are two of the most widely adopted short-range wireless communication technologies in IoT.

- Wireless fidelity (Wi-Fi) operates on the IEEE 802.11 standard and provides high-speed data transmission over local area networks (LANs). It is commonly used in smart home automation, IIoT, and smart healthcare applications where high data throughput is required. However, Wi-Fi consumes significant power, making it less suitable for battery-powered IoT devices.
- Bluetooth is a low-power wireless technology that operates in the 2.4 GHz frequency band. It is widely used in wearable devices, smart home automation, and healthcare monitoring systems. Bluetooth low energy (BLE) is a power-efficient variant designed specifically for IoT applications, enabling devices to operate for extended periods on minimal power [23].

- **Zigbee and Z-Wave**

Zigbee and Z-Wave are two popular wireless mesh networking technologies designed for IoT applications that require low power consumption and secure communication.

- **Zigbee** operates on the IEEE 802.15.4 standard and supports low-power, low-data-rate communication. It is widely used in home automation, smart lighting,

and industrial monitoring systems due to its ability to form mesh networks, which extend coverage and improve reliability.
- **Z-Wave** is another low-power wireless communication protocol that operates in the sub-1 GHz frequency band. It is primarily used in home automation systems, providing efficient connectivity between smart devices such as security sensors, thermostats, and lighting controls.

- **LoRaWAN and NB-IoT**

Long-range communication technologies like LoRaWAN and NB-IoT are designed for large-scale deployments where devices need to transmit data over long distances with minimal power consumption.

- **Long-Range Wide Area Network (LoRaWAN)** is a low-power, long-range communication protocol that operates in unlicensed frequency bands. It is ideal for IoT applications such as smart agriculture, asset tracking, and environmental monitoring. LoRaWAN provides high penetration capabilities, enabling communication even in remote or underground locations.
- **Narrowband IoT (NB-IoT)** is a cellular-based communication standard developed by 3GPP that operates within licensed frequency bands. It is optimized for deep indoor coverage, low-cost deployment, and extended battery life, making it suitable for smart metering, industrial monitoring, and smart city applications.

- **5G and its Impact on IoT**

The advent of 5G technology is revolutionizing IoT by providing ultra-low latency, high data rates, and massive device connectivity. With enhanced network slicing capabilities, 5G enables dedicated communication channels for different IoT applications, ensuring optimized performance [14]. The impact of 5G on IoT includes:

- *Massive IoT connectivity*: 5G can support a significantly higher number of connected devices per square kilometer compared to previous cellular technologies.
- *Ultra-reliable low-latency communication (URLLC)*: This is critical for applications like autonomous vehicles, industrial automation, and remote health care.
- *Enhanced mobile broadband (eMBB)*: 5G provides high-speed data transmission, enabling real-time analytics and high-definition video streaming in IoT applications.

1.7.2 IoT Network Protocols

In addition to wireless communication technologies, IoT devices rely on specialized network protocols to ensure reliable and efficient data exchange. These protocols determine how devices communicate, manage data transmission, and optimize network resources.

- **Message Queuing Telemetry Transport (MQTT)**

MQTT is a lightweight publish–subscribe messaging protocol designed for low-bandwidth, high-latency networks. It is widely used in IoT applications due to its efficiency and reliability. Key features of MQTT include:

- *Low overhead*: Minimal packet size reduces bandwidth consumption.
- *QoS levels*: Ensures reliable message delivery with three quality of service (QoS) levels.
- *Scalability*: Supports large-scale IoT deployments by efficiently handling multiple client connections.

- **Constrained Application Protocol (CoAP)**

CoAP is an application-layer protocol optimized for resource-constrained IoT devices. It follows a request–response model like HTTP but is lightweight and efficient. Key characteristics of CoAP include:

- *UDP-based communication*: Reduces network overhead compared to TCP-based protocols.
- *Support for RESTful APIs*: Enables easy integration with web services.
- *Efficient message exchange*: Supports multicast communication, reducing network congestion.

- **Advanced Message Queuing Protocol (AMQP)**

AMQP is a message-oriented protocol designed for secure and reliable message exchange. It is commonly used in IIoT and enterprise applications. Features of AMQP include:

- *Message queuing and routing*: Supports complex message distribution mechanisms.
- *Interoperability*: Ensures seamless communication between different IoT platforms.
- *Security*: Provides encryption and authentication mechanisms for secure data transmission.

- **HTTP/HTTPS in IoT**

Hypertext transfer protocol (HTTP) and hypertext transfer protocol secure (HTTPS) are widely used for communication between IoT devices, particularly in cloud-based applications. While HTTP is efficient for sending and receiving data to and from web servers, HTTPS provides encrypted communication, enhancing data security in IoT ecosystems. However, due to their high bandwidth and power consumption, HTTP/HTTPS are unsuitable for low-power IoT devices.

1.7.3 Challenges in IoT Networking and Connectivity

Despite the advancements in IoT communication technologies and protocols, several challenges persist in ensuring seamless connectivity and data exchange. These include:

- *Scalability*: IoT networks must handle an exponential increase in connected devices without compromising performance.
- *Interoperability*: Different communication standards and protocols create compatibility issues between IoT devices from various manufacturers.
- *Energy efficiency*: Many IoT applications require ultra-low power consumption, necessitating energy-efficient communication protocols.
- *Security threats*: IoT devices are vulnerable to cyberthreats such as data breaches, denial-of-service (DoS) attacks, and unauthorized access.
- *Network congestion*: The increasing number of connected devices can lead to congestion and reduced network performance, especially in urban areas.

Addressing these challenges requires continuous advancements in IoT communication technologies, including the adoption of AI-driven network management, improved security frameworks, and the development of more efficient protocols.

1.8 Data Management and Analytics in IoT

The IoT ecosystem generates an enormous amount of data, necessitating robust and efficient data management and analytical techniques to derive meaningful insights. IoT data management involves data acquisition, storage, processing, and analysis to enhance decision-making, automation, and predictive capabilities. The integration of advanced technologies such as cloud computing, edge computing, AI, and ML further enhances the efficiency of IoT systems. Given the scale, velocity, and sensitivity of IoT data, ensuring security, privacy, and regulatory compliance is essential [9]. This section explores key aspects of IoT data management and analytics, covering data collection and processing mechanisms, storage solutions, security challenges, and AI-driven analytics.

1.8.1 Data Collection and Processing in IoT Systems

IoT systems rely on a wide range of sensors, devices, and communication networks to collect and process data in real time. The effectiveness of an IoT system largely depends on how efficiently data is gathered, transmitted, processed, and stored for further analytics. The process of data collection and processing is distributed across multiple layers, including sensor nodes, edge computing devices, and cloud-based platforms.

Sensor-Based Data Acquisition

Sensors play a fundamental role in IoT by capturing various environmental and operational parameters such as temperature, humidity, motion, pressure, and energy consumption. These sensors are embedded in IoT devices and continuously monitor conditions in real time. The data collected by these sensors is typically transmitted to a gateway or processing unit via communication protocols such as Wi-Fi, Bluetooth, Zigbee, LoRaWAN, or cellular networks. One of the challenges in sensor-based data acquisition is managing the vast volume of raw data generated. Efficient data preprocessing techniques such as filtering, noise reduction, and compression are often employed to improve data quality before further analysis. Additionally, energy efficiency in sensor nodes is crucial, particularly in remote or battery-operated IoT applications.

Edge Computing for Real-Time Processing

Edge computing has emerged as a crucial paradigm in IoT data management, enabling real-time data processing closer to the source. Unlike traditional cloud-based models where data is sent to centralized servers for analysis, edge computing processes data at the network edge—on IoT devices, gateways, or local servers. Advantages of edge computing in IoT include:

- *Reduced Latency:* Since data is processed locally, there is minimal delay in obtaining actionable insights. This is critical for applications like autonomous vehicles, industrial automation, and healthcare monitoring.
- *Bandwidth Optimization:* By filtering and aggregating data before sending it to the cloud, edge computing reduces network congestion and lowers communication costs.
- *Enhanced Security:* Sensitive data can be processed locally, reducing exposure to cyberthreats and unauthorized access.

Edge computing is particularly beneficial for time-sensitive IoT applications, such as predictive maintenance in manufacturing, smart grids, and telemedicine, where quick decision-making is essential.

Cloud-Based IoT Data Management

While edge computing enhances real-time processing, cloud computing remains indispensable for large-scale IoT data management. The cloud provides scalable storage, computational resources, and analytical capabilities to handle the vast amount of data generated by IoT devices. Features of cloud-based IoT data management include:

- *Scalability*: Cloud platforms such as AWS IoT, Microsoft Azure IoT, and Google Cloud IoT offer flexible storage and computing resources to accommodate fluctuating data loads.
- *Advanced Analytics*: Cloud-based AI and ML tools facilitate deep data analysis, predictive modeling, and automation.
- *Interoperability*: Cloud services integrate with multiple IoT devices and protocols, ensuring seamless data exchange and device management.

Despite its advantages, cloud computing poses challenges such as high latency, dependency on Internet connectivity, and security concerns. Therefore, a hybrid model combining edge and cloud computing is often adopted to optimize IoT data management [4].

1.8.2 IoT Data Storage and Security

With the exponential growth of IoT data, ensuring secure and efficient storage is a major challenge. IoT-generated data is highly diverse, ranging from structured sensor readings to unstructured video streams, necessitating specialized storage solutions. Additionally, security measures must be robust to protect against cyberthreats and ensure compliance with data privacy regulations.

Big Data Challenges in IoT

The IoT ecosystem produces massive datasets that pose several challenges, including:

- *Volume*: IoT systems generate petabytes of data daily, requiring scalable storage solutions.
- *Velocity*: Real-time data streams demand rapid ingestion, processing, and retrieval mechanisms.
- *Variety*: IoT data comes in multiple formats, including structured (databases), semi-structured (JSON, XML), and unstructured (videos, images).
- *Veracity*: Data quality and accuracy are crucial, as IoT data is prone to noise, duplication, and inconsistency.

Big data technologies such as Hadoop, Apache Spark, and NoSQL databases (MongoDB and Cassandra) are widely used to handle IoT data efficiently [20].

Secure Data Storage Methods

Ensuring secure data storage in IoT systems is critical due to the risk of cyberattacks, data breaches, and unauthorized access. Common methods for secure data storage include:

- *Encryption*: Advanced encryption techniques such as advanced encryption standard (AES) and Rivest–Shamir–Adleman (RSA) are used to protect data both in transit and at rest.
- *Access Control Mechanisms*: Role-based access control (RBAC) and multi-factor authentication (MFA) limit unauthorized access to sensitive data.
- *Blockchain for IoT Security*: Blockchain technology offers decentralized, tamper-resistant storage for IoT transactions, ensuring data integrity and authenticity.

1.8.3 Data Privacy Regulations (GDPR, HIPAA, etc.)

To address concerns regarding user privacy and data protection, various regulatory frameworks have been established:

- *General Data Protection Regulation (GDPR)*: Ensures that personal data is collected and processed with explicit user consent, granting individuals the right to access and delete their data.
- *Health Insurance Portability and Accountability Act (HIPAA)*: Mandates strict data protection measures for healthcare-related IoT applications.
- *California Consumer Privacy Act (CCPA)*: Provides consumers with transparency and control over their personal data.

IoT organizations must comply with these regulations to avoid legal penalties and build user trust.

1.8.4 Machine Learning and AI in IoT Data Analytics

The integration of AI and ML in IoT enhances the ability to derive insights, automate processes, and improve decision-making. ML algorithms analyze large datasets, identify patterns, and predict future trends, making IoT systems more intelligent and self-learning. Applications for AI and ML in IoT include:

- *Predictive Maintenance*: ML models analyze sensor data to detect anomalies and predict equipment failures before they occur.
- *Anomaly Detection*: AI-powered IoT systems can identify unusual behavior in cybersecurity applications, such as detecting unauthorized access attempts.
- *Smart Automation*: AI-driven IoT devices, such as smart thermostats and autonomous vehicles, adapt to user behavior and optimize performance.

A simple ML-based predictive maintenance model can be represented as:

$$P(F) = \sum_{i=1}^{n} w_i X_i + \epsilon$$

where

$P(F)$ is the probability of failure,

X_i represents sensor variables such as temperature, vibration, and pressure,

w_i are the learned weights, and

ϵ is the error term.

These AI-driven techniques significantly improve the efficiency and reliability of IoT applications.

1.9 IoT Applications in Environmental Monitoring

The IoT has emerged as a transformative technology in environmental monitoring, enabling real-time data collection, analysis, and response to ecological changes. By integrating smart sensors, wireless communication networks, and cloud-based analytics, IoT facilitates proactive decision-making in areas such as air and water quality monitoring, precision agriculture, and disaster management. These advancements enhance sustainability efforts by providing accurate and timely information that can mitigate environmental degradation and support conservation initiatives [15]. This section explores the role of IoT in environmental monitoring, with a detailed discussion on its applications in air quality assessment, water resource management, precision farming, and disaster mitigation strategies.

1.9.1 Smart Air Quality Monitoring

Air pollution is a significant global concern, affecting human health, ecosystems, and climate patterns. Traditional air quality monitoring systems rely on fixed stations that provide limited spatial coverage. IoT technology, through wireless sensor networks (WSNs) and cloud computing, has revolutionized air pollution monitoring by offering real-time, continuous, and widespread data collection.

IoT-Based Air Pollution Detection

IoT-enabled air quality monitoring systems employ low-cost gas sensors, microcontrollers, and cloud-based platforms to measure pollutants such as particulate matter (PM2.5, PM10), carbon monoxide (CO), nitrogen dioxide (NO_2), and sulfur dioxide (SO_2). These systems use wireless communication protocols like LoRaWAN, Zigbee, and NB-IoT to transmit data to centralized servers for real-time analysis.

Advanced data analytics and machine learning models process this data to detect pollution trends, identify sources of contamination, and provide actionable insights for regulatory agencies.

Mathematically, an IoT-based air quality index (AQI) can be expressed as:

$$\text{AQI} = \sum_{i=1}^{n}\left(\frac{C_i}{C_{\text{standard},i}} \times W_i\right)$$

where C_i is the concentration of pollutant i, $C_{\text{standard},i}$ is its regulatory threshold, and W_i is the weight assigned to the pollutant based on its health impact.

Wireless Sensor Networks for Urban Air Monitoring

Wireless sensor networks (WSNs) play a crucial role in urban air quality monitoring by deploying a distributed network of sensors across cities. These networks transmit pollutant data in real time to cloud-based platforms, where spatial interpolation techniques generate a comprehensive pollution map. Smart air quality monitoring systems also integrate AI-based predictive models to forecast pollution trends and suggest mitigation measures, such as traffic rerouting and industrial emissions control [8]. A typical IoT-based air monitoring system consists of:

- Sensor nodes (for pollutant detection)
- Communication modules (Wi-Fi, LoRa, or 5G networks)
- Cloud analytics platforms (for data processing and visualization)
- User interfaces (mobile applications and web dashboards).

1.9.2 Water Quality Monitoring and Management

Water pollution and scarcity are critical challenges in environmental sustainability. IoT solutions enable real-time monitoring of water quality parameters such as pH, dissolved oxygen (DO), turbidity, and chemical contaminants, ensuring efficient water resource management.

IoT-Enabled Real-Time Water Quality Assessment

IoT-based water quality monitoring systems use smart sensors to collect and transmit water parameter data to cloud-based analytics platforms. These systems detect contaminants such as heavy metals, nitrates, and microbial pathogens, ensuring safe drinking water and sustainable aquatic ecosystems.

The water quality index (WQI) is often calculated as:

$$\text{WQI} = \sum_{i=1}^{n}\left(\frac{P_i}{P_{\text{standard},i}} \times W_i\right)$$

where P_i is the measured concentration of pollutant i, $P_{\text{standard},i}$ is its permissible limit, and W_i is its relative weight.

Remote Sensing and Smart Irrigation Systems

IoT-based remote sensing technology integrates satellite imagery, ground-based sensors, and AI models to monitor soil moisture, nutrient levels, and crop water requirements. Smart irrigation systems leverage this data to optimize water usage through automated irrigation scheduling. A typical smart irrigation system includes:

- Soil moisture sensors (for water requirement assessment)
- Weather stations (for real-time climatic data)
- Actuators (for controlling irrigation valves)
- Cloud analytics (for decision-making and automation).

These systems reduce water wastage, enhance crop yields, and promote sustainable farming practices.

1.9.3 Smart Agriculture and Precision Farming

IoT has revolutionized agriculture by introducing precision farming techniques that optimize resource utilization and improve productivity. IoT-driven solutions enable real-time crop monitoring, automated irrigation, and livestock management, as shown in Fig. 1.1, which depicts a smart agriculture system using sensors, Arduino, data preprocessing, and machine learning to predict optimal crop selection.

– IoT-Driven Crop Monitoring

IoT sensors monitor soil moisture, temperature, humidity, and nutrient levels to provide farmers with actionable insights for optimizing crop growth. These data

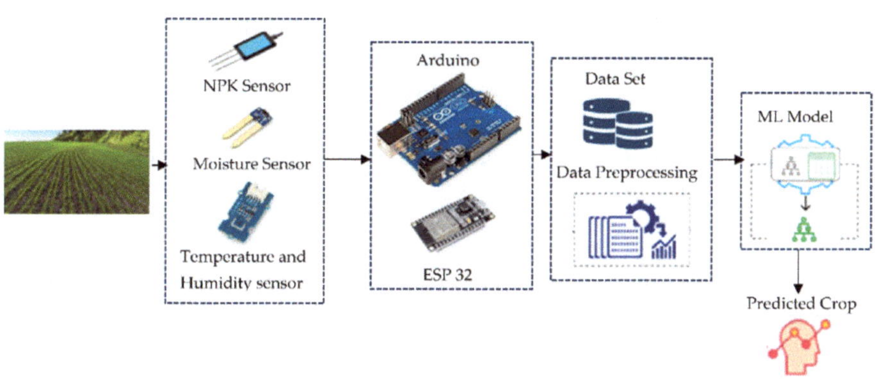

Fig. 1.1 Smart agriculture system using sensors, Arduino, data preprocessing, and machine learning to predict optimal crop selection (*Source* [3])

points are transmitted to cloud platforms, where AI-driven models analyze trends and recommend interventions such as fertilization schedules and pest control measures.

– Automated Irrigation Control

Automated irrigation systems use IoT-enabled controllers to regulate water supply based on soil moisture levels and weather conditions. This approach ensures optimal water usage, preventing over-irrigation and water stress in crops. Smart irrigation can reduce water consumption by up to 30% and improve crop yield [7].

– Livestock Tracking and Health Monitoring

IoT-based livestock management employs wearable sensors, GPS tracking, and RFID technology to monitor animal movement, health parameters, and reproductive cycles. These systems detect diseases early, optimize feeding schedules, and reduce mortality rates, enhancing farm productivity.

1.9.4 Disaster Management and Climate Monitoring

IoT technology plays a critical role in disaster management by providing early warnings and real-time monitoring of extreme weather events. By integrating satellite data, remote sensing, and AI models, IoT systems enhance disaster preparedness and response.

IoT-based flood and drought prediction systems utilize hydrological sensors, weather stations, and satellite imagery to assess water levels, soil moisture, and precipitation patterns. AI-driven models analyze historical and real-time data to forecast potential flooding or drought conditions. These systems help governments and communities take preventive actions, such as reservoir management and drought mitigation strategies.

IoT-driven early warning systems integrate seismic sensors, ocean buoys, and meteorological data to detect natural disasters such as earthquakes, tsunamis, and hurricanes. These systems provide real-time alerts through mobile applications, automated sirens, and emergency response networks, reducing casualties and economic losses.

A general equation for disaster prediction using IoT-based data analytics is:

$$R_{\text{risk}} = f(S, T, H, M)$$

where S represents sensor data, T is temporal patterns, H denotes historical trends, and M is meteorological parameters.

1.10 Challenges and Limitations of IoT Adoption

The adoption of IoT has been accelerating across various industries, transforming operational efficiency, decision-making processes, and service delivery. However, despite its vast potential, IoT adoption faces numerous challenges and limitations that hinder its seamless implementation and widespread use [24]. These challenges can be categorized into technical barriers, economic and infrastructure constraints, and regulatory and standardization issues. Addressing these challenges is critical for ensuring the sustainable and scalable growth of IoT ecosystems.

1.10.1 Technical Challenges in IoT Implementation

– **Scalability and Interoperability Issues**

One of the foremost technical challenges in IoT adoption is scalability, which refers to the ability of an IoT system to accommodate an increasing number of connected devices without significant degradation in performance. As the number of IoT-enabled devices grows exponentially, networks and cloud infrastructures must handle massive amounts of data, increasing the demand for bandwidth, processing power, and storage capacity. Managing such a vast and heterogeneous network requires robust architectures capable of dynamically allocating resources and balancing workloads.

Another critical challenge is interoperability, which arises due to the lack of uniform communication protocols and device standards across IoT manufacturers. IoT devices, sensors, and platforms are developed by different vendors using proprietary technologies and communication standards. This fragmentation creates difficulties in seamless data exchange and integration between heterogeneous systems. Standardization efforts such as those by IEEE, ITU-T, and IETF have sought to address these issues, but achieving universal interoperability remains an ongoing challenge [10].

To overcome these challenges, researchers are developing middleware solutions, open-source IoT platforms, and standardized communication protocols such as message queuing telemetry transport (MQTT), constrained application protocol (CoAP), and open platform communications unified architecture (OPC-UA). However, widespread adoption of such solutions still requires industry collaboration and regulatory frameworks.

– **Energy Efficiency and Power Consumption**

IoT devices, particularly those deployed in remote or resource-constrained environments, must operate on low power while maintaining high efficiency. Many IoT applications, such as environmental monitoring, smart agriculture, and industrial automation, rely on battery-powered sensors and edge devices. However, power consumption remains a significant bottleneck due to continuous data transmission, complex computations, and communication overheads.

Energy efficiency in IoT systems is influenced by various factors, including sensor design, data processing methods, and communication protocols. Low-power communication technologies such as long-range wide area network (LoRaWAN), narrowband IoT (NB-IoT), and Zigbee are being developed to reduce power consumption in IoT networks [5]. Additionally, energy harvesting techniques, such as solar, kinetic, and radio frequency energy harvesting, offer promising solutions for extending the operational lifetime of IoT devices.

To further enhance energy efficiency, researchers are focusing on edge computing and fog computing to process data closer to the source, reducing the need for continuous cloud communication and minimizing energy consumption. Integrating machine learning algorithms for intelligent power management also plays important role in optimizing energy usage in IoT systems.

1.10.2 Economic and Infrastructure Constraints

– **High Costs of Deployment**

The cost of deploying IoT solutions remains a significant barrier, particularly for small and medium enterprises (SMEs) and organizations in developing regions. The high costs stem from several factors, including:

- The hardware costs of IoT-enabled sensors, actuators, gateways, and networking components.
- The software development is required for data processing, analytics, and cloud integration.
- The installation, maintenance, and security expenses are associated with managing a large-scale IoT ecosystem.

Although the cost of IoT hardware has been decreasing due to advances in semiconductor technology and mass production, the costs associated with connectivity, storage, and data processing remain high [17]. Moreover, subscription-based cloud services and real-time data analytics platforms increase costs, making it difficult for cost-sensitive industries to adopt IoT solutions. Governments and technology firms are exploring strategies to reduce these costs, including public–private partnerships, subsidies for IoT adoption, and open-source IoT solutions. Additionally, advancements in edge computing can help reduce cloud service costs by minimizing data transmission and processing requirements.

– **Infrastructure Gaps in Developing Regions**

In many developing regions, the successful deployment of IoT solutions is hindered by poor network infrastructure, unreliable electricity supply, and limited access to advanced computing resources. IoT devices rely on stable, high-speed Internet connectivity to transmit data; however, in many rural or remote areas, broadband access is either unavailable or prohibitively expensive. The lack of infrastructure extends to cloud and edge computing resources, which are essential for processing and analyzing IoT-generated data. Without reliable infrastructure, organizations and industries struggle to implement real-time monitoring, predictive analytics, and automation, limiting the potential benefits of IoT. To address these challenges, governments and international organizations are investing in 5G and low-power wide area network (LPWAN) technologies, satellite-based IoT solutions, and decentralized computing frameworks. Initiatives such as Project Loon by Alphabet and Starlink by SpaceX aim to provide high-speed Internet access to underserved areas, potentially enabling broader adoption of IoT [13].

1.10.3 Regulatory and Standardization Barriers

– **Need for Global IoT Regulations**

As IoT adoption grows, there is an increasing need for global regulatory frameworks to ensure security, privacy, and ethical deployment of IoT solutions. However, the lack of standardized international regulations leads to fragmentation and compliance challenges across different regions. IoT devices collect vast amounts of data, including personally identifiable information (PII), financial records, and sensitive industrial data. Without stringent regulations, issues such as data breaches, cyberthreats, and unauthorized surveillance become significant risks. Existing regulatory frameworks, such as the General Data Protection Regulation (GDPR) in Europe and the California Consumer Privacy Act (CCPA) in the USA, provide guidelines for data privacy, but a unified global standard for IoT security and governance remains lacking [11]. Organizations such as International Organisation for Standardisation (ISO) and International Telecommunication Union (ITU) are working toward creating global IoT standards, but achieving worldwide consensus remains a challenge. Governments and industry leaders must collaborate to develop regulations that balance innovation, security, and privacy concerns while ensuring cross-border compliance [1].

– **Compliance Challenges for Industries**

Different industries face unique compliance challenges when integrating IoT technologies. For instance, in health care, IoT-enabled medical devices must comply with strict regulations, such as the Health Insurance Portability and Accountability Act (HIPAA) in the USA. Similarly, in IIoT, sectors like manufacturing, energy, and transportation must comply with safety and operational standards to ensure reliable and secure IoT deployments. Compliance with multiple regional and industry-specific regulations adds complexity and increases costs for IoT adoption [16]. Companies must invest in certification processes, security audits, and legal expertise to ensure regulatory compliance, which can deter smaller enterprises from integrating IoT solutions. To mitigate these challenges, industries are adopting blockchain-based security frameworks, decentralized identity management, and AI-driven compliance monitoring systems to automate regulatory adherence. Governments and regulatory bodies must also work toward harmonizing IoT standards across industries to reduce compliance burdens [18].

1.11 Conclusion

IoT has emerged as a transformative technology, revolutionizing industries through its ability to connect devices, process vast amounts of data, and enhance decision-making. The global IoT market was valued at approximately USD 662 billion in 2023 and is projected to reach USD 3.3 trillion by 2030, driven by advancements in 5G, AI, and edge computing. IoT applications span diverse sectors, including health care, agriculture, environmental monitoring, and smart cities, and offer solutions that enhance efficiency and sustainability. For instance, precision agriculture powered by IoT has increased crop yields by up to 25%, while smart grid implementations have reduced energy consumption by 20% in some urban areas. Despite its rapid growth, IoT adoption remains constrained by several challenges, including security vulnerabilities, data privacy concerns, and infrastructure limitations, particularly in developing regions. Studies indicate that nearly 57% of IoT devices are susceptible to cyberthreats, underscoring the need for robust security frameworks. Furthermore, interoperability issues and the lack of universal IoT standards hinder seamless integration across platforms. Addressing these challenges through AI-driven analytics, blockchain-enabled secure transactions, and sustainable IoT practices will be crucial to future advancements. As technology evolves, the integration of 6G networks and autonomous IoT systems is expected to further enhance connectivity, automation, and intelligence, paving the way for a more interconnected and efficient digital ecosystem.

References

1. Alsubaei F, Abuhussein A, Shiva S (2018) An overview of enabling technologies for the Internet of Things. Internet Things A-Z:77–112
2. Balaji S, Nathani K, Santhakumar R (2019) IoT technology, applications and challenges: a contemporary survey. Wirel Pers Commun 108:363–388
3. Bouni M, Hssina B, Douzi K, Douzi S (2024) Integrated Iot approaches for crop recommendation and yield-prediction using machine-learning. IoT 5(4):634–649. https://doi.org/10.3390/iot5040028
4. Ding J, Nemati M, Ranaweera C, Choi J (2020) IoT connectivity technologies and applications: a survey. IEEE Access 8:67646–67673
5. Domínguez-Bolaño T, Campos O, Barral V, Escudero CJ, García-Naya JA (2022) An overview of IoT architectures, technologies, and existing open-source projects. Internet Things 20:100626

6. Dudhe PV, Kadam NV, Hushangabade RM, Deshmukh MS (2017) Internet of Things (IOT): an overview and its applications. In: 2017 International conference on energy, communication, data analytics and soft computing (ICECDS). IEEE, pp 2650–2653
7. Gokhale P, Bhat O, Bhat S (2018) Introduction to IOT. Int Adv Res J Sci Eng Technol 5:41–44
8. Govinda K, Saravanaguru RAK (2016) Review on IoT technologies. Int J Appl Eng Res 11:2848–2853
9. Gupta BB, Quamara M (2020) An overview of Internet of Things (IoT): architectural aspects, challenges, and protocols. Concurr Comput Pract Exp 32:e4946
10. Hassan R, Qamar F, Hasan MK, Aman AHM, Ahmed AS (2020) Internet of Things and its applications: a comprehensive survey. Symmetry 12:1674
11. Haxhibeqiri J, De Poorter E, Moerman I, Hoebeke J (2018) A survey of LoRaWAN for IoT: from technology to application. Sensors 18:3995
12. Khanna A, Kaur S (2020) Internet of things (IoT), applications and challenges: a comprehensive review. Wirel Pers Commun 114:1687–1762
13. Landaluce H, Arjona L, Perallos A, Falcone F, Angulo I, Muralter F (2020) A review of IoT sensing applications and challenges using RFID and wireless sensor networks. Sensors 20:2495
14. Li C, Wang J, Wang S, Zhang Y (2024) A review of IoT applications in healthcare. Neurocomputing 565:127017
15. Mishra H, Mishra D (2023) Artificial intelligence and machine learning in agriculture: transforming farming systems. Res Trends Agric Sci 1:1–16
16. Mishra H, Mishra D (2024) AI for data-driven decision-making in smart agriculture: from field to farm management. In: Artificial intelligence techniques in smart agriculture. Springer Nature Singapore, Singapore, pp 173–193
17. Mouha RARA (2021) Internet of things (IoT). J Data Anal Inf Process 9:77
18. Perwej Y, Haq K, Parwej F, Mumdouh M, Hassan M (2019) The internet of things (IoT) and its application domains. Int J Comput Appl 975:182
19. Porkodi R, Bhuvaneswari V (2014) The internet of things (IOT) applications and communication enabling technology standards: an overview. In: 2014 International conference on intelligent computing applications. IEEE, pp 324–329
20. Ramson SJ, Vishnu S, Shanmugam M (2020) Applications of internet of things (IoT)–an overview. In: 2020 5th international conference on devices, circuits and systems (ICDCS). IEEE, pp 92–95
21. Shah SH, Yaqoob I (2016) A survey: Internet of Things (IOT) technologies, applications and challenges. In: 2016 IEEE smart energy grid engineering (SEGE), pp 381–385
22. Tiwari AK, Mishra H, Nishad DC, Pandey A (2023) Sustainable water management in agriculture: irrigation techniques and water conservation. Dr. Ajay B. Jadhao 53
23. Tripathy BK, Anuradha J (eds) (2017) Internet of things (IoT): technologies, applications, challenges and solutions. CRC Press
24. Villamil S, Hernández C, Tarazona G (2020) An overview of internet of things. Telkomnika (Telecommun Comput Electron Control) 18:2320–2327

Chapter 2
Historical Context and Evolution of Environmental Monitoring Technologies

Manish Baboo Agarwal, Manu Mehrotra, Manish Kumar Panday, Vinay Jadon, Seema Agarwal, and Pankaj Mittal

Abstract The historical record and development series of the tools used in environmental monitoring show a clear progression from initial experimental methods to modern, digitized, and satellite-based systems. In the early stages, environmental monitoring was conducted primarily through observation and documentation, using fundamental instruments such as thermometers and rain gauges. The Industrial Revolution of the eighteenth–nineteenth centuries heralded a new era in which pollution from factories and urbanization necessitated the development of more systematic methods of monitoring. More structured chemical and biological monitoring of air and water quality emerged in the early twentieth century in response to growing concerns about industrial waste and public health. The environmental movements that emerged in the mid-twentieth century greatly increased public concern over pollution and, combined with the establishment of regulatory frameworks such as the U.S. Environmental Protection Agency (EPA), led to improvements in monitoring methods. Remote sensing, satellite imagery, and automated sensor networks developed in the late twentieth century increased the rate at which environmental data was collected. The incorporation of geographic information systems (GIS), Internet of things (IoT) sensors, and artificial intelligence (AI) in the twenty-first century has revolutionized monitoring effectiveness by enhancing its accuracy, accessibility, and speed. These improvements are essential for monitoring biodiversity loss, pollution, natural disasters, and climate change. These materials exhibit magnetic and electric properties, enabling efficient use of electricity. These materials are synthesized via thin-film deposition and additive manufacturing, thereby expanding their applications across fields such as supercool sensors, actuators, and smart materials.

M. B. Agarwal (✉) · M. Mehrotra
Sharda School of Basic Sciences and Research, Sharda University, Agra, Agra, Uttar Pradesh, India
e-mail: mba.mems@gmail.com

M. K. Panday · V. Jadon · P. Mittal
Department of Applied Sciences, Anand Engineering College, Agra, Uttar Pradesh, India

S. Agarwal
Department of Mathematics, SRMIST, Ghaziabad, Uttar Pradesh, India

S. Gautam et al. (eds.), *Smart Environmental Monitoring*,
https://doi.org/10.1007/978-981-92-0294-2_2

Keywords Environmental monitoring · Remote sensing · IoT sensors · GIS · Artificial Intelligence · Sustainability · Regulatory frameworks

2.1 Introduction

Environmental monitoring is the active, systematic process of gathering, interpreting, and analyzing environmental data, including air and water pollution, soil health, ecological diversity, and related indicators. This practice supports policy formulation for the preservation and sustainable development of available resources by enabling real-time detection of socio-environmental changes.

In general, environmental monitoring helps identify socio-environmental problems, develop solutions, and assess the effectiveness of regulatory measures. For air quality monitoring (AQM) systems, technologies enable the calculation and monitoring of pollutant emission rates [1, 2]. In ecological systems, water quality indicators are crucial for determining the concentration and condition of water droplets that can be consumed by citizens. Assured progress toward any socio-environmental goal is impossible without accurate data, which can be obtained by utilizing new technologies.

In the current era of technological advancement, comprehensive environmental monitoring systems (CEMS) are among the most essential tools for environmental monitoring, as they offer greater accuracy than outdated manual sampling approaches.

For environmental monitoring, manual sampling, along with laboratory analysis, was the foundational cornerstone. Laboratory work on water samples provided numerous treatment options. The problem was time-consuming and did not yield accurate, rapid results.

In the context of environmental monitoring, the development and integration of sensors and data loggers marked a new technological frontier for automatic data collection. Now, data can be retrieved without manual effort and with much greater accuracy, especially through long-term automated data-logging systems. An example of such sensors is those that measure air pollutants, such as sulfur dioxide and nitrogen oxides, enabling real-time monitoring of air quality for timely responses to pollution events [3, 4].

The invention of remote sensing and satellite technologies was a steppingstone toward enabling large-scale monitoring of Earth's environment. Satellites equipped with multispectral and hyperspectral sensors can track changes in land use, deforestation, and even ocean conditions. This was important because it enabled monitoring of global trends worldwide, including regions previously considered remote and inaccessible. According to a Nova Space report, earth observation (EO) satellites are estimated to monitor ecosystems on more than 70 billion occasions annually by 2033 [5–7].

We can now explore the world without restrictions, aided by advanced technologies such as satellites and drones. The Internet of things (IoT) has revolutionized satellite systems by integrating sensor tags that enable concurrent data collection and transmission, thereby enabling real-time exploration [8].

Various other devices can be used to monitor sound, air, and water pollution, soil moisture levels, and to gather data from a single central system. IoT devices have also been integrated into urban air quality monitoring to support effective public health decision-making [9, 10].

These technologies offer unprecedented opportunities for analyzing pertinent issues including climate change, energy, and environmental degradation. Environmental strategies are implemented based on the collected data, risks, and environmental impacts. The integration of AI, IoT, and big data tools permits advanced analysis and synthesis [1, 11].

AI algorithms could recognize patterns and outliers within datasets, which enable predictive modeling and early warning systems. For example, AI models can predict future pollution levels, which allows relevant authorities to put mitigation measures ahead of time into place. The integration of AI with big data analytics is currently improving the efficiency and accuracy of ecosystem monitoring [12, 13].

Innovative technologies have developed methods such as scraping out DNA from the fluid which enables species identification through environmental samples like water, soil, or dust. This is formally called environmental monitoring of DNA or eDNA monitoring. This technique presents an easier, non-destructive, accurate and affordable approach to traditional biological diversity monitoring methods. A startup company based out of Guildford, Nature Metrics uses eDNA monitoring to assess biodiversity and assists businesses with compliance using nature-related regulations [14, 15].

The introduction of low-cost sensors has expanded engagement with citizen-science programs. This technology enables individual and community participation in data collection, which in turn enhances public understanding of environmental problems. For example, lower-cost air quality sensors allow citizens to track their region's pollution levels, thus increasing local participation and advocating for the environment.

The history of monitoring the environment shows advancement in copious streams, focusing on efficiency, precision with a broader inclusivity. All the changes, including the shift from sampling by hand to IoT technologies like AI and eDNA monitoring, have helped us further comprehend our planet and protect it. While these technological advancements are quite encouraging, but infrastructure, data security, interoperability, and the monitoring system's infrastructure costs pose challenges that need to be addressed. Full potential of these technologies cannot be utilized if such obstacles exist.

2.2 Early Ways of Tracking Environmental Data

Like many other things, environmental monitoring has come a long way over the centuries, going from simple observation-based methods to highly advanced technological systems. Studying history helps in grasping the foundational techniques of today's environmental monitoring systems.

2.2.1 Observation and Manual Recording Techniques

Monitoring of the environment in ancient cultures was mainly qualitative using the human anatomy to perform manual observations. Ancient Egyptians had structures called nilometers, which were used to measure the water height of the Nile River in order to predict agricultural yield. These readings had to be interpreted manually, and they gave an estimate of when flooding would occur so that farming could be initiated.

Early sailors also used some floating object to estimate ocean current and their direction by observing how they moved in the water.

2.2.2 Basic Instruments (Thermometers, Rain Gauges, etc.)

The invention of simple instruments facilitated the deepening of environmental monitoring studies. In the seventeenth century, several key instruments were invented:

- *Rain Gauge*: Used in estimating volume of rainfall, rain gauges have greatly assisted in the scientific study of volume of water in the hydrological cycle and planning in agriculture.
- *Thermometer*: In 1714, Daniel Gabriel Fahrenheit invented a mercury-in-glass thermometer that streamlined the measurement of temperature.
- *Anemometer*: The measurement of wind speed was made easier by the invention of the 4-cup anemometer in 1846 by Thomas Romney Robinson.
- *Barometer*: The measurement of atmospheric pressure with a barometer was pioneered by Evangelista Torricelli in 1643.
- The creation of these instruments made the collection of environmental data precise and quantitative. This marked the birth of the science of meteorology and hydrology.

2.2.3 Gaps in the Initial Monitoring Techniques

While their impact was felt, the early methods of monitoring the environment had a number of gaps which include:

Query Based Collection: As data collection relied on a person's observations, biases and inconsistencies were easily introduced. For instance, water quality checks and its accompanying metadata notes were prone to inaccurate manual entries and therefore lacked reliability.

Geographical Distance and Time Gaps: Early techniques often provided data of certain locations and intervals which hindered measuring advanced environmental activities. For example, deposit gauges designed to measure atmospheric particulates were static and could not gather data over extensive places.

Slow Automated Processing: The lack of automated systems meant that data analysis took more than a set time which negatively impacted the ability to make timely decisions. In water quality monitoring, the use of standard methods of collection of a sample, followed by work in the lab, required more time for analysis.

Single Parameter Measurement: Limited sophisticated multipurpose tools meant there was insufficient coverage to all the environmental parameters hence a narrow view of the environment. For example in early times, rain gauges subset for measuring rainfall simply recorded the amount of rainfall, without measuring factors such as the intensity and even the chemicals that make up the rain.

The data accuracy and reliability factors of the research were assessed in relation to the accuracy of the primary instruments used in the study. Calibration or operational issues of the measuring instruments, as well as the environmental influences on the apparatus, were prominent to study with accuracy. One example is the potential horizontal movement of wind over rain gauges, which could affect the precipitation data recorded over a defined period. All of these constraints reinforce the requirement of automated and more sophisticated environmental monitoring technologies that exist nowadays.

Any contemporary environmental monitoring rests on the simple science of primary observations and basic monitoring instruments. Primary methods are of great importance, despite their numerous shortcomings. All of them have contributed tremendously toward the formulation of a proper understanding of the processes taking place in the environment and have emphasized the pressing need for modern technological tools.

2.3 Industrial Revolution and the Rise of Systematic Monitoring

The Industrial Revolution began in the late 1800s and continued into the 1900s. It was a crucial period in human history that transformed the economy into an industrialized one. During this period, inventions and technology underwent massive changes which improved the economic state of the countries. However, the increased economic activities worsened the country's environmental problems. Hence, the systematic methods of environmental monitoring came into existence.

2.3.1 *Impact of Industrialization on Environmental Pollution*

The period of industrialization caused significant environmental devastation. Air, water, and soil pollution reached critical levels as factories spewed uncontrollable amounts of toxins. Industrial cities were covered with dark smog caused by intolerable factory emissions that exposed the population to serious health challenges including breathing problems. Industrial waste also polluted the rivers which resulted in ecological destruction, alongside public health that used this water.

A notable example of industrial-associated crimes against nature is the Bhopal incident of 1984. In Bhopal, India, a gas leak from a pesticide factory led to thousands of fatalities and countless others suffering from chronic illnesses with no accessible medical aid. This incident brought attention toward a need for powerful policies to check and control the environment to stop future tragedies (Fig. 2.1) [16].

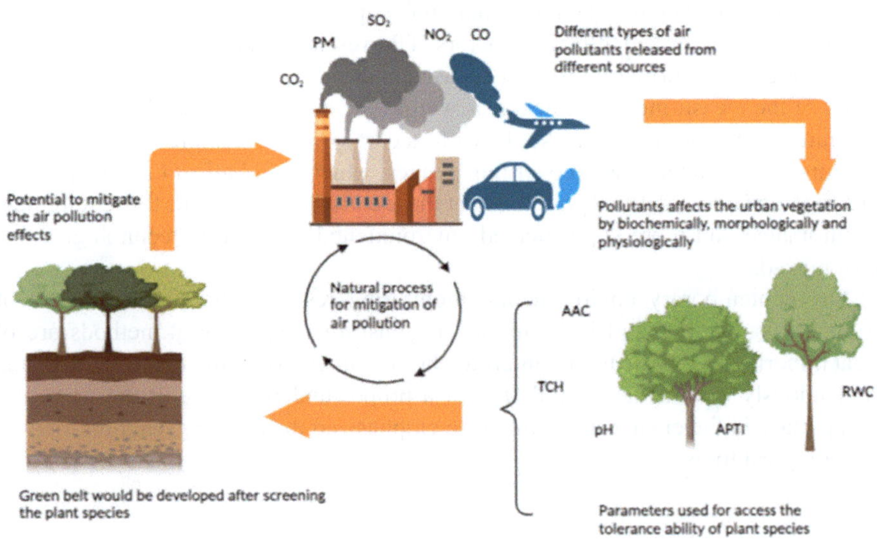

Fig. 2.1 Effect of industrialization on environmental pollution [17]

2.3.2 *Development of Chemical and Biological Monitoring Techniques*

To combat the growing concern over pollution, some sort of chemical and biological monitoring processes were developed with the sole purpose of detecting and measuring these pollutants as well as evaluating the health of the environment.

Chemical Monitoring

Under chemical monitoring, investigations of the environment involve sampling and determining the level of pollutants. The development of instruments such as gas chromatography-mass spectrometry (GC-MS) contributed significantly in this area. By 1979, GC-MS was the 'go to' instrument for water and waste water pollutant analysis at the US EPA. It was now possible to diagnose many organic compounds and their derivatives, which strengthened the capability of regulation by the EPA [18, 19].

Biological Monitoring

Living organisms are indicators of the health status of ecosystems, and biological monitoring examines these organisms ecosystem. For instance, environmental DNA (eDNA) methods enable one to detect some species by analyzing DNA fragments extracted from water, soil, and other samples. This method is now widely used because it does not disturb the organisms being studied and enhances assessment of biodiversity and pollution impacts on a wide range of species [20, 21].

2.3.3 *Early Regulatory Responses to Environmental Concerns*

The primary regulatory responses designed to control health damaging pollution and other effects of industrialization were prompted by the analysis of these problems.

- **Legislative Measures**

To address pollution and preserve the environment, governments started enacting laws. For example, in the United States, the Toxic Substances Control Act (TSCA) of 1976 allowed the Environmental Protection Agency (EPA) to control the currently used chemicals. This legislation sought to mitigate the unreasonable risks allocated by the chemicals business on human life and the environment, thereby marking the initial steps for the comprehensive policy on chemicals.

- **International Agreements**

On a global level, international agreements were put in place to mitigate cross-border environmental concerns. For example, certain protocols were designed to curtail the release of substances that lead to acid rain, signaling the knowledge of the need for some form of coordinated environmental management [22, 23].

– **Technological Innovations**

Advances in technology have made environmental monitoring and enforcement much easier. For example, remote sensing has enabled large-scale environmental surveillance, allowing remote monitoring of deforestation, land-use change, and pollution. These measures facilitated the detection of environmental crimes and the implementation of corrective measures.

The environmental ramifications of the Industrial Revolution necessitated more effective, systematic control and monitoring. The biological and chemical monitoring advancements, along with other ecological policies, have significantly resolved some of the environmental issues that stemmed from industrialization. There remains a need for innovation and robust political oversight to protect the environment and human health during industrial activities.

2.4 Middle of the Twentieth Century: Regulatory Movements and Environmental Policy

A new chapter in the environmental history of the world opened in the middle of the twentieth century. It was characterized by a heightened level of concern for the environment, policy formation, and the development of environmental monitoring technologies.

2.4.1 Escalation of Environmental Concern and Action

The modern environmental movement is attributed to the ecosystem conservation movement and environmental activism that sprang forth from the publication of Rachel Carson's Silent Spring in 1962. Her work highlighted the use of pesticides like DDT that were used indiscriminately without appreciating the destructive effect they have on the environment and human life.

This heightened mass concern was aided by a few other catastrophe events. An example of this is the Cuyahoga River Fire. The river engulfed in flames in 1969 as an outcome of industrial waste into water bodies of Ohio which further signified the neglect toward the preservation of natural resources. These things mobilized people to unite for the environmental cause and brought about activism.

Substantial awareness was noticeable and brought to action with the establishment of the first commemorative Ceremony also known as 'Earth Day', observed on April 22, 1970. This initiative was taken by an American congressman Gaylord

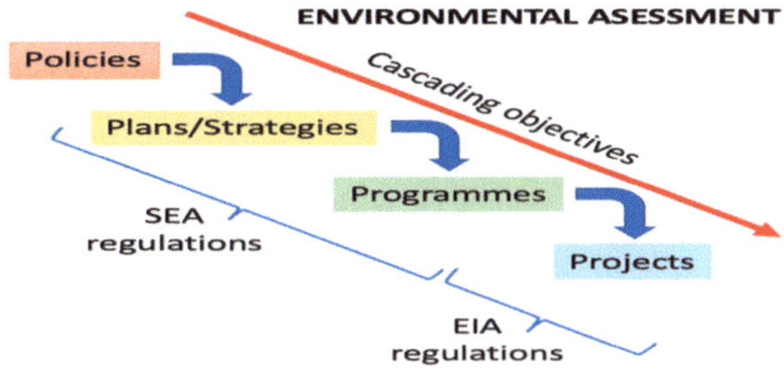

Fig. 2.2 Strategic environmental assessment (SEA) versus environmental impact assessment (EIA) [24]

Nelson and resulted in more than 20 million voicing their outrage through celebration, proclaiming a ban on environmental modification without scrutiny.

2.4.2 *Creation of Environmental Implementation Agencies (EIA)*

Prior to the first Earth Day, congress initiated a nationwide poll to use mobile air pollution control devices to be placed on automobiles. This gave birth to the formation of the environmental implementation agencies for the accomplishment of active structured governance on the environmental impact assessment. Thus in November 1970, they were successfully able to do so by creating The United States Environmental Protection Agency (EPA) which was able to amalgamate federal research and monitory functions for the enforcement of protective laws.

The EPA greatly contributed toward the enforcement of important ecologic laws such as the Clean Air Act of 1970 and the Clean Water Act of 1972 where the latter defined the criteria of accepted standards for the quality of air and water within the United States.

Globally, the establishment of the United Nations Environment Programme UNEP in 1972 after the Stockholm Conference on the Human Environment was a landmark event. UNEP played a critical role in coordinating global efforts on the environment and sustainable development (Fig. 2.2).

2.4.3 *Advances in Air and Water Quality Monitoring*

During the mid-twentieth century period, there were also some major improvements in the monitoring of the environment. Environmental data collection till this point was irregular and devoid of uniformity. With the foundation of environmental agencies, they set up organization monitoring systems.

The introduction of the Clean Air Act permitted the federal government to trim industrial and mobile source emissions, which spurred the creation of air quality monitoring networks. These technologies that allowed the measurement of air pollutants like sulfur dioxide and particulate matter, which greatly affected the air quality. The Clean Water Act similarly required monitoring of water bodies to enforce the set standards, leading to the creation of advanced water sampling and quality analysis methods which enhanced pollutant detection and the safeguarding of aquatic life [25].

Technology improvement greatly increased the provision of data needed to support environmental policies which in turn resulted in enviable improvements in the environment during the last few decades of the twentieth century.

To sum up, the mid-twentieth century marked a turning point in the evolving history of the environment with an increased focus on examination and legislation, along with technology that provided the groundwork of contemporary conservation activities.

2.5 Monitoring Technology Advances: Late Twentieth Century

The late twentieth century is marked by a period of profound change in environmental monitoring that came with the adoption of new technologies which improved methods of information collection, evaluation, and understanding. Such period saw the advent of remote sensing and satellite imaging, automated sensor networks, and the ability to collect and analyze information in a digital form.

2.5.1 The Scope of Remote Sensing and Satellite Imagery

Remote sensing is defined as obtaining information from the surfaces of the Earth without coming in direct contact with the surface, and this information is usually obtained through sensors mounted on satellites and aircrafts. In the second half of the twentieth century, environmental satellite technologies expanded in breadth and depth owing to their ability to monitor the Earth longitudinally and latitudinally. A major innovation came in 1972 with NASA's launch of the first Earth observation satellite, Landsat-1. It came equipped with the multichannel scanner system (MSS), which allowed for imaging in several different spectral bands. This capability made it possible to assess land use, vegetation, and geological features at a region and analyze them over time to solve them. The success of Landsat-1 led to several follow-on missions, creating the longest continuous record of Earth's land surfaces from space.

SPOT-1 (Satellite Pour l'Observation de la Terre) was launched in 1986 by the French Space Agency CNES, which made available for use the first High Resolution Optical Imager with Stereoscopic View Capability. This enabled very accurate topographic mapping and metropolitan spatial planning. In the 1990s, commercial remote sensing companies began to emerge, including DigitalGlobe and GeoEye, which launched satellites equipped with high-resolution imaging systems for defense, agriculture, and disaster monitoring.

In addition, the European Space Agency's ERS-1 satellite, launched in 1991, carried synthetic aperture radar (SAR) sensors that operate day and night regardless of cloud cover; these features improved monitoring of ocean and land surface waves.

2.5.2 Development of Automated Sensor Networks

The late twentieth century marked a shift from manual environmental data collection to automated sensor networks, which improved data precision and frequency. The development of microelectronics has led to the miniaturization and increased energy efficiency of sensors used to measure environmental parameters, such as temperature, humidity, and pollutants. In addition, such sensors could be used in deployed networks that provide continuous, real-time data streams [26–28].

A clear example of this innovation is automated weather stations that use various sensors to measure and record meteorological variables without any human involvement. This data is automatically uploaded to central databases, where it can be combined with other data to improve the accuracy of weather forecasts and climate research. Monitoring systems such as the Silvanet Wildfire Sensor from Dryad Networks can detect incendiary gases, such as hydrogen and carbon monoxide, and warn authorities before serious damage occurs.

2.5.3 Growth of Digital Data Collection and Analysis

Digital technologies enhanced information collection and analysis of environmental data by replacing analogue systems with more efficient digital systems in the latter half of the twentieth century. Detailed datasets can be processed and visualized because of the creation of personal computers and specialized software. GIS systems, or geographic information systems, became essential for managing spatial data by integrating diverse datasets to assess environmental patterns and trends. Digital data loggers replaced manual data collection by automatically recording sensor data over time. This automation enhanced data reliability and facilitated long-term environmental studies. The interpretation of remote sensing data has been enhanced by the development of digital image processing techniques. Environmental assessments became more accurate and detailed due to algorithms developed for image classification, change detection, and feature extraction. Shifts in the collection of data to a digital format integrated with upcoming technologies of the Internet, to support the establishment of real-time monitoring systems for the environment. Data from satellite, radiofrequency, or cellular communication-enabled sensor networks can be interconnected, allowing data to be accessed instantly and enabling timely decision-making [29].

Advances in technology in the latter half of the twentieth century transformed environmental monitoring. High-resolution images and scans of the Earth were obtained through satellite imagery and remote sensing. Environmental monitoring can now be conducted 24/7 in real time due to automated sensor networks. Environmental policies can now be developed with greater certainty owing to the ease of data compilation and analysis. These advancements formed the basis for the modern sophisticated environmental monitoring systems.

2.6 Twenty-First Century: The Digital and Smart Technology Revolution

In the twenty-first century, environmental monitoring experienced a revolutionary change due to the application of digital and smart technology. Advanced tools have emerged for biosphere observation, analysis, and management using GIS, IoT, and AI technology which have resulted in well-informed decisions alongside proactive environmental preservation.

2.6.1 Role of Geographic Information Systems (GIS)

In environmental monitoring, the role of geographic information systems (GIS) extends from providing spatial information to the acquisition, processing, and presentation of information, which makes it crucial. GIS allows the integration of multiple datasets for phenomena to be understood at different scales (spatially and thematically) which leads to sensible strategies for management as well as planning [30].

According to Jack Dangermond, founder and president of Esri, addressing environmental problems begins with trusting the data available, which is critical in solving climate situations. With details mapping and analytic capabilities, Emissions Policy Decisions are facilitated through Esri's GIS technology. Esri uses GIS along with partnerships obtained from the U.S National Climate Assessment and Bezos Earth Fund to spearhead biodiversity hot spot together with innovative climate change works. Most recently, Esri's work has included New York City and Southern California's environmental justice issues mapping, which demonstrates the urgency of placing data on maps to better understand the implications of climate risk and foster evidence-based action.

Environmental monitoring is enhanced with combined technologies and GIS. For example, it is possible to monitor deforestation, urban encroachment, and loss of natural habitats in real time by integrating GIS with remote sensing data. These tools allow stakeholders to spot emerging patterns, evaluate conservation programs, and derive precise solutions.

2.6.2 Integration of Internet of Things (IoT) Sensors

The scope, accuracy, and efficiency of environmental monitoring have been greatly improved by IoT, in which a network of connected devices collects environmental data and transmits it in real time. New IoT sensors are being used to monitor different aspects of the environment. An example is Townsville's technology company, LiXiA, which developed cost-effective, energy-efficient IoT sensors for flood-level monitoring and power-line safety. Their sensors capture relevant data at a reasonable price, significantly less than conventional flood sensors, allowing local councils to respond quickly. Following positive feedback from pilots, the sensors are now compliant with the latest flood-warning regulations, enabling broader deployment and improving community resilience to floods.

In forest conservation, sophisticated IoT devices ranging from drones to micro sensors are utilized to observe forest ecosystems, monitor wildlife, and identify potential threats. For example, Dryad Networks' Silvanet Wildfire Sensors identify early warning signs of forest fires, allowing prompt intervention. Treevia's digital dendrometers perform similar functions by relaying information about tree growth and reading climate changes, giving insight into sustainable management of forests (Fig. 2.3).

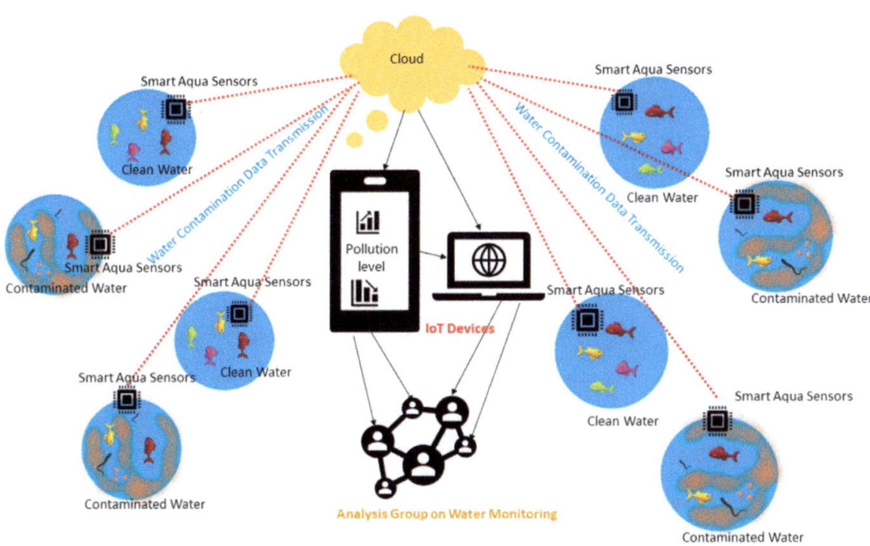

Fig. 2.3 Internet of things (IoT) sensors for environmental monitoring [31]

2.6.3 *The Role of Artificial Intelligence (AI) in Monitoring Environmental Resources*

AI can be a robust tool for environmental monitoring because it enables improved data analysis, flexible modeling, and autonomous decision-making. Having AI integrated with IoT and GIS technologies ensures that environmental resources are addressed promptly and effectively.

Across the world, AI technologies are being integrated into the monitoring and response infrastructure for natural disasters, yielding improved outcomes for densely populated cities. Some applications AI technology covers include public alert systems, real-time response to disasters, data collection, and even forecasting accuracy. One specific use AI has is predicting the landfall of hurricanes. This can allow authorities to take proactive steps to mitigate harmful impacts. Moreover, AI can facilitate post-disaster assessments by analyzing satellite imagery, thereby informing funding and resource allocation and recovery processes (Fig. 2.4).

In biodiversity and environmental conservation, AI enables monitoring of ecosystems, detection of illegal activities, and support for biodiversity measures. For example, AI can address multiple problems simultaneously, such as proactively mitigating pollution by predicting pollutant levels in advance. Additionally, the integration of AI with big data analytics improves the overall accuracy and efficiency of these monitoring efforts.

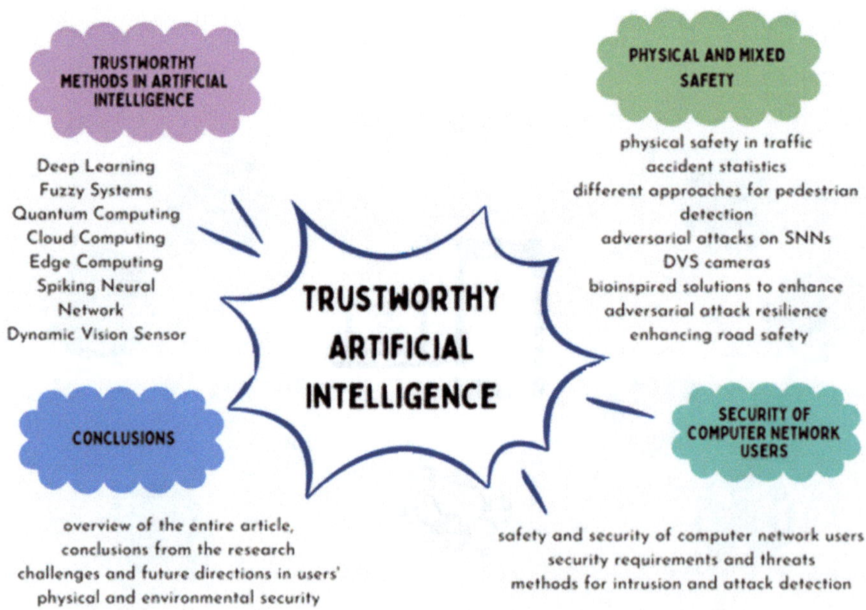

Fig. 2.4 Artificial intelligence (AI) in monitoring environmental resources [32]

Advanced monitoring systems have emerged from the integration of AI and IoT. One example is the use of AI-powered cameras in Northwest Tasmania, which improved the early detection of bushfires. They have successfully pinpointed the exact location of the fire, thereby protecting businesses from the threat of bushfires in the region. This protects the forestry industry, local businesses, and public safety from the ever-increasing risk of bushfires. In addition, environmental intelligence companies are using AI technology to monitor greenhouse gases [32, 33].

As an example, Kayrros has created an AI application Methane GPT, which allows for the civic tracking of methane-producing sites. This application uses a large language model in conjunction with the company's Methane Watch map, providing information on methane emissions from individual facilities at national and global scales. Such initiatives help improve citizens' climate literacy while providing mechanisms to hold emitters accountable, thereby aiding the mitigation of global warming.

2.7 Applications for Modern Environmental Monitoring Technologies

The twenty-first century is marked by unprecedented advancements in digital and smart technologies that have transformed environmental monitoring. The application of GIS, IoT sensors, and AI has significantly improved the accuracy and efficiency of environmental management processes, enabling real-time problem detection and remediation rather than post-hoc responses. These technologies support the real-time collection and analysis of data, the timely identification of potential environmental threats, and effective decision-making, thereby improving the protection and sustainability of natural ecosystems.

With the continued advancement of these technologies, their integrated application has the potential to address complex environmental problems more efficiently. The application and continuous innovation of new tools and systems will play an important role in sustaining the environment and combating climate change.

Contemporary technological resources for environmental monitoring have fundamentally changed our ability to observe, evaluate, and respond to changes in surrounding ecosystems. These are highly beneficial in situations involving pollution, climate change, biodiversity loss, and natural disasters, among others.

2.7.1 Tracking Pollution and Climate Change

The advancement of monitoring pollution and analyzing the impacts of climate change is augmented by satellite remote sensing and AI. For example, environmental intelligence companies have created AI programs that analyze satellite images to track greenhouse gas emissions, such as methane, enabling the public to monitor emissions from methane-reduction facilities. Such information is useful for garnering public support, enhancing accountability, and informing policy decisions.

The use of the Internet of things (IoT) in cities has enabled the deployment of large-scale networks of air quality monitoring systems that measure pollutant concentrations. These systems provide real-time monitoring of air pollution, with responses to hazardous levels implemented immediately. The information aids in planning to reduce emissions [35, 37].

2.7.2 Monitoring Biodiversity and Ecosystem Health

Innovations in technology have simplified biodiversity monitoring and the assessment of ecosystem health. Nature Metrics and other companies that use eDNA methods offer cost-effective biodiversity monitoring, which supports monitoring of their environmental impacts and helps ensure compliance with environmental laws through evidence-based reporting.

In addition, environmental sensing technologies, such as microsensors and drones, are used to monitor forests and wildlife and to detect threats. These instruments provide accurate and detailed information, which consistently improves our understanding of ecosystem processes and supports effective conservation.

2.7.3 Natural Disaster Prediction and Mitigation

Modern environmental monitoring technologies have a significant impact on natural disaster prediction and mitigation. IoT-based environmental monitoring systems that incorporate sensors and AI technologies can issue timely warnings to relevant authorities and the public for disasters such as wildfires or floods upon detecting early signs of such events. For instance, certain fire detection systems are designed to use environmentally monitoring heat sensors that can detect fires in their earliest stages, enabling timely action to control and reduce damage [1, 36].

In flood-prone areas, AI-based flood forecasting projects analyze historical rainfall and monsoon records, and simulate flood events, to generate timely warnings. Such systems assist the relevant stakeholders in minimizing the risk of loss of life and damage. These systems support disaster preparedness and ecosystem management. Advanced monitoring techniques, such as acoustic monitoring, are also used to track marine animals and detect environmental changes in ocean ecosystems. Overall, modern environmental monitoring technologies have improved pollution tracking, biodiversity monitoring, and the prediction of natural disasters, making these tasks much easier.

These innovations are necessary for global environmental management in the face of persistent, continuous challenges.

2.8 Emerging Trends and Issues Related to Environmental Monitoring Technologies

Some upcoming trends and issues are developing with the advancement of these technologies for environmental monitoring.

2.8.1 Disruptive Technologies and Their Novel Uses

The adoption of new technologies fundamentally changes how we monitor the environment.

Artificial Intelligence (AI) and Machine Learning (ML): AI and ML algorithms make it possible to analyze data in a more sophisticated manner by making predictive models, performing real-time anomaly detection and more. For instance, California uses AI satellites to predict and keep track of wildfires so that response efforts can be carried out quickly.

Internet of Things (IoT): IoT devices, such as sensors and smart collars, are actively helping to track environmental and wildlife health parameters. For instance, dairy IoT wearables on farms help diagnose illness within the herd and assist with prompt intervention.

Remote Sensing and Drones: Drones equipped with advanced sensors are being used to acquire comprehensive data on deforestation, forest health, and biodiversity. Additionally, some devices can be used to detect signs of forest fires or to monitor tree growth, thereby improving forest conservation.

Environmental DNA (eDNA) Monitoring: eDNA methods work by collecting gene fragments from the target ecosystem to detect which species are present and monitor diversity. Nature Metrics is one of the businesses that uses eDNA to assess species presence and ecosystem health. This alternative is non-invasive and cost-effective compared to conventional methods.

2.8.2 Environmental Data Collection Ethical and Privacy Issues

The use of monitoring technology is raising further ethical and privacy-related problems.

Surveillance and Abuse of Data: Cameras on garbage trucks raise surveillance and privacy concerns because they can capture identifiable information in real time, which can be misused.

Approval and Openness: Trust and ethical considerations demand proper communication regarding the collection of public or private environmental data by interfaces and communities. Communities' trust needs proper consent.

Sensitive Information: IoT technologies and sensors generate large volumes of data that require robust cybersecurity to protect sensitive and private information from unauthorized access.

2.8.3 Policies and Worldwide Cooperation for Continuous Monitoring

In order for policies to effectively deal with the problems concerning the environment, international collaboration needs to be put in place.

Data Policy: For environmental policies to properly address relevant issues, an effective stratified approach, based on the prioritization of identified and verifiable issues, that cut across various ecologies and economies on a global scale through localized collaborative data collection, analysis, and issue assessment, would ensure the proper resolution of environmental challenges from a cumulative perspective.

Capacity building: The investment in infrastructure and training of environment monitoring in developing countries enhances international global environmental monitoring participation in data collection and analysis.

Public–Private Enterprises: Private sectors, governments, and nongovernmental organizations can drive large-scale monitoring projects through collaborations which include resource sharing and innovation.

Harnessing emerging technologies, global collaboration to address ethical and privacy issues, and fostering global cooperation is what will ensure sustainable and effective stewardship, thus marking the future of monitoring environmental technologies.

2.9 Conclusion

Environmental monitoring technologies have advanced significantly over the decades. There has been a shift from manual sampling methods to sophisticated systems powered by AI and integrated with IoT. These innovations have enhanced the monitoring capabilities of real-time environmental assessments. The availability of remote sensing technologies, along with the capabilities of big data analytics and environmental DNA (eDNA), has widened the reach of environmental assessments and done so more efficiently and economically. As environmental challenges become more complex, technology plays an increasingly crucial role in mitigating environmental risks.

The advances in technologies for monitoring the environment have evolved in the following phases:

1. **Manual and Laboratory-Based Monitoring**—Environmental monitoring started with monitoring via sample taking and laboratory examination of the sample, resulting in very slow movement both in time and region.
2. **Automation and Sensor-Based Monitoring**—Introduction of automated sensors led to increased real-time effort and reduced manual work on non-stop data accumulation of air and water, noise pollution, and even radiation sensors.
3. **Satellite and Remote Sensing**—With the innovation in imaging satellites, monitoring the state of the environment like climate change, deforestation, or the health of the oceans has become possible. Such changes were viewed globally due to its sights.
4. **Internet of Things (IoT) and Cloud-Based Systems**—The use of smart sensors within the IoT framework enables the automatic collection of data. The cloud computing paradigm allows for easy and instant sharing, processing and even real-time analysis of data. With timely and accurate information available, prompt, well-informed decisions can be made about pollution levels and resource use.
5. **Advanced Technologies and Big Data**—The recent surge in applying AI and ML technologies greatly improves the processing of environmental datasets. Pollution, natural disasters, and shifts in biological phenomena are now predicted with these models.
6. **Innovative Methods like eDNA and Citizen Science**—Environmental DNA (eDNA) analysis enables biologists to estimate species diversity with DNA remnants present in water and soil. Environmental assessments can be conducted by the public using portable sensors and smartphone applications. This new development has increased the amount of monitoring and data collection done.

More automated approaches will use AI and self-driving features. Significant improvements to monitoring systems will be achieved using AI models, especially in forecast aspects. Anticipatory triggers for change will engage proactively responding programs for pollution, contamination, and natural disasters, pre-initiated triggers.

Having data protection and monitoring integrity with blockchain technology-environmental data collection and reporting discrepancies have raised concerns over data issues. Consequently, blockchain may provide protective measures for trackability, transparency, and reporting reliability in the collection of sensitive information.

Emerging Wearable Sensors and Nanoscale Technologies—Due to advancements in ultra-sensitive, nano-sized sensors, accurate and comprehensive surveillance will be much easier. Individuals could be provided with real-time situational data through portable and wearable environmental sensors.

Greater Integration with Climate Action and Policy—Future environmental monitoring systems will play a crucial role in shaping environmental policies and ensuring compliance with international agreements such as the Paris Climate Accord.

Citizen Science and Community Engagement—The increasing availability of user-friendly monitoring tools will encourage more public participation, making environmental monitoring a collective responsibility.

References

1. Olawade DB, Wada OZ, Ige AO, Egbewole BI, Olojo A, Oladapo BI (2024) Artificial intelligence in environmental monitoring: advancements, challenges, and future directions. Hyg Environ Health Adv 12:100114. https://doi.org/10.1016/j.heha.2024.100114
2. Samad A, Garuda S, Vogt U, Yang B (2023) Air pollution prediction using machine learning techniques—an approach to replace existing monitoring stations with virtual monitoring stations. Atmos Environ 310:119987. https://doi.org/10.1016/j.atmosenv.2023.119987
3. Wani AK, Akhtar N, Mir TUG, Chopra C, Singh R, Hong JC, Kadam US (2024) CRISPR/Cas12a-based biosensors for environmental monitoring and diagnostics. Environ Technol Innov 34:103625. https://doi.org/10.1016/j.eti.2024.103625
4. Bhatia D, Paul S, Acharjee T, Ramachairy SS (2023) Biosensors and their widespread impact on human health. Sens Int 5:100257. https://doi.org/10.1016/j.sintl.2023.100257
5. Janga B, Asamani G, Sun Z, Cristea N (2023) A review of practical AI for remote sensing in earth sciences. Remote Sens 15(16):4112. https://doi.org/10.3390/rs15164112
6. Lassalle G, Ferreira MP, La Rosa LEC, Scafutto RDM, De Souza Filho CR (2022) Advances in multi- and hyperspectral remote sensing of mangrove species: a synthesis and study case on airborne and multisource spaceborne imagery. ISPRS J Photogramm Remote Sens 195:298–312. https://doi.org/10.1016/j.isprsjprs.2022.12.003
7. Abdulraheem MI, Zhang W, Li S, Moshayedi AJ, Farooque AA., Hu J (2023). Advancement of remote sensing for soil measurements and applications: a comprehensive review. Sustainability 15(21):15444. https://doi.org/10.3390/su152115444
8. Ullah I, Adhikari D, Su X, Palmieri F, Wu C, Choi C (2024) Integration of data science with the intelligent IoT (IIoT): current challenges and future perspectives. Digit Commun Netw. https://doi.org/10.1016/j.dcan.2024.02.007
9. Bobulski J, Szymoniak S, Pasternak K (2024) An IoT system for air pollution monitoring with safe data transmission. Sensors 24(2):445. https://doi.org/10.3390/s24020445
10. Kumar V, Sharma KV, Kedam N, Patel A, Kate TR, Rathnayake U (2024) A comprehensive review on smart and sustainable agriculture using IoT technologies. Smart Agric Technol 8:100487. https://doi.org/10.1016/j.atech.2024.100487
11. Santos MRC, Carvalho LC (2025) AI-driven participatory environmental management: innovations, applications, and future prospects. J Environ Manage 373:123864. https://doi.org/10.1016/j.jenvman.2024.123864
12. Kuppusamy S, Meivelu M, Praburaman L, Alam MM, Al-Sehemi AG, Anbarasu K (2024) Integrating AI in food contaminant analysis: enhancing quality and environmental protection. J Hazard Mater Adv 100509. https://doi.org/10.1016/j.hazadv.2024.100509
13. Dharmarathne G, Abekoon A, Bogahawaththa M, Alawatugoda J, Meddage D (2025) A review of machine learning and internet-of-things on the water quality assessment: methods, applications and future trends. Res Eng 26:105182. https://doi.org/10.1016/j.rineng.2025.105182
14. Lu S, Zeng H, Xiong F, Yao M, He S (2024) Advances in environmental DNA monitoring: standardization, automation, and emerging technologies in aquatic ecosystems. Sci China Life Sci 67(7):1368–1384. https://doi.org/10.1007/s11427-023-2493-5

15. Shu L, Ludwig A, Peng Z (2020) Standards for methods utilizing environmental DNA for detection of fish species. Genes 11(3):296. https://doi.org/10.3390/genes11030296
16. Broughton E (2005) The Bhopal disaster and its aftermath: a review. Environ Health 4(1). https://doi.org/10.1186/1476-069x-4-6
17. Mehmood Z, Yang H-H, Awan MUF, Ahmed U, Hasnain A, Luqman M, Muhammad S, Sardar AA, Chan T-Y, Sharjeel A (2024) Effects of air pollution on morphological, biochemical, DNA, and tolerance ability of roadside plant species. Sustainability 16(8):3427. https://doi.org/10.3390/su16083427
18. Kadadou D, Tizani L, Alsafar H, Hasan SW (2024) Analytical methods for determining environmental contaminants of concern in water and wastewater. MethodsX 12:102582. https://doi.org/10.1016/j.mex.2024.102582
19. Duff D, Lennard C, Li Y, Doyle C, Edge KJ, Holland I, Lothridge K, Johnstone P, Beylerian P, Spikmans V (2023) Portable gas chromatography–mass spectrometry method for the in-field screening of organic pollutants in soil and water at pollution incidents. Environ Sci Pollut Res 30(40):93088–93102. https://doi.org/10.1007/s11356-023-28648-w
20. Rishan ST, Kline RJ, Rahman MS (2023) Applications of environmental DNA (eDNA) to detect subterranean and aquatic invasive species: a critical review on the challenges and limitations of eDNA metabarcoding. Environ Adv 12:100370. https://doi.org/10.1016/j.envadv.2023.100370
21. Sahu A, Singh M, Amin A, Malik MM, Qadri SN, Abubakr A, Teja S, Dar SA, Ahmad I (2025) A systematic review on environmental DNA (eDNA) Science: an eco-friendly survey method for conservation and restoration of fragile ecosystems. Ecol Ind 173:113441. https://doi.org/10.1016/j.ecolind.2025.113441
22. Escobar-Pemberthy N, Ivanova M (2020) Implementation of multilateral environmental agreements: rationale and design of the environmental conventions index. Sustainability 12(17):7098. https://doi.org/10.3390/su12177098
23. Li X, Shen X, Jiang W, Xi Y, Li S (2024) Comprehensive review of emerging contaminants: detection technologies, environmental impact, and management strategies. Ecotoxicol Environ Saf 278:116420. https://doi.org/10.1016/j.ecoenv.2024.116420
24. Rodrigo-Ilarri J, González-González L, Rodrigo-Clavero M-E, Cassiraga E (2020) Advances in implementing strategic environmental assessment (SEA) techniques in Central America and the Caribbean. Sustainability 12(10):4039. https://doi.org/10.3390/su12104039
25. Hubbell JA, Ryan JC (2021) Environmental history. In: Routledge eBooks, pp 89–108. https://doi.org/10.4324/9781351200356-6
26. Shahid S, Brown DJ, Wright P, Khasawneh AM, Taylor B, Kaiwartya O (2025) Innovations in air quality monitoring: sensors, IoT and future research. Sensors 25(7):2070. https://doi.org/10.3390/s25072070
27. Aarif MKO, Alam A, Hotak Y (2025) Smart sensor technologies shaping the future of precision agriculture: recent advances and future outlooks. J Sens 2025(1). https://doi.org/10.1155/js/2460098
28. Anwer AH, Saadaoui M, Mohamed A, Ahmad N, Benamor A (2024) State-of-the-Art advances and challenges in wearable gas sensors for emerging applications: innovations and future prospects. Chem Eng J 502:157899. https://doi.org/10.1016/j.cej.2024.157899
29. Da Costa TP, Da Costa DMB, Murphy F (2024) A systematic review of real-time data monitoring and its potential application to support dynamic life cycle inventories. Environ Impact Assess Rev 105:107416. https://doi.org/10.1016/j.eiar.2024.107416
30. Bill R, Blankenbach J, Breunig M, Haunert J, Heipke C, Herle S, Maas H, Mayer H, Meng L, Rottensteiner F, Schiewe J, Sester M, Sörgel U, Werner M (2022) Geospatial information research: state of the art, case studies and future perspectives. PFG J Photogram Remote Sens Geoinf Sci 90(4):349–389. https://doi.org/10.1007/s41064-022-00217-9
31. Ullo SL, Sinha GR (2020) Advances in smart environment monitoring systems using IoT and sensors. Sensors 20(11):3113. https://doi.org/10.3390/s20113113
32. Szymoniak S, Depta F, Karbowiak Ł, Kubanek M (2023) Trustworthy artificial intelligence methods for users' physical and environmental security: a comprehensive review. Appl Sci 13(21):12068. https://doi.org/10.3390/app132112068

33. Shivaprakash KN, Swami N, Mysorekar S, Arora R, Gangadharan A, Vohra K, Jadeyegowda M, Kiesecker JM (2022) Potential for artificial intelligence (AI) and machine learning (ML) applications in biodiversity conservation, managing forests, and related services in India. Sustainability 14(12):7154. https://doi.org/10.3390/su14127154
34. Akter S, Babu MM, Hani U, Sultana S, Bandara R, Grant D (2024) Unleashing the power of artificial intelligence for climate action in industrial markets. Ind Mark Manage 117:92–113. https://doi.org/10.1016/j.indmarman.2023.12.011
35. Gu Z, Zeng M (2023) The use of artificial intelligence and satellite remote sensing in land cover change detection: review and perspectives. Sustainability 16(1):274. https://doi.org/10.3390/su16010274
36. Albahri A, Khaleel YL, Habeeb MA, Ismael RD, Hameed QA, Deveci M, Homod RZ, Albahri O, Alamoodi A, Alzubaidi L (2024) A systematic review of trustworthy artificial intelligence applications in natural disasters. Comput Electr Eng 118:109409. https://doi.org/10.1016/j.compeleceng.2024.109409
37. Popescu SM, Mansoor S, Wani OA, Kumar SS, Sharma V, Sharma A, Arya VM, Kirkham MB, Hou D, Bolan N, Chung YS (2024) Artificial intelligence and IoT driven technologies for environmental pollution monitoring and management. Front Environ Sci 12:1336088. https://doi.org/10.3389/fenvs.2024.1336088

Chapter 3
Introduction to IoT-Based Environmental Monitoring

Amitabh Padmanabhan and Varun Kumar Gupta

Abstract People, in the past few years, have a much deeper understanding of the environment they live in. With that heightened consciousness, there is a great need for accurate environmental monitoring systems. The application of such systems to air quality, water quality, farming, and wildlife habitats monitoring, which is important for health and has considerable industrial relevance, is very imperative. Air contamination from hazardous gases can lead to serious dangers to the safety of workers in industries such as mining and heavy industries. Monitoring in such dangerous surroundings can be a matter of life-and-death. Systems for implementing large-scale sensor for environmental monitoring suffer from problems of data collection, management, connectivity, and power consumption. These challenges, IoT technology is well-suited to meeting. By storing the collected data on a web server, it is easily accessible for a user anywhere with Internet access or cloud architecture enables seamless monitoring and visualization of the data. It leads to proactive policymaking, early environmental risk detection, and evidence-based formulation toward mitigating issues related to climate change, biodiversity conservation, and sustainable resource management.

Keywords Sensor · Environmental · IoT · Climate · Biodiversity · Hazardous

3.1 Introduction

Environmental monitoring is the cornerstone of environmental management and ecological research. It offers the data to help us know how ecosystems work, how pollutants are changing, and how natural resources are changed—what we are using or what we are not using. The systematic advancement of environmental monitoring developed over the past 50 years, especially because industrial development further exacerbated environmental degradation [1].

A. Padmanabhan · V. K. Gupta (✉)
Anand School of Engineering and Technology, Sharda University Agra, Agra, India
e-mail: varungupta.rkg@gmail.com

S. Gautam et al. (eds.), *Smart Environmental Monitoring*,
https://doi.org/10.1007/978-981-92-0294-2_3

In the case of biodiversity monitoring, population and species distribution-related risk indicators, and ecosystem stability-related indicators, these are important for planning conservation. Climate change awareness also—temperature trends and sea-level rise—enable the context of trends in global warming. Evaluation of the water quality may also have an effect, as water quality plays a determinant role in human health, the ecological and biological balance of aquatic environments, and the sustainable management of agriculture [2].

For much of the past few decades, environmental monitoring involved manual sampling methods but also relied on sensors. Although these methods laid the groundwork for contemporary environmental science, they did have clear shortcomings. Instead, data tended to be collected on the fly—not from an ongoing basis. This meant that a lot of large differences, particularly those driven by rare environmental occurrences that had large effects, were sometimes neglected [3].

Data gaps in long-range environmental interpretations can distort, even slight data. A number of traditional monitoring systems encounter the problems as follows:

- Monitoring data is collected at times, which prevents one from noting some of the fast changing environmental conditions.
- Manual processes both cost money and demand human resources.
- Hazard detection, however, if it is delayed, can attenuate response performance.
- Variability in sampling methods may also contribute to measurement difference.

This landscape was drastically changed with the advent of embedded systems and IoT technologies. Intelligent sensing devices enable uninterrupted, automatic, and remote (at-a-glance) monitoring, rendering most of the structural limitations of traditional technologies obsolete [1].

The Internet of Things (IoT) refers to an ecosystem of interconnected physical devices equipped with sensors, communication interfaces, and processing capabilities. These instruments can collect environmental data and send it over networks for future analyses. For environmental sensors, the IoT is used for the real-time monitoring of air pollutants, water chemistry, atmospheric conditions, wildlife movement, and urban heat patterns [4]. Continuity is one of the key benefits of IoT-based systems. Unlike traditional monitoring setups, IoT networks operate persistently, monitoring environmental changes as they occur. This real-time capability improves situational awareness and enables prompt action when thresholds get crossed. The key benefits include:

- Continuous rather than sporadic sampling,
- Reduced reliance on physical site visits,
- More consistent measurements with less human error,
- Automatic alerts in case of unusual environmental conditions,
- Analytics driven by AI that is predictive.

IoT moves monitoring from being responsive to proactive by embedding intelligence into environmental sensing infrastructure. Instead of reacting to environmental damage, authorities and industries can intervene early based on live data streams and predictive models.

3.2 IoT for Environmental Monitoring

IoT-based environmental monitoring systems are deployed with distributed sensors across various operating environments, including urban, industrial, agricultural, forest, and marine ecosystems. These sensors continuously record environmental data and forward it to a central location, such as the cloud or a database. Once collected, data can be processed, analyzed, and visualized to support decision-making. This information is relayed over communication networks to various cloud platforms or control centers, which can be analyzed to:

- Detect pollution levels,
- Monitor weather patterns,
- Track deforestation,
- Detect forest fires,
- Monitor water bodies,
- Enable early warning systems for natural disaster (Table 3.1).

3.2.1 IoT Architecture for Environmental Monitoring

1. **Perception Layer (Sensing Layer)**

Function: Responsible for collecting raw environmental data.

Components:

- Sensors (e.g., temperature, humidity, air quality, CO_2, pH, noise, etc.),
- Actuators (optional, for responding to conditions, e.g., turning on alarms or sprinklers).

Example: A PM2.5 sensor measures air pollution in real time.

2. **Network/Transport Layer**

Function: Transmits the collected data from the perception layer to the processing systems or cloud.

Communication Technologies:

- Wi-Fi, Bluetooth,
- ZigBee, LoRaWAN,
- NB-IoT, LTE/5G,
- Ethernet, RFID.

Example: A GSM module sends temperature data from a rural station to a central server.

3. **Data Processing Layer (Middleware Layer)**

Table 3.1 Key components of IoT for environmental monitoring

S. No.	Component	Role	Examples/ technologies
1	Sensors/actuators	Collect environmental data such as temperature, humidity, air quality, noise, light, radiation, pH	MQ-135 (air quality) – DHT11/DHT22 (temperature and humidity) – Water pH sensors – Ultrasonic sensors (water level)
2	Microcontroller/ microprocessor	Acts as the brain of the system, controlling sensors and processing collected data	– Arduino – Raspberry Pi – ESP8266/ESP32 (with Wi-Fi)
3	Communication module	Transfers data from sensors/ microcontrollers to the cloud or a remote server	– Wi-Fi – GSM/3G/4G – LoRaWAN – Zigbee – NB-IoT
4	Cloud platform/ server	Stores and processes the data received from sensors	– AWS IoT core – Google Cloud IoT – ThingsBoard – Blynk/ThingSpeak
5	Data analytics and visualization	Analyzes collected data and provides dashboards for trends, anomalies, and insights	– Power BI – Grafana – Custom dashboards (web/mobile apps)
6	Power source	Powers the IoT device/sensors	P-Batteries – Solar panels (for remote/off-grid use)
7	End-User Interface	Allows stakeholders to view and interact with environmental data	Mobile apps – Web portals – SMS/Email alerts

Function: Processes and filters the data, manages devices, and ensures secure data flow.

Components:

- IoT gateways,
- Edge computing devices,
- Data aggregation and preprocessing tools.

Example: An edge device filters out redundant data before uploading to the cloud to save bandwidth.

4. **Application Layer**

Function: Interfaces with users and provides environmental insights and services.

Features:

- Dashboards,
- Reports,
- Alerts and notifications,
- Analytics and visualization.

Example: A web app showing real-time air quality index (AQI) with graphs and location data.

5. **Business/Management Layer (Optional)**

Function: Manages the overall IoT ecosystem including:

- Decision-making,
- Security and privacy,
- Policy enforcement,
- Cost optimization.

Example: The system alerts authorities if pollution exceeds permissible levels (Fig. 3.1).

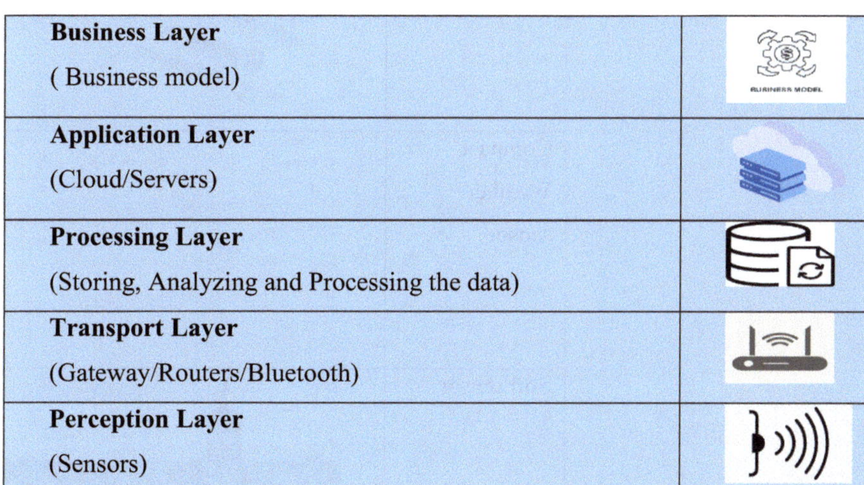

Fig. 3.1 Layer architecture of IoT for monitoring environmental parameter

3.2.2 Sensors and Actuators in Environmental Monitoring

Sensors and actuators in environmental monitoring systems are the primary operating components. Sensors determine and measure environmental parameters, and actuators act on such measurements. This feedback loop, in other words, describes how environmental information results in guided system actions. Actuators operate by receiving commands from control units based on predetermined thresholds or algorithmic decisions. For example, when the air quality deteriorates beyond acceptable levels, an actuator will trigger an alarm or activate ventilation equipment. To ensure safe water flow, soil moisture readings may automatically initiate water flow through controlled valves in irrigation systems. Therefore, actuators are the "action layer" of the monitoring framework.

IoT systems rely on different categories of sensors depending on the environmental domain being monitored. Each sensor type is designed to measure specific parameters with suitable sensitivity and precision (Fig. 3.2).

a. Air Quality Sensors

Fig. 3.2 Various type of sensor

<table>
<tr><td>Air sensor</td><td></td></tr>
<tr><td>Water sensor</td><td>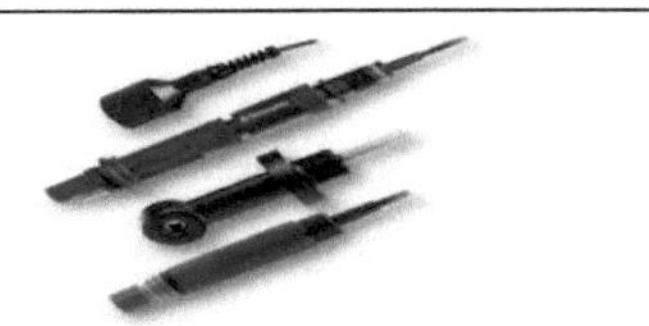</td></tr>
<tr><td>Compact Weather sensor</td><td>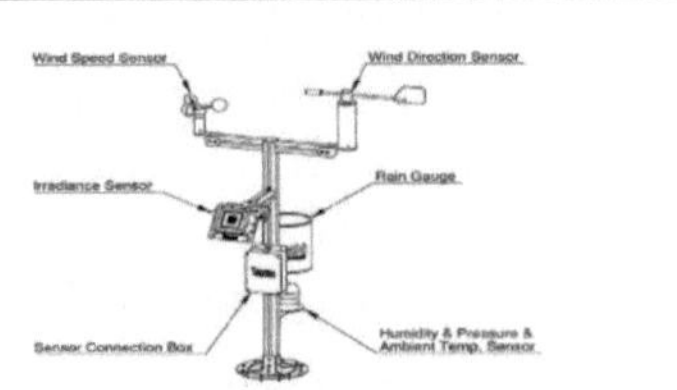
</td></tr>
<tr><td>Soil sensor</td><td></td></tr>
</table>

Air quality sensors are widely used in both indoor and outdoor monitoring systems. They detect pollutants and gaseous compounds that influence air quality and public health.

- **CO_2 Sensors**: Measure carbon dioxide concentration, often used in indoor air quality assessments and greenhouse gas studies.
- **NO_2 Sensors**: Detect nitrogen dioxide, typically associated with vehicle emissions and industrial exhaust.
- **PM2.5 Sensors**: Monitor fine particulate matter capable of penetrating deep into the respiratory system.
- **VOC Sensors**: Identify volatile organic compounds emitted from industrial processes, solvents, paints, and cleaning agents.

These sensors contribute significantly to air quality index (AQI) calculations and regulatory monitoring frameworks.

b. Water Quality Sensors

Water quality monitoring is essential for ensuring ecosystem health and safe water supply. IoT-enabled water sensors provide continuous assessment of chemical and physical properties.

- **pH Sensors**: Determine the acidity or alkalinity of water.
- **Turbidity Sensors**: Measure water clarity by detecting suspended particles.
- **Dissolved Oxygen (DO) Sensors**: Evaluate oxygen availability for aquatic organisms.
- **TDS Sensors**: Measure total dissolved solids to assess overall water purity.

Such sensors are commonly deployed in municipal water systems, irrigation networks, and wastewater treatment plants.

c. Soil Monitoring Sensors

Soil monitoring plays a critical role in precision agriculture and land management. IoT-based soil sensors enable farmers and researchers to monitor field conditions continuously.

- **Moisture Sensors**: Assess soil water content to optimize irrigation scheduling.
- **Temperature Sensors**: Track soil temperature, influencing seed germination and plant growth.
- **Nutrient Sensors**: Measure levels of nitrogen, phosphorus, and potassium, which are vital for crop health.

These sensors enhance agricultural productivity while reducing unnecessary resource consumption.

d. Weather Sensors

Weather monitoring sensors collect atmospheric data necessary for forecasting and climate analysis.

- Temperature and Humidity Sensors: Provide basic atmospheric condition measurements.
- Wind Speed Sensors: Useful in renewable energy planning and storm monitoring.
- Precipitation Sensors: Record rainfall, snow, or hail accumulation.

By integrating these sensors into IoT networks, meteorological data can be collected at localized levels, improving forecasting precision.

e. Sensor Calibration and Accuracy Considerations

Accurate environmental data is fundamental for reliable analysis and decision-making. However, sensor readings may vary over time due to environmental exposure, aging components, or external interference. Calibration is therefore necessary to maintain measurement reliability. Calibration involves adjusting sensor output to align with established reference standards. The frequency of calibration depends on sensor type, deployment conditions, and sensitivity requirements. For example, gas sensors operating in industrial zones may require more frequent recalibration than temperature sensors in controlled environments.

Several factors influence sensor accuracy:

- Environmental Conditions: Extreme temperatures, humidity, and dust accumulation may affect readings.
- Sensor Drift: Long-term exposure can gradually shift sensor output from its baseline, requiring periodic correction.
- Cross-Sensitivity: Some sensors respond to multiple stimuli, potentially reducing specificity.

Maintaining sensor accuracy ensures that environmental monitoring systems deliver trustworthy data, ultimately supporting effective interventions and policy decisions.

3.3 IoT-Based Environmental Monitoring System Design

Designing an IoT-based environmental monitoring system requires thoughtful integration of hardware components, software platforms, communication technologies, and data processing mechanisms. Unlike isolated sensing devices, IoT systems function as coordinated ecosystems where each layer supports the next. The ultimate objective is to create a system that is reliable, scalable, energy-efficient, and capable of delivering meaningful insights in real time. To understand how such systems operate, it is useful to examine their hardware foundation first, followed by the software and data architecture that enables intelligent monitoring.

3.3.1 Hardware Components

The physical infrastructure forms the backbone of any IoT-based monitoring solution. These hardware elements interact directly with environmental parameters and ensure seamless data transmission to processing units.

At the core of the hardware architecture lies the processing unit. Microcontrollers and microprocessors act as the system's control center, collecting signals from sensors, performing initial computations, and managing communication protocols.

- Arduino is widely used for small-scale or prototype deployments due to its simplicity and flexibility.
- Raspberry Pi offers more advanced processing capabilities and can handle local analytics tasks.
- ESP32 is a low-cost, energy-efficient option with built-in Wi-Fi and Bluetooth, making it suitable for wireless sensor networks.

These devices serve as the bridge between raw environmental sensing and intelligent data handling.

Beyond controllers, a complete hardware setup includes:

- Sensors, which capture environmental data,
- Actuators, which enable physical responses,
- Power sources, such as batteries or solar panels,
- Communication modules, including Wi-Fi, LoRa, Zigbee, or GSM.

Together, these components create a cohesive sensing and transmission infrastructure capable of operating in diverse environmental conditions.

3.3.2 Software Frameworks and Platforms

While hardware collects data, software systems transform it into actionable intelligence. Without appropriate platforms for programming, integration, and visualization, environmental data remains underutilized.

IoT software frameworks generally fall into two categories: open-source platforms and industrial-grade cloud solutions.

Open-Source Platforms

Open-source environments are particularly valuable for research, experimentation, and academic implementations.

- Node-RED enables flow-based visual programming for IoT applications.
- ThingSpeak supports data collection and analysis with MATLAB integration.
- OpenHAB and Home Assistant allow integration of environmental sensors within smart ecosystems.

These platforms provide flexibility and lower entry barriers for developers and researchers.

Proprietary Platforms

For large-scale or industrial deployments, enterprise-grade platforms are often preferred.

- AWS IoT, Google Cloud IoT, and Azure IoT Hub offer secure device management, large-scale storage, and advanced analytics capabilities.
- Blynk provides a user-friendly interface for real-time monitoring via mobile dashboards.

The choice of platform typically depends on scalability requirements, security standards, and long-term deployment goals.

3.3.3 Data Collection, Processing, and Transmission

It is essential that a system be able to efficiently handle continuous data access once the hardware and software elements are installed. Environmental data starts as analog or digital signals from sensors. Microcontrollers take these signals and convert them into formatted data. In advance of transmission, basic preprocessing at the device level is possible; examples include filtering noise or applying thresholding. It avoids unnecessary data flow and improves system operation. The data is conveyed over specific protocols such as MQTT, HTTP, or CoAP, depending on the network architecture and available bandwidth. The selected communication protocol significantly impacts latency, reliability, and power usage. Through this systematic method, raw environmental observations are transformed into linear data streams.

3.3.4 Cloud Integration and Data Analytics

Cloud computing plays a central role in storing, analyzing, and visualizing environmental data at scale. Environmental monitoring scales from region to region and sector to sector, so centralized data management is imperative. Cloud computing is the key to storing, analyzing, and visualizing large volumes of environmental information efficiently. In the cloud, you perform on the data storage, cloud platforms perform the following:

- Storage data to preserve not only current but also past and live data,
- Advanced analytics, including trend analysis and anomaly detection,
- Dashboards and reporting tools and dashboards or reporting visualization,

- Automatic alerts that go off when the system has crossed a threshold for data that is considered environmental, serving the user immediately with no time delay at all with an alarm system.

With the help of machine learning algorithms, cloud systems can allow predictive environmental modeling to take the place of descriptive reports. This integrated monitoring turns normal systems into an effective agent for a decision-support system that actively comprehends information, enabling better utilization of real-time data for decision-making. Cloud systems, by means of machine learning algorithms, can enable the use of predictive environmental modeling to supersede descriptive reports. This integrated monitoring transforms conventional systems into an effective decision-support agent that actively interprets information and can use real-time data to make decisions.

3.3.5 *Edge and Fog Computing for Real-Time Monitoring and Processing*

Cloud-based frameworks provide advanced analytical capabilities, where cloud processing alone may cause latency and bandwidth bottlenecks through centralized processing alone. Nowadays, edge and fog computing architectures are increasingly promoted for environmental monitoring systems worldwide. Edge computing entails processing data at the edge, typically by local processing at or near the sensor, e.g., an ESP32 or a Raspberry Pi might analyze ambient data and deploy an actuator at the next-level without having the cloud confirm it, i.e., very handy in near real-time but not necessarily for a fire awareness, for example, or leaks from hazardous gas. Fog computing, which occurs between edge devices and the cloud, is hybrid processing. A local server or gateways will aggregate and filter out crucial information and then forward it back to the cloud. This reduces energy usage and enhances system reliability (Fig. 3.3).

By combining edge intelligence with cloud analytics, IoT-based environmental systems achieve both responsiveness and scalability.

3.4 Applications of IoT in Environmental Monitoring

The true impact of IoT in environmental monitoring becomes evident when examining its practical implementations. Across sectors such as urban governance, agriculture, water management, and waste control, IoT systems are no longer experimental—they are actively shaping environmental decision-making processes. By enabling continuous data acquisition and automated alerts, these systems allow stakeholders to move from reactive correction to proactive prevention. Below are

Fig. 3.3 IoT and real-time environment monitoring

key domains where IoT-based environmental monitoring is making measurable contributions.

Air Quality Monitoring

Air pollution remains one of the most pressing environmental challenges worldwide. Industrial emissions release harmful gases such as carbon monoxide, nitrogen oxides, and various hydrocarbons, while vehicular exhaust and agricultural methane further degrade air quality [10]. Continuous exposure to these pollutants affects respiratory health, contributes to climate change, and reduces overall quality of life. IoT-enabled air monitoring systems deploy distributed sensors across cities, industrial zones, and residential areas. These sensors measure parameters such as CO_2 concentration, particulate matter (PM2.5 and PM10), nitrogen dioxide, sulfur dioxide, and volatile organic compounds.

Practical implementations include:

- Installation of air sensors at traffic intersections to monitor emission hotspots,
- Real-time indoor air monitoring in offices and hospitals,
- Continuous methane detection in agricultural storage and waste facilities,
- Integration with smart city dashboards displaying live air quality index (AQI) data.

Municipal authorities can use this data to regulate traffic flow, restrict industrial emissions during peak pollution periods, and design healthier urban infrastructure. In this way, IoT systems support both regulatory compliance and public awareness.

Water Resource Management

Water quality and availability are central to environmental sustainability and human well-being. Pollution, population growth, and climate variability have increased pressure on freshwater resources. Traditional water testing methods often involve manual sampling and laboratory analysis, which may delay contamination detection. IoT-based water monitoring systems provide continuous oversight of water bodies and supply systems. Sensors installed in pipelines, reservoirs, and rivers measure parameters such as pH levels, turbidity, dissolved oxygen, conductivity, and total dissolved solids.

Practical applications include:

- Monitoring municipal water treatment plants for contamination,
- Supervising groundwater and storm water systems,
- Detecting leaks in urban distribution pipelines,
- Ensuring irrigation water quality in agricultural fields.

For example, if abnormal pH levels are detected in a municipal pipeline, the system can automatically trigger alerts before contaminated water reaches households. Such early warning capabilities significantly reduce health risks.

Soil and Agriculture Monitoring

The agricultural sector has seen increasing integration of Internet of Things (IoT) technologies, with applications expanding each year. IoT-based systems enable the monitoring of environmental parameters, allowing access to valuable farm data that can support improvements in both the quality and quantity of agricultural output [7]. To optimize plant development and mitigate adverse effects from extreme weather, data on soil moisture, temperature, and humidity (using sensors such as the DHT11) are gathered and analyzed. In conditions such as high temperature, the agricultural sector has increasingly adopted IoT technologies to improve productivity and resource efficiency. Environmental conditions such as soil moisture, temperature, and humidity directly influence crop growth cycles [8]. IoT-enabled smart farming systems use distributed soil sensors to collect field-level data. Farmers can access this information through mobile dashboards, enabling data-driven irrigation and fertilization decisions.

Practical implementations include:

- Automated irrigation systems that activate when soil moisture drops below threshold levels,
- Monitoring temperature and humidity during flowering and vegetative crop stages,
- Nutrient-level tracking to optimize fertilizer application,
- Environmental data analysis for predicting pest outbreaks.

By reducing water wastage and improving yield quality, IoT systems contribute to sustainable agriculture and food security.

Weather and Climate Monitoring

As Mishra and Singh [5] noted, this real-time data acquisition enables a proactive approach to weather forecasting and environmental monitoring. Unlike traditional weather systems, which often involve delayed updates and limited reach, IoT-powered solutions provide instantaneous insights, making them invaluable in critical scenarios such as storm warnings and flood predictions. Gupta and Pal [6] highlighted that the integration of IoT with existing weather monitoring frameworks allows for enhanced scalability, reliability, and coverage, ensuring that even remote areas can access accurate environmental data.

- IoT weather stations collect data on temperature, humidity, rainfall, wind speed, etc.
- Real-time and hyper-local climate data support forecasting and disaster management.
- Urban microclimate mapping,
- Early warning systems for floods, storms, and droughts,
- Integration with AI for climate pattern analysis.

Waste Management

Improper waste handling presents environmental and public health risks. Urban waste collection systems often operate on fixed schedules, regardless of actual bin fill levels. This results in inefficiency, overflow, or unnecessary resource use. IoT-based smart waste systems incorporate sensors within waste bins to measure fill levels and transmit data to central dashboards. Municipal authorities can optimize collection routes based on real-time need.

- Smart bins that notify collection teams when capacity reaches a threshold,
- Tracking segregation efficiency at the source,
- Monitoring industrial waste disposal practices,
- Generating analytics to support recycling initiatives.

These systems reduce operational costs, minimize environmental contamination, and improve urban sanitation management. Across all these domains, IoT-based environmental monitoring systems share a common strength: they transform environmental data into actionable intelligence. Rather than relying on periodic manual assessments, stakeholders can make continuous, evidence-based decisions grounded in real-time observations.

3.5 Challenges and Limitations in IoT-Based Environmental Monitoring

Although IoT-based environmental monitoring systems offer significant advantages, their implementation is not without obstacles. Technical limitations, operational complexities, economic constraints, and human factors can affect long-term sustainability. However, most of these challenges are manageable when approached strategically. Rather than viewing them as barriers, they can be seen as areas requiring thoughtful system design and policy support. Below is a structured examination of these challenges, followed by practical mitigation approaches.

3.5.1 Technical Challenges

Power Supply Constraints

Many environmental monitoring systems are deployed in remote or off-grid areas such as forests, coastal zones, or agricultural lands. Ensuring uninterrupted power supply under such conditions is a major challenge. Battery depletion or inconsistent energy availability may lead to data gaps. Integrating renewable energy solutions—particularly solar-powered sensor nodes—significantly enhances the feasibility of deployment. Energy-efficient microcontrollers (e.g., low-power ESP modules) and sleep-mode programming strategies can further extend battery life.

Network Connectivity Issues

Reliable data transmission depends on stable network connectivity. In rural or geographically isolated areas, Internet coverage may be weak or inconsistent. Relying solely on high-bandwidth communication systems can increase operational costs. Low-power wide-area networks (LPWANs) such as LoRaWAN and NB-IoT offer extended range with minimal energy consumption. Hybrid communication architectures—combining local gateways with cellular or satellite uplinks—can improve reliability.

Data Accuracy and Sensor Reliability

Sensors may degrade over time due to environmental exposure. Factors such as humidity, dust, chemical contamination, or extreme temperatures can influence measurement precision. Additionally, sensor drift can compromise long-term accuracy. Regular calibration schedules, automated drift-detection algorithms, and redundancy through multiple sensor deployments can enhance reliability. Implementing periodic validation against reference instruments ensures sustained accuracy.

Interoperability and Standardization

Environmental monitoring networks often incorporate devices from different manufacturers. Without standardized communication protocols, system integration becomes complex and scalability is limited. Adopting standardized protocols such as MQTT or CoAP improves compatibility across devices. Open architecture frameworks and API-based integrations further enhance interoperability.

Security and Privacy Risks

IoT systems transmit large volumes of environmental data across networks, making them vulnerable to cyber threats. Unauthorized access, data manipulation, or interception can compromise system integrity. End-to-end encryption, secure authentication mechanisms, firmware updates, and network segmentation significantly reduce cybersecurity risks. Incorporating blockchain-based validation frameworks is an emerging approach for enhancing data integrity.

3.5.2 Operational and Maintenance Challenges

Beyond technical design, practical deployment introduces operational complexities.

Scalability and Network Expansion

As environmental monitoring networks grow, managing thousands of sensor nodes becomes increasingly complex. Maintenance scheduling, firmware updates, and failure detection require systematic coordination. Cloud-based device management platforms with automated health monitoring and remote firmware update capabilities streamline large-scale operations.

Harsh Environmental Conditions

Sensors deployed in forests, oceans, deserts, or industrial zones face extreme environmental stress. Wildlife interference, corrosion, and mechanical wear can shorten device lifespan. Using ruggedized enclosures, weather-resistant materials, and protective housing extends equipment durability. Periodic inspection schedules should be incorporated into operational planning.

Data Management Overload

Continuous monitoring generates large volumes of data. Without efficient filtering and processing strategies, organizations may struggle with storage costs and analysis complexity. Edge and fog computing architectures reduce unnecessary cloud traffic by filtering redundant data locally. Advanced analytics tools and AI-based pattern recognition streamline data interpretation.

3.5.3 Economic and Policy Limitations

The cost of hardware, infrastructure setup, connectivity, and software licensing can be substantial, particularly for developing regions or small municipalities. Open-source platforms, modular deployment strategies, and phased scaling approaches reduce initial financial burden. Government incentives and public–private partnerships can further support adoption. In some regions, clear legal frameworks governing the use, sharing, and protection of IoT data may be absent. Developing standardized regulatory guidelines and promoting transparent data governance policies enhance stakeholder trust and system sustainability.

3.5.4 Human and Knowledge Barriers

Effective IoT deployment requires interdisciplinary expertise spanning electronics, networking, data analytics, and environmental science. A shortage of trained personnel can limit system performance. Capacity-building initiatives, technical training programs, and collaboration between academia and industry help bridge knowledge gaps. Public skepticism or limited awareness can hinder the acceptance of environmental monitoring technologies. Community participation, transparent data-sharing practices, and awareness campaigns increase public trust and encourage cooperative environmental governance. By addressing these challenges proactively through technological refinement, policy development, and stakeholder collaboration, IoT-based environmental monitoring systems can achieve long-term sustainability and scalability.

3.5.5 Future Trends and Research Directions in IoT-Based Environmental Monitoring

As environmental pressures intensify and technological capabilities expand, IoT-based environmental monitoring systems are entering a transformative phase. The next generation of monitoring solutions will not simply collect data—they will interpret, predict, secure, and autonomously respond to environmental changes. Emerging research is increasingly focused on making these systems more intelligent, decentralized, energy efficient, and resilient. Several innovation pathways are expected to redefine the landscape of environmental monitoring in the coming years.

3.5.6 Artificial Intelligence and Machine Learning Integration

The integration of artificial intelligence (AI) and machine learning (ML) with IoT platforms represents one of the most significant advancements in environmental monitoring. Traditional monitoring systems primarily report real-time conditions; AI-enabled systems, in contrast, can identify patterns and forecast future environmental states.

By analyzing historical and live data streams, AI algorithms can:

- Detect anomalies in pollution levels or temperature fluctuations,
- Predict wildfire risks based on humidity and wind conditions,
- Forecast air quality deterioration in high-traffic zones,
- Model long-term climate variations using multi-parameter datasets.

For example, an AI-driven monitoring system could anticipate hazardous air quality several hours before it reaches critical thresholds, allowing authorities to issue preventive advisories rather than reactive warnings. Future research is focused on developing lightweight ML models that operate directly on edge devices. This approach reduces latency, lowers cloud dependency, and enables immediate decision-making even in connectivity-constrained environments.

3.5.7 Blockchain for Secure and Transparent Data Management

As environmental data becomes increasingly valuable for policymaking and compliance verification, ensuring its integrity and authenticity becomes essential. Blockchain technology offers a decentralized mechanism for securing IoT-generated data.

By storing environmental records in immutable ledgers:

- Data tampering becomes significantly more difficult.
- Transparency in reporting improves.
- Trust between institutions, industries, and the public increases.

Smart contracts could automate regulatory compliance—for instance, triggering automatic notifications when emissions exceed permissible limits. Current investigations explore lightweight blockchain architectures compatible with resource-constrained IoT devices. Integrating blockchain with AI-based analytics may create secure and autonomous environmental governance frameworks.

3.5.8 5G and Next-Generation Connectivity

The rollout of 5G networks—and future ultra-fast communication technologies—will dramatically enhance IoT deployment capabilities. Environmental monitoring systems often require large-scale, simultaneous data transmission from distributed sensors. High bandwidth and ultra-low latency networks are critical for such operations.

With advanced connectivity:

- Real-time streaming from thousands of sensors becomes feasible.
- Smart city-scale environmental dashboards can operate seamlessly.
- Remote or previously inaccessible regions gain improved monitoring coverage.

Network slicing in 5G environments may allow dedicated bandwidth allocation for environmental monitoring services, ensuring reliability even during peak network demand. Researchers are examining the integration of 5G edge computing with environmental IoT nodes to enable near-instantaneous data processing and decision support.

3.5.9 Green IoT and Sustainable Infrastructure

Ironically, while IoT systems aim to promote environmental sustainability, large-scale deployment may increase energy consumption and electronic waste. The concept of Green IoT addresses this paradox by emphasizing environmentally responsible system design.

Future innovations will focus on:

- Ultra-low-power sensors and communication modules,
- Renewable energy-powered IoT nodes (solar, wind, kinetic),
- Biodegradable or recyclable hardware components,
- Energy-harvesting technologies that eliminate external power reliance.

Self-sustaining environmental sensor nodes capable of harvesting ambient energy represent a promising research frontier. Such systems would significantly extend operational lifespans while reducing maintenance costs. The development of self-powered, adaptive IoT sensors capable of dynamically adjusting energy usage based on environmental conditions is an emerging area of exploration.

3.5.10 Autonomous Environmental Monitoring Ecosystems

Beyond individual technological enhancements, the broader vision involves creating autonomous environmental monitoring ecosystems. These systems would integrate sensing, analytics, secure data management, and automated response mechanisms into a unified infrastructure.

In such a framework:

- Sensors detect environmental changes.
- AI models analyze trends and predict risks.
- Blockchain ensures secure data validation.
- Edge devices execute immediate corrective actions.
- Cloud systems coordinate large-scale analytics.

This convergence of technologies has the potential to transform environmental governance from reactive regulation to predictive and preventive management.

The future of IoT-based environmental monitoring lies not merely in expanding sensor networks, but in deepening system intelligence, enhancing security, and prioritizing sustainability. Continued interdisciplinary collaboration among environmental scientists, engineers, data analysts, and policymakers will be essential to unlock the full potential of these innovations.

3.6 Conclusion

IoT-based environmental monitoring represents a significant evolution in the way environmental data is collected, processed, and utilized. What once depended on periodic manual observations has now transformed into continuous, interconnected, and intelligent sensing ecosystems. This shift from isolated monitoring instruments to integrated digital infrastructures marks a fundamental technological advancement in environmental management.

At its core, IoT introduces real-time visibility into environmental conditions. However, the true innovation lies in its convergence with complementary technologies. Artificial intelligence enhances data interpretation and predictive modeling. Edge and fog computing reduce latency and enable immediate localized responses. Cloud platforms ensure scalability and long-term storage, while blockchain introduces new dimensions of data security and transparency. Meanwhile, 5G connectivity expands deployment capacity, allowing environmental systems to operate at unprecedented scale.

As these technologies continue to mature, environmental monitoring systems are moving beyond simple measurement toward autonomous decision-support frameworks. Instead of merely reporting environmental degradation, next-generation systems will anticipate risks, optimize resource use, and initiate corrective actions with minimal human intervention.

The evolution of IoT-based monitoring also reflects a broader technological transition—from reactive environmental governance to predictive and data-driven sustainability strategies. By integrating sensing, analytics, connectivity, and automation, IoT platforms are reshaping environmental oversight into a dynamic and intelligent process.

Future progress will depend on continued innovation in energy-efficient hardware, secure communication architectures, adaptive analytics, and scalable system design. With interdisciplinary collaboration and responsible technological deployment, IoT-based environmental monitoring has the potential to become a foundational pillar of smart cities, resilient agriculture, sustainable industries, and climate adaptation frameworks.

In essence, the integration of IoT with emerging digital technologies is not simply enhancing environmental monitoring—it is redefining it.

References

1. Lu T, Chen X, Bai W (2019) Research on environmental monitoring and control technology based on intelligent Internet of Things perception. J Eng. https://doi.org/10.1049/joe.2018.9154
2. Katie B (2024) Internet of Things (IoT) for environmental monitoring. Int J Comput Eng 6(3):29–42. https://doi.org/10.47941/ijce.2139
3. Biber E (2013) The challenge of collecting and using environmental monitoring data. Ecol Soc 18(4):68. https://doi.org/10.5751/ES-06117-180468
4. Radha C, Madheswaran M, Lokesh M, Mohammad Althaf M (2024) Environmental monitoring in Internet of Things. Int J Res Appl Sci Eng Technol 12(IV). https://doi.org/10.22214/ijraset.2024.60086
5. Mishra D, Singh A (2021) Design and implementation of IoT-based weather monitoring system. Int J Adv Res Comput Sci 12(1):100–105
6. Gupta P, Pal SK (2020) A study of weather monitoring using IoT systems. J Emerg Technol Innov Res 7(8):102–110
7. Bhagyashree M, Tapakire A, Patil MM (2019) IoT based smart agriculture using Thingspeak. www.ijert.org
8. Nievola CC, Carvalho CP, Carvalho V, Rodrigues E (2017) Rapid responses of plants to temperature changes. https://www.ncbi.nlm.nih.gov/pmc/articles/PMC5800372/
9. Ansari S, Deshmukh RR (2017) Estimation of soil moisture content: a review
10. Zhang H, Srinivasan R (2020) A systematic review of air quality sensors, guidelines, and measurement studies for indoor air quality management. Sustainability 12:9045. [CrossRef]

Part II
Core Systems and Intelligence

Chapter 4
A Comprehensive Review of Smart Environmental Monitoring Systems Using IoT and Sensors for a Greener Future

G. Jegadeeswari, D. Lakshmi, Sudheer Choudari, Nidal Al Said, Deepak Hajoary, and B. Kirubadurai

Abstract Nowadays, pollution from water, radiation, and air quality are very crucial environmental challenges that need a proper monitoring system to enhance public health and improve sustainable development. In the modern era, the growth of environmental monitoring systems has led to the advancement of smart environmental monitoring systems (SEMs), driven by modern sensor technologies and the advancement of IoT. This chapter provides a comprehensive review of key research contributions in SEMs, with a focus on water and air quality, radiation pollution, and agricultural monitoring. This chapter organizes SEM applications by their specific objectives and delves into the types of machine learning techniques, classification methods, and sensors employed. A detailed analysis of the existing literature highlights important insights, recommendations, and evolving trends in SEM research. This study critically evaluates how technological advancements in sensors, IoT, and machine learning have contributed to more efficient and intelligent environmental monitoring. Finally, the chapter proposes a framework that combines modern machine learning

G. Jegadeeswari
Electrical and Electronics Engineering, Saveetha Engineering College, Chennai, India

D. Lakshmi (✉)
Electrical and Electronics Engineering, AMET Deemed to be University, Chennai, India
e-mail: lakshmi.d@ametuniv.ac.in

S. Choudari
Civil Engineering, Centurion University of Technology and Management, Vizianagaram, Andhra Pradesh, India

N. Al Said
College of Mass Communication, Ajman University, Ajman, UAE

D. Hajoary
Department of Management Studies, Bodoland University, Kokrajhar, Assam, India

B. Kirubadurai
Aeronautical Engineering, VelTech Rangarajan Dr. Sagunthala R&D Institute of Science and Technology, Chennai, India

S. Gautam et al. (eds.), *Smart Environmental Monitoring*,
https://doi.org/10.1007/978-981-92-0294-2_4

methods, denoising techniques, and standards for wireless sensor networks (WSNs) to enhance the overall effectiveness of SEMs.

Keywords Smart environment monitoring (SEM) · Sustainable development · Internet of Things · Environment pollution · Smart sensor · Wireless sensor networks

4.1 Introduction

Globally, there are 17 sustainable developments which rely on various aspects like quality education, economy, agriculture, industry, and many other, with the environment playing a critical role in this equation. The health and hygiene of a population are key components of national progress and societal sustainability, both of which depend on maintaining a hazard and pollution-free, clean environment. Therefore, environmental monitoring is very much essential for ensuring the health and well-being of citizens. EM involves the effective management of disasters, pollution control, and addressing challenges arising from unhealthy environmental conditions. This includes monitoring air and water pollution, weather changes, hazardous radiation, earthquakes, and many other ecological concerns. Pollution sources can be done by both human as well as natural made, and the role of environmental monitoring (EM) is to tackle these issues specially to ensure healthy world and society and to protect the environment.

Modern advancement in science and technology, mainly machine learning and artificial intelligence, have renovated conventional EM into a more efficient intelligent system like smart environmental monitoring systems (SEMs). These technologies allow for more accurate result in monitoring the environmental factors such as mitigating unwanted effects and optimizing pollution control [1]. The development of smart cities, for example, replacing an older method of urban planning by engaging wireless networks to monitor vehicular pollution levels in cities. The Internet of Things devices integrated into WSNs are used for tasks such as pollution control, vehicle tracking, waste management, and temperature control. As a result, advanced environmental monitoring is increasingly referred to as SEMs, by integrating IoT, wireless sensors, and AI [2].

Studies mentioned in the current literature highlight the various applications of SEMs, such as assessing burned areas using remote sensing and multispectral satellite imaging, IoT-based environmental monitoring and health systems, smart environment systems in marine with multimodal sensing networks, and more [3–5]. Additionally, the use of wireless devices in WSNs has prompted the development of protocols and standards for the effective implementation of IoT-based SEM systems.

The global effort to safeguard the environment is very crucial for achieving healthy society, sustainable economic growth, and agriculture. SEM's primary goal is to address environmental challenges by enabling smart monitoring system, ensuring the major key factors such as public health and growth are well-regulated [6–8]. SEM systems are applied in various sectors, including weather forecasting, air pollution control, crop damage monitoring, and water quality monitoring. The adoption of IoT and wireless networks has simplified and optimized environmental monitoring through AI-based systems. SEM systems use a range of smart sensors, WSNs, and IoT devices to provide accurate and dynamic environmental monitoring, enabling systems that adapt to adjustable circumstances.

Nowadays, smart sensors, IoT, and WSNs are the main pillars for SEM systems, which enable data connectivity for environmental criteria like air and water quality, and temperature [9–11]. For instance, a cloud-based SEMs can monitor water contamination and its control by integrating IoT devices and appropriate sensors. These devices, equipped with ML and AI capabilities, enable real-time monitoring and quality analysis, ensuring that water sources are safe and clean [12, 13]. The monitoring organization can access the cloud-based system, where data from various sensors, such as aqua sensors, is analyzed, enabling effective decision-making and pollution control (Fig. 4.1).

Another example of a smart environmental monitoring (SEM) system, with a broader scope and general-purpose capabilities, is depicted in Fig. 4.2. This system addresses various environmental factors, including humidity, temperature, radiation, dust, and UV radiation. This setup exemplifies the concept of a "smart city" [15–17], where a SEM system is employed to ensure a healthy and sustainable environment for its citizens. By integrating diverse environmental monitoring factors through a WSN, the system offers a comprehensive approach to maintaining the well-being of the urban population.

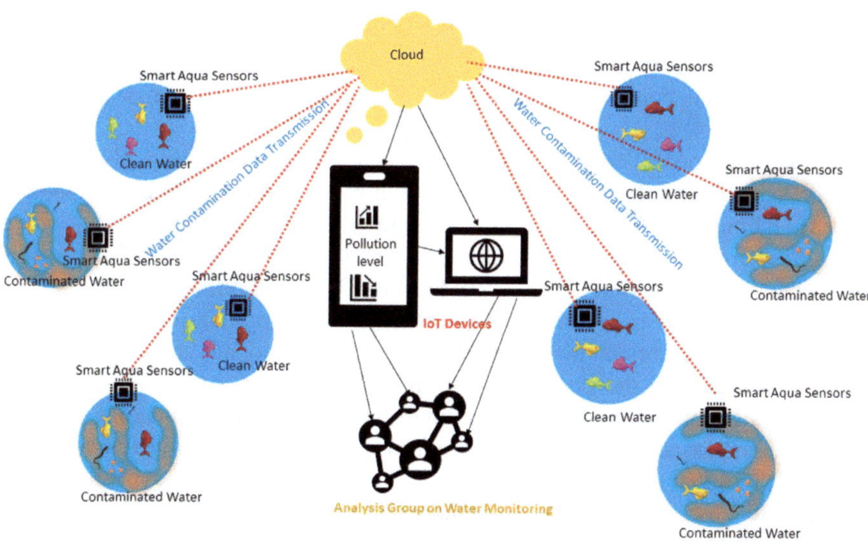

Fig. 4.1 Smart environment monitoring system using the cloud connecting IoT and sensors [14]

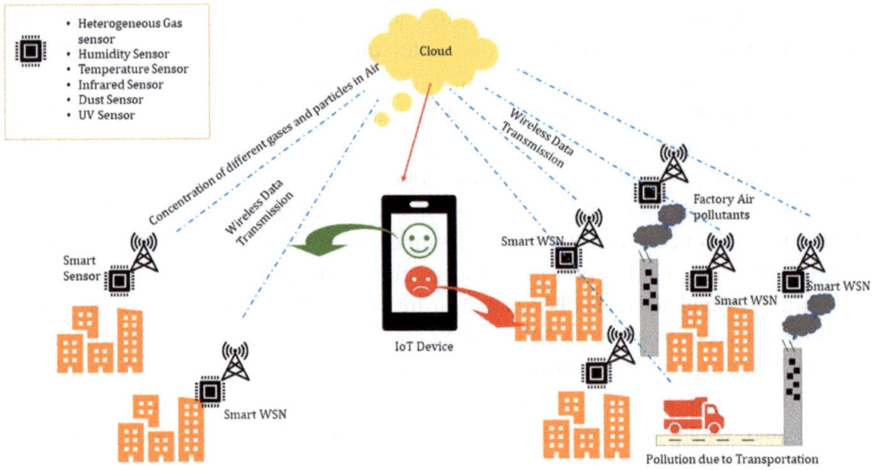

Fig. 4.2 SEM system addressing various issues in the environment using wireless sensor networks (WSNs) and IoT devices

Focusing on agriculture, a key driver of national growth, it is evident that smart environmental monitoring (SEMs) can play a crucial role in fostering "smart or green agriculture" [18–21], addressing the major challenges and factors that contribute to sustainable growth and productivity within the agricultural sector. In this scenario, factors such as soil health, moisture levels, water contamination, and water quantity are critically monitored to ensure sustainable agricultural productivity [22, 23]. By effectively tracking and managing these elements, SEM systems provide vital insights that help optimize farming practices, conserve resources, and enhance overall agricultural output, contributing significantly to the sector's long-term sustainability (Fig. 4.3).

Although various SEM approaches have been discussed, our critical analysis of prior contributions indicates that few thorough reviews provide in-depth coverage of SEM techniques [25, 26]. Comprehensive reviews of SEM techniques are scarce in the literature. Although these studies cover various facets of SEM, there is not a comprehensive review that covers every facet of SEM method.

Although there are many causes of environmental contamination, the most important ones in the research at the moment are radiation, sound, air, and water pollution [27, 28]. This lack of thorough research on SEM encourages us to present a thorough analysis that addresses the main environmental factors influencing health as well as the main strategies, like IoT and sensor technologies, employed to lessen the difficulties these factors present.

The main concerns regarding environmental monitoring, SEM, and the part that IoT, AI, and WSNs play in the deployment of SEM systems have been described in this section. The chapter's subsequent sections are arranged as follows: A comparative analysis of SEM system advancements is presented in Sect. 4.3, relevant

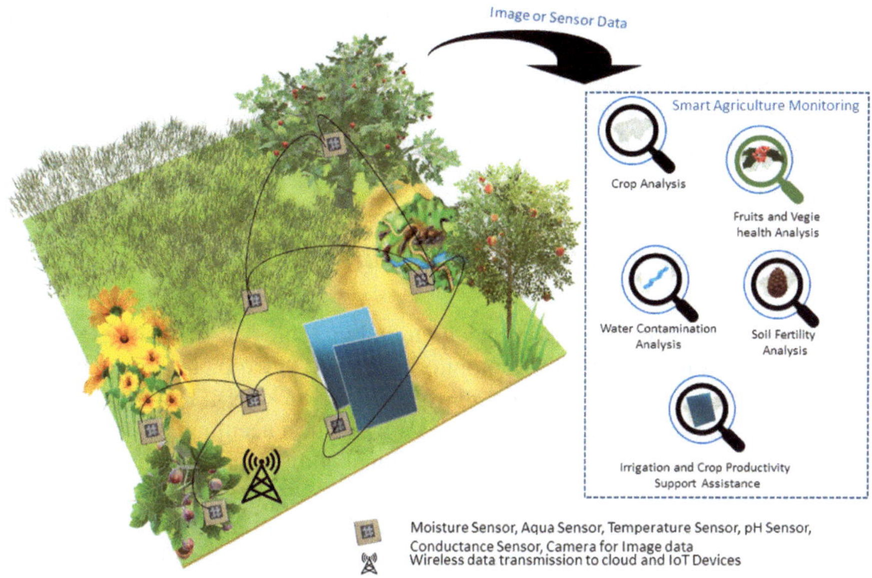

Fig. 4.3 Smart agriculture monitoring system using sensors and IoT devices [24]

research and studies are discussed in Sect. 4.2, and the study's importance and recommendations are emphasized in Sect. 4.4.

4.2 Related Research and Study

Recent studies indicate that smart environmental monitoring systems are increasingly used for a range of objectives and employing a variety of techniques. A substantial body of research on SEM is organized by its goals and techniques. The author's goal in this chapter is to critically present the main conclusions and constraints of recent SEM research. A thorough examination of several SEM application areas is also included in the chapter, including radiation monitoring [29, 30], air quality monitoring [31–34], water quality monitoring [35, 36], soil monitoring [37–39], and ocean and marine environment monitoring.

We found that a comprehensive review of SEM techniques, particularly developments in IoT and sensor technologies, has not yet been conducted [40, 41]. Although there is a wealth of literature on research topics addressing environmental issues such as radiation, air pollution, water pollution, and smart agriculture, there is a lack of synthesis and consolidation.

The study's conclusions are intended to help researchers, scientists, and policymakers develop frameworks and techniques for efficiently monitoring environments affected by issues such as radiation, water pollution, and poor air quality. The foundation of both developed and developing economies, agricultural productivity, is at risk due to these factors, which also affect the environment [42–44]. Therefore, this section also examines smart agriculture monitoring systems, highlighting SEM's contribution to improving agricultural resilience and sustainability.

4.2.1 Study Based on Smart Agriculture (SAM)

A summary of key research studies can be found in Table 4.1. One notable work is the implementation of plant growth monitoring named "gCrop" [45], which utilizes IoT, machine learning, and wireless sensor networks (WSN). This system uses a third-degree regression model and achieves a prediction accuracy of 98% [46, 47]. However, it is worth noting that the computational complexity is high, which could limit its scalability.

In terms of crop quality assessment, a study [48, 49] utilized synthetic aperture radar (SAR) data to monitor the quality of paddy rice. Additionally, leaf area and dimensions are essential determinants of crop growth. One study [50] measured leaf area index using SVMs combined with a Gaussian process model [51], achieving 89% accuracy with a limited sample size, consistent with prior work on crop quality assessment.

Other notable studies have explored various aspects of SEM systems, including smart farming [52], pest monitoring [53], and crop area monitoring [54], thereby further expanding the applications of SEM in agriculture. These studies illustrate the growing role of IoT, machine learning, and sensor technologies in advancing agricultural practices, with a focus on improving crop health, optimizing resource use, and enhancing overall farm productivity.

Table 4.1 IoT and machine learning; IoT and WSN; and an extensive number of research contributions

Year	Research using IoT and WSN	Research using IoT and machine learning
1990–1995	24	2
1995–2000	21	2
2001–2005	7	7
2006–2010	22	2
2010–2015	541	175
2015–2020	6181	3004

4.2.2 Study Based on Smart Water Pollution Monitoring Systems

Smart water pollution monitoring techniques and systems that make use of wireless sensors, machine learning, and the Internet of Things have been the subject of numerous studies. One such study used machine learning and remotely sensed images to forecast lagoon water pollution levels, which are important for farming [55–57]. Although a standard neural network-based machine learning model was employed in this work, the prediction outcomes were not very good. Another study used IoT devices and machine learning techniques to classify water contamination in order to distinguish between clean and polluted water [58]. The data collected was restricted to a small area, despite the fact that the paper presented a real-time contamination monitoring system.

Real-time drinking water quality assessment and classification into drinkable and non-drinkable water using AI-SVM-based techniques are examples of other noteworthy works [59, 60]. In order to help stop man-made sources of contamination, another study [61] investigated video-based surveillance of water quality and pollutants. This system used machine learning to determine whether the water was clean or polluted, and IoT tools for video surveillance. Additionally, in [62], a drinking water quality prediction model was put forth, utilizing a feature-based methodology to assess drinking water and forecast its suitability for human consumption. Last but not least, a study on measuring the amount of chlorophyll-A in lake water [63, 64] used various machine learning models and suggested it as a technique for managing lake water cleaning in real time.

These contributions highlight the significance of real-time data collection, prediction models, and classification techniques for efficient water quality management by illuminating the various uses of machine learning, the Internet of Things, and wireless sensors in smart water pollution monitoring systems [65, 66].

4.2.3 Study Based on Smart Air Quality Monitoring

There has been considerable recent research on smart air quality monitoring systems and methods. One noteworthy study [67, 68] used machine learning techniques in conjunction with heterogeneous sensors to characterize the air quality. Although this method effectively tracked and described air quality, interoperability problems arose because different sensor types were used.

Using both stationary and mobile sensor nodes, another approach to assessing air quality was investigated [69–71]. With compatible sensors deployed as mobile nodes that operate efficiently under dynamic conditions, this system monitors air quality in both stationary and mobile environments.

Gas sensors were used in a study on air quality control in [72] that used IoT and machine learning techniques to collect air particles and analyze air pollutants.

Similarly, Imen et al. [73] employed wireless sensor networks and mobile sensor nodes to monitor air quality within moving cars. The sensor data was processed using machine learning techniques. Furthermore, infrared sensors were employed in [74] to assess air quality, specifically by spectroscopically analyzing volatile organic compounds and applying machine learning techniques to the data.

In [75], advanced machine learning methods were used to forecast PM2.5, a key air quality indicator, from spatiotemporal data. This technique provided important insights into air quality forecasting by analyzing the pollutant levels over various distances and time periods. Additional research on urban air quality was presented in [76], where several forecasting models were proposed to assess pollutants such as O_3, SO_2, and NO_2, and model comparisons were conducted.

To forecast pollution levels, an RFID- and gas-sensor-based air quality control system was implemented in [77]. RFID was mainly used for pollutant detection and communication with WSNs via an IoT-connected architecture, while IoT devices were utilized in this work to analyze sensory data from gas sensors.

Additional research [78] focused on applying machine learning techniques to spatiotemporal data to detect multiple pollutants, including PM2.5, PM10, CO, O_3, SO_2, lead, NO_x, and benzene. Deep learning techniques were incorporated to enable a more thorough examination of O_3 components.

Finally, a study in [79] investigated the use of heterogeneous sensors in air quality monitoring, estimating air quality by analyzing sensor data using support vector machine (SVM). This demonstrates the variety of techniques employed in SAQM systems, all of which integrate IoT, machine learning, and cutting-edge sensor technologies to provide more accurate and efficient air quality monitoring.

4.3 Discussion, Analysis, and Recommendation

Based on a thorough literature review of numerous smart environmental monitoring (SEM) systems, this section offers an analysis, discussion, and some important recommendations. The primary focus of the SEM systems examined in this review is agricultural monitoring systems, water pollution monitoring systems, and air quality assessment systems, along with the related subsidiary applications. A few significant studies from the previous 20 years have also been included, even though the main emphasis is on recent research contributions. With a strong focus on air quality assessment [1, 4, 7, 9, 11, 80, 81] water pollution monitoring [82], radiation monitoring [81], and smart agriculture monitoring systems [1, 14, 28, 54, 60, 62, 63], the study encompasses a wide range of SEM techniques used for various purposes. The following are some important findings from this thorough investigation of SEM techniques:

Goal and Use: Although there is a wide range of research on SEM, smart air quality monitoring, smart water pollution monitoring, and smart agriculture monitoring receive a lot of attention. Research on soil moisture, humidity, air quality, and water pollution offer important information for creating wholesome environmental systems, which can help smart agriculture and promote long-term economic growth.

Machine Learning Methods: A predominant trend in SEM, particularly in SAM and SWPM, is the use of convolutional neural networks (CNN)-based deep learning methods. Other deep learning models are less frequently applied in these studies.

Data Variability: Sensory data varies significantly across SEMs applications, and there is no single, robust dataset that supports the majority of research methods. The types of data and regions of interest differ across various studies, which complicates generalizing findings.

Classification and Prediction: Most SEMs methods are focused on classification (e.g., determining whether water is clean or polluted) or prediction (e.g., forecasting air and water quality degradation).

Challenges Observed in SEMs Research:

Although every SEM system has different difficulties based on its use, a number of typical problems were found:

Heterogeneous Sensors: Because data collected from various sensor types may not be readily compatible, the use of heterogeneous sensors frequently results in interoperability problems in data analysis.

Limited Sample Sizes: A lot of research has a small sample size, which can compromise the findings' robustness and generalizability.

Noisy Data: Both internal and external factors can contribute to noise, which is a persistent issue in sensor data. This noise impairs the accuracy of the results and makes the data analysis process more difficult.

Conventional Machine Learning Techniques: Neural networks and support vector machines are two examples of techniques commonly employed in SEMs. Although these techniques have been used broadly for classification and training data, more recent methods are less common.

Computational Complexity and Big Data: Although some studies have examined deep learning and fuzzy-based techniques, these approaches often struggle with big data or require substantial processing power.

Absence of Robust Approaches: To date, no single, reliable machine learning technique has been found that can successfully address all issues raised by environmental monitoring. A universal approach that can adjust to various monitoring goals, data kinds, and sensor types is required.

Research Trends and Insights: A trend analysis was conducted using publication data from the Science Direct database, spanning the years 1990 to 2020, to assess the growth of SEMs research. The findings, summarized in Table 4.1, reveal a steady increase in SEMs research over this period, particularly in studies involving IoT and WSN technologies. However, the integration of modern machine learning techniques remains relatively underdeveloped compared to traditional research approaches that do not incorporate machine learning. Notably, artificial intelligence (AI) plays a crucial role in enhancing SEMs capabilities, making it an essential factor when deploying IoT devices in WSNs.

These findings suggest that despite significant progress in SEM research, issues with data variability, sensor interoperability, and the uptake of more sophisticated machine learning techniques still exist. This points to an opportunity to advance the development of more resilient and flexible SEM systems that can integrate various sensors and data types while reducing computational complexity.

By classifying the studies into two main groups SEM integrating IoT with wireless sensor networks (WSN) and SEM incorporating IoT with machine learning, Fig. 4.4 shows research trends on SEMs. The adoption of SEM technologies has been steadily increasing year over year, as shown in the figure. It also emphasizes how little machine learning is currently being used in SEM systems for tasks like training, classification, and prediction. Although SEM development has been significantly impacted by IoT and WSN technologies, the integration of machine learning still receives little research attention and is not widely used. In addition to showing a significant chance to further enhance SEM performance by incorporating advanced machine learning techniques, the figure also emphasizes the growing importance and impact of WSN and IoT technologies within SEM systems.

The following suggestions are put forth for creating better, more reliable, and more intelligent environmental monitoring systems (SEMs) in light of the discussion and analysis above:

Creation of a Machine Learning Framework: To manage different environmental monitoring tasks, a thorough framework for machine learning techniques should be created. This framework would improve the classification and predictive capabilities of SEMs and make it easier to integrate machine learning into them.

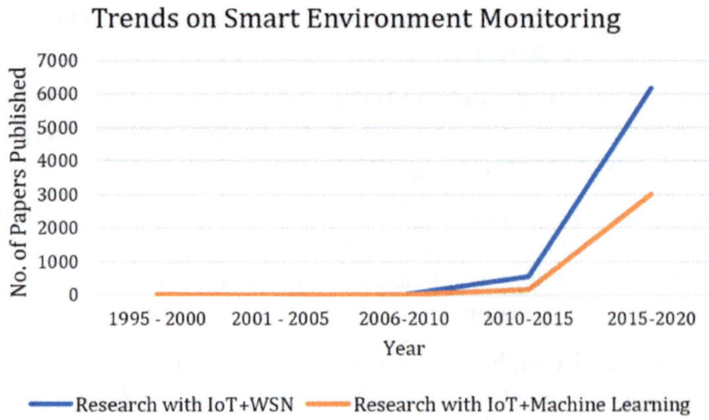

Fig. 4.4 Trends of SEM methods

Design of Robust Models: Regardless of the SEMs particular goal, a robust collection of prediction, classification, and forecasting models that can function on any data must be created. These models should be sufficiently adaptable to address a range of environmental monitoring challenges across diverse fields.

Application of Denoising Techniques: As a preprocessing step in SEM systems, appropriate denoising techniques ought to be applied. The quality of analysis has been impacted by the improper handling of data noise in numerous studies. Denoising techniques would ensure cleaner data and enhance the overall functionality of SEM systems.

Data Deduplication and Big Data Management: Methods for managing big data problems and data deduplication must be investigated. Large datasets pose challenges for many studies; addressing this problem would improve data management and analysis in SEMs.

Infrastructure Issues in Developing Areas: SEM systems are essential to a country's sustainable development, especially those that use sensors, WSN and other IoT. Nonetheless, a major obstacle to implementing IoT and sensor-based systems in rural regions of underdeveloped and developing nations is the lack of requisite infrastructure. For these technologies to be available and reasonably priced, local and international governments must get involved.

Resolving Interoperability Issues: By developing protocols and standards, the interoperability challenges encountered when organizing various types of sensors can be addressed. For efficient analysis and collection, these standards would ensure that data from multiple sensors are well-suited and can be integrated into SEMs.

A thorough, in-depth review of SEMs is lacking, despite efforts to incorporate observations from key review articles on the subject. This study, which focused on important research addressing environmental challenges driven by key factors, was motivated by this gap. The suggestions made considering the review's conclusions are meant to direct the creation of a more reliable and efficient SEMs that can handle the difficulties associated with environmental monitoring using sensor and AI technologies.

4.4 Conclusions and Future Scope of Work

This chapter presents a critical, comprehensive review of several environmental monitoring systems used for various purposes and applications. The discussion and analysis of the literature review have led to various key recommendations. Remarkably, there is clear evidence that extensive research is needed on big data, methods for handling deep learning techniques, and issues with noisy data. Furthermore, the advancement of an outline for robust classification methods in SEMs has been recognized as critical for refining their performance.

The main aim of this chapter is to focus on air and water quality monitoring systems, with an emphasis on smart cultivation systems, with the aim of addressing several environmental challenges. Most of the challenges associated with implementing these systems, including AI, smart sensors, and wireless sensor networks, should be addressed to guarantee sustainable growth through SEMs.

The contribution of regulatory bodies, environmental organizations, and efforts to increase public consciousness will further reinforce SEMs initiatives. Additionally, by employing appropriate filters and signal-processing techniques during data preprocessing, the low quality of sensory data in many SEMs can be mitigated. As a result, the data will be better suited for later SEMs tasks. To provide a more comprehensive approach to environmental monitoring, the future scope of this research will investigate additional environmental factors, such as the integration of SEMs, natural disasters, and sound pollution.

References

1. Sayed E, Ahmed A, Yousef ME (2019) Internet of things in smart environment: concept, applications, challenges, and future directions. World Sci News 134:1–51
2. Jamil MS, Jamil MA, Mazhar A, Ikram A, Ahmed A, Munawar U (2015) Smart environment monitoring system by employing wireless sensor networks on vehicles for pollution free smart cities. Procedia Eng 107:480–484. [CrossRef]
3. Bhoomika KN, Deepa C, Rashmi RKS (2016) Internet of Things for environmental monitoring. Int J Adv Netw Appl 497–501
4. Cicala L, Angelino CV, Parrilli S, Fiscante N, Liberata S, Addabbo P (2018) Unsupervised post-fire assessment of burned areas with free and open multispectral data using OBIA, pp 1–21. https://hal.univ-reunion.fr/hal-01957184. Accessed 17 Dec 2018
5. Gaglio S, Re GL, Martorella G, Peri D, Vassallo SD (2014) Development of an IoT environmental monitoring application with a novel middleware for resource constrained devices. In: Proceedings of the 2nd conference on mobile and information technologies in medicine (MobileMed 2014), Prague, Czech Republic
6. Zhang D, Eng B, Prof S, Connor NEO, Regan PF (2015) Multi-modal smart sensing network for school of electronic engineering. PhD thesis, Dublin City University, Dublin, Ireland
7. Tadejko P (2017) Environmental monitoring systems using internet of things-standards and protocols Pawel Tadejko Environmental policy and management. Ekon. I Środowisko, vol 4
8. Kulkarni PH, Kute PD (2016) Internet of Things based system for remote monitoring of weather parameters and applications. Int J Adv Electron Comput Sci 3:68–73
9. Kamal R (2017) Lesson 11 internet connected environment (weather, air pollution and forest fire) monitoring, pp 1–41
10. Air Quality Monitoring Using IoT and Big Data. GSMA 2018. https://www.gsma.com/iot/wp-content/uploads/2018/02/iot_clean_air_02_18.pdf. Accessed 31 May 2020
11. Jovanovska EM, Davcev D (2020) No pollution smart city sightseeing based on WSN monitoring system. In: Proceedings of the 2020 sixth international conference on mobile and secure services (MobiSecServ), Miami Beach, FL, USA, pp 1–6
12. Arco E, Boccardo P, Gandino F, Lingua A, Noardo F, Rebaudengo M (2016) An integrated approach for pollution monitoring: smart acquirement and smart information. ISPRS Ann Photogramm Remote Sens Spat Inf Sci 4:67–74. [CrossRef]
13. Pavithra G (2018) Intelligent monitoring device for agricultural greenhouse using IOT. J Agric Sci Food Res 9:2–5

14. Pathak A, Uddin MA, Jainal Abedin M, Andersson K, Mustafa R, Hossain MS (2019) IoT based smart system to support agricultural parameters: a case study. Procedia Comput Sci 155:648–653. [CrossRef]
15. Sivakannu G, Balaji S (2017) Implementation of smart farm monitoring using IoT. Int. J. Curr. Eng. Sci. Res. 4:21–27
16. Dhas YJ, Jeyanthi P (2017) Environmental pollution monitoring system using internet of things (IoT). J Chem Pharm Sci 10:1391–1395
17. Ullo S, Vaccaro A, Velotto G (2010) The role of pervasive and cooperative sensor networks in smart grids communication. In: Proceedings of the 2010 15th IEEE mediterranean electrotechnical conference (Melecon 2010), Valletta, Malta, pp 443–447
18. Morello R, De Capua C, Lugarà M (2013) The design of a sensor network based on IoT technology for landslide hazard assessment. In: Proceedings of the 4th Imeko TC19 symposium on environmental instrumentation and measurements protecting environment, climate changes and pollution control, Lecce, Italy, pp 99–103
19. Gardner J (2014) Smart sensors in mobile phones for environmental monitoring. In: Proceedings of the core-group meeting at Eurosensors—2014 conference, Brescia, Italy
20. Shahzadi R, Ferzund J, Tausif M, Asif M (2016) Internet of Things based expert system for smart agriculture. Int J Adv Comput Sci Appl 7. [CrossRef]
21. Carminati M, Kanoun O, Ullo SL, Marcuccio S (2019) Prospects of distributed wireless sensor networks for urban environmental monitoring. IEEE Aerosp Electron Syst Mag 34:44–52. [CrossRef]
22. Kocakulak M, Butun I (2017) An overview of Wireless Sensor Networks towards internet of things. In: Proceedings of the 2017 IEEE 7th annual computing and communication workshop and conference (CCWC), Las Vegas, NV, USA, pp 1–6
23. LG Uplus Corp (2016) What is IoT? LG Uplus Corp, Seoul, Korea. https://www.uplus.co.kr. Accessed 6 April 2020
24. Gubbi J, Buyya R, Marusic S, Palaniswami M (2013) Internet of Things (IoT): a vision, architectural elements, and future directions. Future Gener Comput Syst 29:1645–1660. [CrossRef]
25. Wong MS, Wang T, Ho HC, Kwok CYT, Lu K, Abbas S (2018) Towards a smart city: development and application of an improved integrated environmental monitoring system. Sustainability 10:623. [CrossRef]
26. Alharbi N, Soh B (2019) Roles and challenges of network sensors in smart cities. In: IOP conference series: earth and environmental science, vol 322, p 012002. [CrossRef]
27. Nayyar A, Puri V (2016) Smart farming: Iot based smart sensors agriculture stick for live temperature and moisture monitoring using arduino, cloud computing & solar technology. In: Proceedings of the international conference on communication and computing systems (ICCCS-2016), Gurgaon, India, pp 673–680
28. Kulkarni MPP (2019) IOT based smart agricultural system. Int J Res Appl Sci Eng Technol 7:2037–2041. [CrossRef]
29. Shweta AM, Nagaveni V (2019) Survey on smart agriculture using IOT. J Comput Progr Multimed 4:6–15
30. Li X, Liu Q, Yang R, Wen J, Zhang J, Cai E, Zhang H (2016) The combination of ground-sensing network and satellite remote sensing in Huailai County. IEEE Sens J 16:3819–3826. [CrossRef]
31. Balakrishnan S, Vasudavan H, Murugesan RK (2018) Smart home technologies: a preliminary review. In: Proceedings of the 6th international conference on information technology: IoT and smart city (ICIT 2018), Hong Kong, China, pp 120–127
32. Mshali H, Lemlouma T, Moloney M, Magoni D (2018) A survey on health monitoring systems for health smart homes. Int J Ind Ergon 66:26–56. [CrossRef]
33. Duisebekova KS, Tuyakova ZN, Amanzholova ST, Sarsenova ZN, Duzbayev NT, Pyagay VT, Aitmagambetov AZ (2019) Environmental monitoring system for analysis of climatic and ecological changes using LoRa technology. In: Proceedings of the 5th international conference on engineering and MIS, Agadir, Morocco, pp 1–6

34. Arora J, Pandya U, Shah S, Doshi N (2019) Survey-pollution monitoring using IoT. Procedia Comput Sci 155:710–715. [CrossRef]
35. Okafor NU, Delaney D (2019) Considerations for system design in IoT-based autonomous ecological sensing. Procedia Comput Sci 155:258–267. [CrossRef]
36. Xu G, Shi Y, Sun X, Shen W (2019) Internet of things in marine environment monitoring: a review. Sensors 19:1711. [CrossRef]
37. Sharma J, John S (2017) Real time ambient air quality monitoring system using sensor technology. Int J Adv Mech Civ Eng 4:72–73
38. Shelestov A, Kolotii A, Lavreniuk M, Medyanovskyi K, Bulanaya T, Gomilko I (2018) Air quality monitoring in urban areas using in-situ ind satellite data within era-planet project Eos Data Analytics, Kyiv, Ukraine National Technical University of Ukraine "Igor Sikorsky Kyiv Polytechnic Institute", Kyiv, Ukraine Space Research Institute. In: Proceedings of the international geoscience and remote sensing symposium (IGARSS 2018), Valencia, Spain, pp 1668–1671
39. Gallah N, Besbes K (2013) Small satellite and multi-sensor network for real time control and analysis of lakes surface waters. In: Proceedings of the RAST 2013: 6th conference on recent advances in space technologies, Istanbul, Turkey, pp 155–158
40. Shaikh SF, Hussain MM (2019) Marine IoT: non-invasive wearable multisensory platform for oceanic environment monitoring. In: Proceedings of the IEEE 5th world forum internet things (WF-IoT 2019), Limerick, Ireland, pp 309–312
41. Durante G, Beccaro W, Peres HEM (2018) IoT protocols comparison for wireless sensors network applied to marine environment acoustic monitoring. IEEE Lat Am Trans 16:2673–2679. [CrossRef]
42. Daniels ET, McPheron BD (2017) A machine learning approach to classifying algae concentrations. In: Proceedings of the 2017 IEEE MIT undergraduate research technology conference (MA URTC 2017), Cambridge, MA, USA, pp 1–4
43. Al Mamun MA, Yuce MR (2019) Sensors and systems for wearable environmental monitoring toward IoT-enabled applications: a review. IEEE Sens J 19:7771–7788. [CrossRef]
44. Shaban KB, Kadri A, Rezk E (2016) Urban air pollution monitoring system with forecasting models. IEEE Sens J 16:2598–2606. [CrossRef]
45. Goodson LH, Jacobs WB, Davis AW (1974) Air pollution monitoring system. In: Pesticides abstracts. United States Environmental Protection Agency, Washington, DC, USA
46. Shinde D, Siddiqui N (2018) IOT based environment change monitoring controlling in greenhouse using WSN. In: Proceedings of the 2018 international conference on information, communication, engineering and technology (ICICET 2018), Pune, India, pp 1–5
47. Dhingra S, Madda RB, Gandomi AH, Patan R, Daneshmand M (2019) Internet of things mobile-air pollution monitoring system (IoT-Mobair). IEEE Internet Things J 6:5577–5584. [CrossRef]
48. Chen WP, Wang LK, Wang TT, Chen YT (2013) An Intelligent management system for aquacultures environmental monitoring and energy conservation. In: Advances in intelligent systems research. Atlantis Press, Paris, France, pp 194–198
49. Mocanu I, Florea AM (2012) A multi-agent supervising system for smart environments. In: Proceedings of the 2nd international conference on web intelligence, mining and semantics, Craiova, Romania
50. Beebi F (2018) Environmental monitoring system using IoT. India Res Tech Org 5:64–68
51. Yamashita K, Ao C, Suzuki T, Xu Y, Li H, Tian J, Kimura K, Kasahara H (2016) Architecture design for the environmental monitoring system over the winter season. In: Proceedings of the 14th ACM international symposium on mobility management and wireless access (MobiWac 2016), Malta, pp 27–34
52. Santos D, Mataloto B, Ferreira JC (2019) Data center environment monitoring system. In: ACM international conference proceeding series, pp 75–81. [CrossRef]
53. Chehri A, Saadane R (2019) Zigbee-based remote environmental monitoring for smart industrial mining. In: ACM international conference proceeding series, pp 2–7. [CrossRef]

54. Kumar S, Chowdhary G, Udutalapally V, Das D, Mohanty SP (2019) GCrop: Internet-of-Leaf-Things (IoLT) for monitoring of the growth of crops in smart agriculture. In: Proceedings of the 5th IEEE international symposium on smart electronic systems (Formerly iNIS) (IEEE-iSES 2019), Rourkela, India, pp 53–56
55. Hosseini M, McNairn H, Mitchell S, Davidson A, Robertson LD (2019) Comparison of machine learning algorithms and water cloud model for leaf area index estimation over corn fields. In: Proceedings of the IGARSS 2019—2019 IEEE international geoscience and remote sensing symposium, Yokohama, Japan, pp 6267–6270
56. Fazai R, Mansouri M, Abodayeh K, Puig V, Selmi M, Nounou H, Nounou M (2019) Multiscale Gaussian process regression-based GLRT for water quality monitoring. In: Conference on control and fault tolerant systems (SysTol), pp 44–49. [CrossRef]
57. Dimitriadis S, Goumopoulos C (2008) Applying machine learning to extract new knowledge in precision agriculture applications. In: Proceedings of the 12th Pan-Hellenic conference on informatics Doryssa seaside resort (PCI 2008), Samos Island, Greece, pp 100–104
58. Amado TM, Cruz JC (2018) Dela development of machine learning-based predictive models for air quality monitoring and characterization. In Proceedings of the TENCON 2018, 2018 IEEE Reg, Jeju, Korea, pp 668–672
59. Saha AK, Saha J, Ray R, Sircar S, Dutta S, Chattopadhyay SP, Saha HN (2018) IOT-based drone for improvement of crop quality in agricultural field. In: Proceedings of the 8th IEEE annual computing and communication workshop and conference (IEEE CCWC), Las Vegas, NV, USA, pp 612–615
60. Di Martini DR, Liesenberg V, Tetila EC, Junior JM, Matsubara ET, Siqueira H, De Castro Junior AA, Araujo MS, Monteiro CH, Pistori H (2019) Machine learning applied to UAV imagery in precision agriculture and forest monitoring in Brazililian Savanah. In: Proceedings of the international geoscience and remote sensing symposium 2019 (IGARSS 2019), Yokohama, Japan, pp 9364–9367
61. Zhou Z, Li S (2017) Peanut planting area change monitoring from remote sensing images based on deep learning. In: Proceedings of the 2017 4th international conference on systems and informatics (ICSAI 2017), Hangzhou, China, pp 1358–1362
62. Boursianis AD, Papadopoulou MS, Diamantoulakis P, Liopa-Tsakalidi A, Barouchas P, Salahas G, Karagiannidis G, Wan S, Goudos SK (2020) Internet of Things (IoT) and agricultural unmanned aerial vehicles (UAVs) in smart farming: a comprehensive review. Internet Things 100187. [CrossRef]
63. Liu L, Wang R, Xie C, Yang P, Sudirman S, Wang F, Li R (2019) Deep learning based automatic approach using hybrid global and local activated features towards large-scale multi-class pest monitoring. In: IEEE international conference on industrial informatics, vol 1, pp 1507–1510
64. Li Y, Wang X, Zhao Z, Han S, Liu Z (2020) Lagoon water quality monitoring based on digital image analysis and machine learning estimators. Water Res 172:115471. [CrossRef]
65. Chen Q, Cheng G, Fang Y, Liu Y, Zhang Z, Gao Y, Horn BKP (2018) Real-time learning-based monitoring system for water contamination. In: Proceedings of the 2018 4th international conference on universal village (UV 2018), Boston, MA, USA, pp 1–5
66. Yan H, Liu Y, Han X, Shi Y (2017) An evaluation model of water quality based on DSA-ELM method. In: Proceedings of the 16th international conference on optical communications and networks (ICOCN 2017), Wuzhen, China, pp 1–3
67. Ragi NM, Holla R, Manju G (2019) Predicting water quality parameters using machine learning. In: Proceedings of the 4th IEEE international conference on recent trends on electronics, information & communication technology (RTEICT-2019), Bengaluru, India, pp 1109–1112
68. Budiarti RPN, Sukaridhoto S, Hariadi M, Purnomo MH (2019) Big Data technologies using SVM (case study: surface water classification on regional water utility company in Surabaya). In: Proceedings of the 2019 international conference on computer science, information technology, and electrical engineering (ICOMITEE 2019), Jember, Indonesia, vol 1, pp 94–101
69. Jalal D, Ezzedine T (2019) Toward a smart real time monitoring system for drinking water based on machine learning. In: Proceedings of the 27th international conference on software, telecommunications and computer networks (SoftCOM 2019), Split, Croatia, pp 1–5

70. Bouamar M, Ladjal M (2007) Evaluation of the performances of ANN and SVM techniques used in water quality classification. In: Proceedings of the IEEE international conference on electronics, circuits and systems, pp 1047–1050. [CrossRef]
71. Pang Z, Jia K, Feng J (2014) A water environment security monitoring algorithm based on intelligent video surveillance. In: Proceedings of the 2014 tenth international conference on intelligent information hiding and multimedia signal processing (IIH-MSP 2014), Kitakyushu, Japan, pp 191–194
72. Liu J, Zhang Y, Qian X (2009) Modeling chlorophyll-a in Taihu Lake with machine learning models. In: Proceedings of the 3rd international conference on bioinformatics and biomedical engineering (iCBBE 2009), Beijing, China, pp 1–6
73. Imen S, Chang NB, Yang YJ, Golchubian A (2018) Developing a model-based drinking water decision support system featuring remote sensing and fast learning techniques. IEEE Syst J 12:1358–1368. [CrossRef]
74. Asiful Islam M, Khan RH, Syeed M (2020) A smart and integrated surface water monitor system architecture: Bangladesh perspective. In: ACM international conference proceeding series, pp 8–13. [CrossRef]
75. Mihăiţă AS, Dupont L, Chery O, Camargo M, Cai C (2019) Evaluating air quality by combining stationary, smart mobile pollution monitoring and data-driven modelling. J Clean Prod 221:398–418. [CrossRef]
76. Shetty C, Sowmya BJ, Seema S, Srinivasa KG (2020) Air pollution control model using machine learning and IoT techniques, 1st edn, vol 117. Elsevier Inc., Amsterdam, The Netherlands. ISBN 9780128187562
77. Van Le D, Tham CK (2018) Machine learning (Ml)-based air quality monitoring using vehicular sensor networks. In: Proceedings of the 38th IEEE international conference on distributed computing systems, Vienna, Austria, pp 65–72
78. Liu B, Yan S, Li J, Li Y (2016) Forecasting PM2.5 concentration using spatio-temporal extreme learning machine. In: Proceedings of the 15th IEEE international conference on machine learning and applications (IEEE ICMLA'16), Anaheim, CA, USA, pp 950–953
79. Ayele TW, Mehta R (2018) Air pollution monitoring and prediction using IoT. In: Proceedings of the 2nd international conference on inventive communication and computational technologies (ICICCT 2018), Coimbatore, India, pp 1741–1745
80. Thu MY, Htun W, Aung YL, Shwe PEE, Tun NM (2018) Smart air quality monitoring system with LoRaWAN. In: Proceedings of the 2018 international conference on internet of things and intelligence system (IoTaIS 2018), Bali, Indonesia, pp 10–15
81. Chiwewe TM, Ditsela J (2016) Machine learning based estimation of Ozone using spatio-temporal data from air quality monitoring stations. In: IEEE international conference on industrial informatics 2016, pp 58–63. [CrossRef]
82. Ou CH, Chen YA, Huang TW, Huang NF (2020) Design and implementation of anomaly condition detection in agricultural IoT platform system. In: International conference on information networking, pp 184–189. [CrossRef]

Chapter 5
Lightweight Communication Protocols and AI-Driven Optimization Strategies for Energy-Efficient IoT-Based Environmental Monitoring Systems

Rachmad Andri Atmoko

Abstract Environmental monitoring systems have undergone significant transformation through the integration of Internet of Things (IoT) technologies and artificial intelligence, enabling autonomous real-time sensing across diverse ecological contexts. This chapter examines the critical role of communication protocols and AI-driven optimization in enhancing the performance, sustainability, and scalability of IoT-based environmental monitoring networks. The research explores lightweight communication protocols including MQTT, CoAP, LoRaWAN, and NB-IoT, analyzing their trade-offs in terms of energy efficiency, transmission range, and reliability under various environmental conditions. Multi-objective optimization frameworks utilizing machine learning techniques, particularly reinforcement learning and neural networks, are investigated for their ability to dynamically balance energy consumption, data quality, and network lifetime. The study reveals that MQTT can reduce transmission overhead by 64% while improving sensor battery life by 33%, while hybrid NB-IoT and LoRa strategies effectively address scalability and energy efficiency in wide-area monitoring applications. Artificial intelligence optimization strategies demonstrate significant improvements in system performance through predictive maintenance, demand response optimization, and fault detection mechanisms. Deep reinforcement learning approaches for adaptive sampling rate control achieve up to 43% longer network lifetime compared to fixed-interval sampling, while ANN-based fault detection models attain over 97% accuracy in identifying sensor anomalies. Edge computing frameworks incorporating AI algorithms reduce energy consumption by 55–70% through hierarchical processing and context-aware resource allocation. Sustainable system architectures integrating energy-harvesting technologies, optimized sensor placement, and adaptive edge computing are analyzed for their contribution to long-term deployment viability. Solar-powered LoRaWAN systems with intelligent duty cycling extend operational lifetime to over three years in remote wilderness areas, while advanced time-series compression

R. A. Atmoko (✉)
Faculty of Vocational Studies, Universitas Brawijaya, Malang, Indonesia
e-mail: ra.atmoko@ub.ac.id

S. Gautam et al. (eds.), *Smart Environmental Monitoring*,
https://doi.org/10.1007/978-981-92-0294-2_5

algorithms achieve compression ratios exceeding 100:1 for slowly varying environmental parameters. Real-world case studies spanning air quality monitoring, watershed management, precision agriculture, and biodiversity conservation demonstrate the practical effectiveness of these integrated approaches. The chapter identifies critical research gaps including lifecycle environmental impact assessment, cross-layer protocol optimization, adaptive security frameworks, and semantic interoperability challenges. Future opportunities encompass autonomous grid management systems, explainable AI for environmental decision-making, and federated learning approaches that preserve data privacy while enabling collaborative model improvement across distributed monitoring networks. This comprehensive analysis provides essential insights for researchers, practitioners, and policymakers developing next-generation environmental monitoring systems that balance technological innovation with ecological responsibility, ensuring sustainable planetary observation capabilities for addressing climate change and environmental conservation challenges.

Keywords IoT environmental monitoring · Communication protocols · Artificial intelligence optimization · Energy harvesting · Edge computing · Sustainable sensing networks · Environmental data analytics

5.1 Introduction

The landscape of environmental monitoring has undergone substantial transformation through Internet of Things (IoT) integration, facilitating autonomous real-time sensing capabilities across varied ecological environments. These range from atmospheric pollution tracking and aquatic quality assessment to forest fire identification and agricultural precision monitoring [1–5]. The foundation of such systems relies on extensive networks of distributed sensing devices and interconnected equipment communicating through constrained, power-sensitive networks. IoT-enabled frameworks encounter intricate balance considerations involving response time, dependability, power consumption, system expansion capabilities, and long-term viability. To tackle these obstacles, researchers increasingly adopt sophisticated communication protocols, efficient networking technologies, and artificial intelligence-based enhancement models [1, 6–13].

Contemporary IoT architectures specifically engineered for environmental applications have enabled remarkable data gathering potential at temporal and spatial resolutions previously considered impractical [2, 5]. These frameworks generally comprise several layers: field-positioned sensor nodes, edge computing devices for initial data processing, and cloud infrastructure for comprehensive analytics and visualization [1, 8]. The diversity of sensing methods—encompassing atmospheric quality sensors, hydrological monitoring devices, acoustic detection systems, and multispectral imaging equipment—requires innovative data fusion and interpretation approaches [4, 14]. Recent developments have shown the effectiveness of combining various sensor streams to create comprehensive environmental models capable of identifying subtle ecological variations and forecasting potential disruptions [6, 7].

Despite these advances, substantial technical obstacles persist in widespread IoT environmental monitoring deployment. Power limitations remain especially problematic in remote ecological locations where traditional energy infrastructure remains unavailable [9, 13]. This situation has encouraged exploration of energy collection mechanisms, ultra-low-power communication protocols, and adaptive duty-cycling algorithms that intelligently control power usage based on environmental circumstances and application needs [15]. Furthermore, network durability becomes essential in harsh environments where sensor nodes may encounter physical damage or communication disruption [9, 16]. Researchers have created robust mesh topologies and self-repairing network architectures that preserve operational integrity despite individual node failures [1, 7].

Data quality and reliability represent another complexity dimension, as environmental sensors face calibration drift, physical contamination, and measurement errors [5, 14]. Machine learning methods have shown potential in identifying anomalous readings, conducting automated calibration, and extracting meaningful signals from noisy environmental data [4, 6]. Additionally, edge computing paradigm integration has enabled sophisticated on-site data analysis, reducing bandwidth needs while providing near-instantaneous actionable intelligence to environmental stakeholders [8, 13]. This distributed intelligence approach proves especially valuable in time-critical applications such as early warning systems for natural disasters or industrial contamination incidents [2].

As environmental IoT deployments expand from localized experimental setups to regional and global monitoring networks, compatibility and standardization become progressively important [1, 9]. Several initiatives have emerged to establish common data models, communication standards, and open interfaces that enable seamless data exchange across heterogeneous environmental sensing platforms [5, 7]. Adopting these standards not only improves system scalability but also promotes data accessibility and reusability across scientific disciplines and policy domains [3].

5.2 Communication Protocols for Environmental IoT Systems

5.2.1 Lightweight and Long-Range Communication Protocols

Communication protocol design represents one of the most crucial aspects affecting IoT system performance and energy consumption. Protocols including MQTT, CoAP, LoRaWAN, NB-IoT, and Wi-Fi exhibit different characteristics regarding bandwidth utilization, transmission range, and energy efficiency. For instance, MQTT demonstrates 64% transmission overhead reduction and 33% sensor battery life improvement in constrained sensor networks [8]. LoRaWAN, featuring long-range and low-power characteristics, supports environmental applications including smart agriculture and air quality monitoring [4, 15]. Hybrid strategies combining NB-IoT and LoRa have been implemented to achieve dual objectives of scalability and low energy consumption in wide-area monitoring [9].

MQTT's publish-subscribe architecture proves especially beneficial in environmental monitoring contexts where multiple subscribers—including researchers, regulatory authorities, and emergency response teams—may need access to identical sensor data streams [8]. Recent implementations have extended MQTT with quality-of-service guarantees to ensure critical environmental alerts reach their destinations reliably even under challenging network conditions [4]. Similarly, CoAP's REST-like interface and inherent support for resource discovery facilitate heterogeneous environmental sensor integration into unified monitoring frameworks [9]. The protocol's integrated congestion control mechanisms help prevent network collapse during high-activity environmental events, such as sudden weather changes that might trigger simultaneous transmissions from multiple sensors [8].

Field studies comparing protocol performance across diverse ecological settings have revealed significant environment-specific variations in reliability and energy efficiency [4]. For example, dense forest canopies substantially reduce signal propagation for most RF technologies, whereas LoRaWAN maintains acceptable packet delivery rates despite these obstructions [9]. In urban environmental monitoring, NB-IoT utilizes existing cellular infrastructure to achieve reliable connectivity despite radio interference and physical obstacles, though at higher energy consumption costs compared to other LPWAN technologies [8, 17]. Researchers have also examined protocol behavior under extreme environmental conditions, documenting how temperature fluctuations, humidity, and precipitation affect transmission reliability and power consumption profiles across different communication technologies [9].

5.2.2 Protocol Trade-Offs and Optimization Modeling

These protocol selections influence key trade-offs. For example, latency must often be compromised for energy efficiency or range. To address such trade-offs, researchers have utilized optimization modeling. Han and Gong applied finite-state Markov chains to model wireless channel behavior together with reinforcement learning algorithms to maintain low latency and timely environmental updates without depleting energy reserves [13]. Other works similarly pursue adaptive protocol tuning using machine learning to enable real-time transmission strategy calibration under changing environmental load conditions [18].

Multi-objective optimization frameworks have been developed to simultaneously address competing metrics including energy efficiency, latency, reliability, and network lifetime [13]. These approaches typically formulate protocol parameter selection as a constrained optimization problem, where environmental requirements (e.g., minimum sampling frequency for valid ecological analysis) establish constraint boundaries [18]. Several studies have implemented dynamic protocol parameter adjustment based on contextual factors such as remaining battery capacity, environmental event frequency, and data priority levels [13]. For instance, during detected environmental anomalies (e.g., sudden pollution spikes or forest fire indicators), the system can temporarily prioritize low latency over energy conservation to provide timely alerts [18].

Game theory has also been applied to model interactions between multiple IoT nodes competing for limited network resources in dense environmental deployments [13]. This approach has yielded distributed decision-making algorithms that achieve near-optimal network performance without requiring centralized control, an important consideration for remote environmental monitoring systems with limited connectivity to central infrastructure [18]. Researchers have further explored semantic information integration to optimize protocol behavior, allowing transmission policies to consider the ecological significance of sensor readings rather than treating all data equally [13]. For example, small temperature variations might be suppressed during transmission when they fall within normal seasonal patterns, while equivalent deviations representing anomalous events would trigger immediate data forwarding [18].

Recent advances in federated learning have enabled collaborative protocol optimization across distributed environmental monitoring networks without requiring raw sensor data centralization [13]. This approach preserves data privacy while allowing the system to learn from collective experience of all deployed nodes, gradually improving communication efficiency based on observed environmental patterns and network conditions [18]. Additionally, digital twin modeling of environmental IoT networks has facilitated extensive simulation-based protocol optimization prior to field deployment, reducing the need for costly trial-and-error adjustments in operational systems [9, 13]. These simulation environments incorporate detailed environmental models to accurately predict how protocol performance will vary across different ecological contexts and seasonal conditions.

5.3 AI-Driven Optimization of IoT Operations

5.3.1 Reinforcement Learning for Update Control

Machine learning techniques are critical in managing trade-offs dynamically. Reinforcement learning, including deep RL, has been used to optimize the timing and energy cost of sensor updates in energy-harvesting networks [9, 13, 19]. These models learn behavior policies that align transmission frequency with environmental conditions and communication channel states.

Recent advances in deep reinforcement learning have demonstrated particular promise in adaptive sampling rate control for environmental monitoring. By modeling the energy-harvesting process as a Markov Decision Process (MDP), researchers have developed policies that maximize information gain while ensuring sustainable operation [13]. These approaches consider both the stochastic nature of energy availability (e.g., solar, wind, or vibration-based harvesting) and the varying information value of environmental measurements [18]. For instance, Han et al. implemented a Q-learning framework that dynamically adjusted sampling rates based on both current battery levels and the rate of environmental parameter change, achieving 43% longer network lifetime compared to fixed-interval sampling approaches [13].

Multi-agent reinforcement learning has further extended these capabilities to collaborative environmental sensing scenarios, where multiple IoT nodes must coordinate their sampling and transmission schedules to maximize coverage while minimizing redundancy [20]. This distributed decision-making approach enables robust adaptation to both spatial and temporal variations in environmental dynamics without requiring constant centralized control [18]. Field deployments in watershed monitoring applications have shown that such collaborative RL approaches can reduce network-wide energy consumption by up to 37% while maintaining equivalent environmental event detection capabilities [13, 19].

Transfer learning techniques have also been applied to accelerate the adaptation of RL policies across different environmental contexts and seasonal conditions [18]. Rather than training models from scratch for each deployment, knowledge transfer from previously optimized networks significantly reduces the learning curve in new environments while preserving domain-specific optimizations [13]. This approach has proven particularly valuable for rapid deployment of environmental monitoring systems in emergency response scenarios such as wildfire monitoring or chemical spill tracking [20].

5.3.2 Fault Detection and Data Validation

To improve system reliability, neural network-based models have been used to detect anomalies in sensor data prior to transmission. For example, an ANN model was able to detect multiple sensor fault types with over 97% accuracy, improving both energy efficiency and data trustworthiness in remote deployments [6]. This reduces retransmission and sensor waste.

Convolutional Neural Networks (CNNs) and Long Short-Term Memory (LSTM) networks have demonstrated particular effectiveness in distinguishing between genuine environmental anomalies and sensor malfunctions by exploiting both spatial and temporal data patterns [6]. These deep learning approaches can identify subtle signatures of sensor drift, calibration errors, and environmental contamination that might otherwise lead to false alarms or missed detection of significant environmental events [18]. Comparative studies have shown that hybrid CNN-LSTM architectures achieve superior fault classification accuracy compared to traditional statistical methods, particularly in noisy environmental settings with multiple interfering variables [6].

Federated learning approaches to fault detection have addressed privacy concerns while enabling collaborative model improvement across distributed environmental monitoring networks [6]. By training local models on device-specific data and aggregating only model updates rather than raw sensor readings, these systems maintain regulatory compliance while benefiting from the diverse fault examples encountered across the network [18]. This approach has proven especially valuable in sensitive environmental applications such as industrial compliance monitoring and drinking water quality assessment [6].

Semi-supervised learning techniques have further reduced the annotation burden for fault detection systems by leveraging large quantities of unlabeled sensor data supplemented with limited expert-verified fault instances [6]. This approach has facilitated the deployment of accurate fault detection in remote environmental monitoring applications where regular physical inspection of sensors is impractical [18]. Researchers have documented how these semi-supervised models continuously improve over time as they encounter new fault patterns, gradually expanding their detection capabilities through operational experience [6].

5.3.3 Predictive Modeling and Anomaly Detection

Predictive systems using machine learning can anticipate sensor behavior and environmental changes, helping reduce redundant measurements and transmitted data. Tatiraju et al. demonstrated improvements of 40% in anomaly detection accuracy and 25% in energy efficiency in real-time WSNs for wildlife and pollution monitoring [1]. These ML-enhanced WSNs can also predict environmental trends and preemptively react to sensor faults.

Ensemble methods combining multiple prediction algorithms have shown superior resilience to the variability inherent in environmental data [1]. These approaches integrate predictions from diverse base models—including gradient boosting machines, random forests, and neural networks—weighted according to their historical performance under similar environmental conditions [18]. Field studies in urban air quality monitoring networks demonstrated that these ensemble approaches reduced false anomaly detections by 53% compared to single-model implementations while maintaining equivalent sensitivity to genuine pollution events [1].

Unsupervised anomaly detection techniques, including autoencoders and isolation forests, have addressed the challenge of identifying novel environmental patterns without requiring extensive labeled training data [1]. These approaches model the normal behavior of environmental parameters and flag significant deviations for transmission, enabling efficient detection of unprecedented events that supervised models might miss [18]. For example, unsupervised anomaly detection deployed in coastal water quality monitoring systems successfully identified previously unknown pollution sources by detecting subtle multivariate parameter deviations that escaped traditional threshold-based monitoring [1].

Time-series forecasting models incorporating domain-specific environmental knowledge have enabled significant reductions in data transmission volume [1]. By transmitting only when measured values deviate significantly from predictions, these systems achieve compression ratios of up to 80% for slowly varying environmental parameters while ensuring rapid notification of unexpected changes [18]. Gaussian process models have proven particularly effective for this purpose, as they provide both predictions and uncertainty estimates that can be used to dynamically adjust transmission thresholds based on confidence levels [1].

Edge computing implementations of these predictive models have overcome latency limitations in traditional cloud-based approaches, enabling real-time anomaly detection even in bandwidth-constrained environments [1, 9]. Model compression techniques, including pruning and quantization, have facilitated the deployment of sophisticated prediction algorithms on resource-constrained IoT devices without sacrificing detection accuracy [1]. Recent advancements in hardware-specific optimization have further reduced the energy footprint of on-device inference, with specialized implementations achieving up to 15× energy efficiency improvements compared to general-purpose model execution [18].

5.4 Sustainable System Architectures and Designs

5.4.1 Energy Harvesting and Autonomous Power Systems

Long-term deployment of IoT monitoring requires sustainable power sources. Several systems integrate solar or RF-based energy harvesting into sensor designs [7, 14, 17]. For instance, a solar-powered system with integrated LoRaWAN enables multi-sensor nodes to function autonomously, eliminating the need for battery replacement in off-grid deployments [4]. Other systems combine optical-electric conversion units to further extend network autonomy [9].

Comprehensive field studies have evaluated the performance of various energy-harvesting technologies across diverse environmental conditions [7]. Solar harvesting demonstrates strong potential in open environments but faces significant challenges in dense forest canopies or frequently overcast regions, where energy yield can decrease by up to 85% [4]. To address these limitations, researchers have developed hybrid harvesting systems that combine multiple energy sources—including solar, wind, thermal gradients, and vibration—to maintain reliable power generation across varying environmental conditions [21]. These multi-modal approaches have shown particular value in coastal and alpine monitoring stations where environmental energy availability fluctuates seasonally [4, 15].

Advanced power management circuits incorporating maximum power point tracking (MPPT) have significantly improved harvesting efficiency in environmental IoT deployments [7]. These systems dynamically adjust harvesting parameters to maximize energy extraction under varying environmental conditions, achieving up to 37% improvement in energy capture compared to static configurations [4]. Additionally, researchers have developed specialized energy storage solutions addressing the unique requirements of environmental monitoring, including wide temperature tolerance ranges and extended cycle life under partial charge conditions [21]. For example, supercapacitor-battery hybrid storage systems have demonstrated superior performance in applications with frequent, small energy-harvesting opportunities, common in vibration or RF-based harvesting scenarios [7, 15].

Novel approaches to RF energy-harvesting leverage ambient radio signals from existing communications infrastructure to power environmental sensors in urban and peri-urban settings [4]. These systems have enabled self-sustaining pollution monitoring networks in metropolitan areas without requiring dedicated power infrastructure or regular maintenance [9]. Researchers have also explored directional energy transfer systems that can remotely power sensors in difficult-to-access environmental monitoring locations, such as forest canopies or underwater habitats [7, 14]. These approaches significantly reduce deployment and maintenance costs while enabling sensing in previously inaccessible environments.

5.4.2 Sensor Deployment and Placement Optimization

One pathway to energy savings is optimizing where and how many sensors are deployed. Ahmad et al. employed QR decomposition techniques to guide energy-efficient sensor placement and gateway configuration, maximizing environmental data collection while minimizing infrastructure overhead [22].

Recent advances in compressive sensing theory have enabled substantial reductions in required sensor density while maintaining environmental monitoring quality [22]. These approaches exploit the inherent spatial and temporal correlation structure of environmental parameters to reconstruct high-resolution data fields from sparse measurements, reducing both equipment costs and energy consumption [7]. Field validations in watershed monitoring applications have demonstrated that optimized sensor placements based on information-theoretic criteria can achieve equivalent detection capabilities with up to 40% fewer nodes compared to uniform grid deployments [22].

Multi-objective optimization algorithms addressing the combined challenges of coverage, connectivity, and energy efficiency have proven particularly valuable for large-scale environmental deployments [22]. These approaches simultaneously consider multiple competing objectives, including sensing coverage, network reliability, energy consumption, and deployment cost [7]. Researchers have developed specialized genetic algorithms and particle swarm optimization techniques tailored to the unique constraints of environmental monitoring contexts, including terrain variability, vegetation interference, and restricted access zones [22].

Mobile sensing platforms, including autonomous drones and robotic surface vessels, have emerged as complementary approaches to static sensor networks, offering adaptive coverage with reduced infrastructure requirements [7]. These systems dynamically adjust their sampling patterns based on environmental conditions and detected anomalies, concentrating measurement resources where they provide maximum information value [22]. Hybrid networks combining fixed sensor infrastructure with mobile sensing elements have demonstrated superior performance in tracking dynamic environmental phenomena such as pollution plumes, algal blooms, and wildlife movements [7, 22].

5.4.3 Adaptive Edge Computing Frameworks

Oliveira et al. and Trihinas et al. independently proposed edge computing frameworks that reduce energy use through adaptive sampling and edge-side prediction [18, 23]. Trihinas's ADMin framework demonstrates over 83% energy savings simply by sending model updates rather than raw monitoring streams. Oliveira's approach combines sampling strategies with local models, yielding over 60% savings on sensor power budgets.

Hierarchical edge computing architectures have further refined these approaches by distributing processing tasks across multiple tiers of computing resources according to their computational intensity and latency requirements [23]. These frameworks allocate simple filtering and aggregation to resource-constrained sensor nodes, intermediate-level processing to field gateways, and complex analytics to cloud infrastructure [24]. Benchmark evaluations across diverse environmental monitoring applications have shown that such hierarchical approaches reduce overall system energy consumption by 55–70% while maintaining or improving response time for critical environmental alerts [23].

Context-aware adaptive computing frameworks dynamically allocate processing resources based on environmental conditions and application requirements [24]. During periods of stable environmental parameters, these systems can reduce sampling rates and processing intensity, entering low-power states to conserve energy [23]. When environmental indicators suggest potential events of interest, the system seamlessly transitions to higher-resolution sampling and more sophisticated edge analytics [24]. Field deployments in wildlife habitat monitoring have demonstrated that such context-aware frameworks can extend system operational lifetime by 2.8× compared to fixed-configuration approaches while maintaining equivalent ecological insight [23].

Federated edge learning approaches enable collaborative model improvement across distributed environmental monitoring networks without requiring centralized data aggregation [24]. By training local models on device-specific data and sharing only model parameters rather than raw measurements, these systems dramatically reduce communication overhead while preserving data privacy and security [23]. Implementation studies in watershed monitoring networks have shown that federated approaches reduce data transmission volume by up to 94% compared to centralized learning while achieving comparable predictive accuracy for environmental parameters [24]. These bandwidth savings translate directly to reduced energy consumption and extended network lifetime [23].

Advanced model compression and acceleration techniques have further expanded the capabilities of edge computing in resource-constrained environmental monitoring systems [24]. Through quantization, pruning, and knowledge distillation, researchers have deployed sophisticated deep learning models on low-power microcontrollers without sacrificing detection accuracy [23]. These optimized models enable complex environmental pattern recognition and anomaly detection directly at the sensing edge, eliminating the need for continuous raw data transmission to more powerful computing resources [24]. Recent implementations in air quality monitoring networks have demonstrated that such optimized edge models can achieve comparable detection performance to cloud-based solutions while reducing energy consumption by over 75% [18, 23].

5.5 Integrated Trade-Off Management: Data, Power, and Communication

5.5.1 Balancing Freshness, Accuracy, and Fidelity

Managing the timeliness ("age of information") and fidelity of sensor data is a key concern in real-time sensing networks. Chen et al. analyzed these trade-offs in selection-based combining transmissions using metrics like MMSE and offered suboptimal sensor node strategies for balancing accuracy with energy efficiency [25].

The age of information (AoI) metric has emerged as a crucial performance indicator in environmental monitoring networks, particularly for time-sensitive applications such as early warning systems and disaster monitoring [25]. Unlike traditional latency metrics that focus solely on transmission delay, AoI captures the staleness of information from the perspective of data consumers, providing a more holistic view of system performance [2]. Researchers have developed analytical frameworks that model the relationship between AoI and environmental parameter dynamics, enabling adaptive sensing strategies that concentrate measurement and transmission resources on rapidly changing variables [25]. Field studies in flood monitoring systems demonstrated that AoI-aware transmission scheduling reduced average information age by 47% compared to periodic reporting while maintaining equivalent energy efficiency [2].

Multi-objective optimization approaches addressing the inherent trade-offs between data freshness, accuracy, and energy efficiency have gained significant traction in environmental IoT research [25]. These frameworks formulate the sensing and transmission scheduling as a constrained optimization problem, with objective functions capturing both the information value of measurements and their resource costs [2]. Pareto-optimal solutions generated by these approaches provide system designers with a spectrum of operating points, allowing application-specific balancing of competing performance metrics [25]. For example, pollution monitoring deployments might prioritize detection accuracy during normal conditions but automatically shift toward timeliness during potential exceedance events [2].

Information-theoretic approaches to sampling rate optimization have further refined these trade-offs by dynamically adjusting measurement frequency based on the information value of collected data [25]. These methods leverage concepts from estimation theory and information theory to quantify the uncertainty reduction achieved by each measurement, enabling resource allocation proportional to information gain [2]. Implementation studies in soil moisture monitoring networks showed that information-theoretic sampling reduced the number of transmitted measurements by up to 76% while maintaining estimation error within predefined tolerance bounds [25]. This substantial reduction in network traffic directly translates to extended battery life and reduced maintenance requirements for remote deployments [2].

5.5.2 Reducing Network Load with AI

Deep learning models were used to detect and filter self-similar sensor data to avoid transmitting redundant or unnecessary updates [20]. Combined with autocorrelation analysis and AI-powered transmission decision-making, these models effectively lower total data traffic and help prolong system life.

Advanced time-series compression algorithms specifically designed for environmental data have achieved remarkable efficiency gains in long-term monitoring applications [20]. These approaches exploit the temporal correlation structure inherent in many environmental parameters to represent measurement sequences with compact models rather than raw data points [2]. For slowly varying parameters like soil temperature or atmospheric pressure, compression ratios exceeding 100:1 have been demonstrated while maintaining reconstruction error below application-specific thresholds [20]. These compression techniques operate across multiple temporal scales, capturing both short-term fluctuations and long-term trends with adaptive resolution [2].

Edge AI approaches that perform local feature extraction and event detection have substantially reduced network load in large-scale environmental deployments [20]. Rather than transmitting raw sensor data for centralized processing, these systems conduct preliminary analysis at the sensing edge, communicating only relevant events or summary statistics [2]. CNN-based acoustic monitoring systems deployed in forest ecosystems, for instance, can identify specific wildlife species locally and transmit only detection events rather than continuous audio streams, reducing data volume by over 99% [20]. This dramatic reduction in transmitted data translates directly to extended network lifetime and increased system scalability [2].

Transfer learning and domain adaptation techniques have addressed the challenge of deploying effective AI models across heterogeneous environmental contexts without requiring extensive local training data [20]. By leveraging pre-trained models and fine-tuning them with limited site-specific measurements, these approaches achieve high accuracy with minimal calibration overhead [2]. Comparative studies across multiple watershed monitoring deployments demonstrated that transfer learning reduced the required calibration period by 73% while maintaining equivalent prediction accuracy for water quality parameters [20]. This accelerated deployment capability is particularly valuable for emergency environmental monitoring scenarios requiring rapid system setup [2].

5.5.3 Secure and Efficient Communication

Despite being critical, secure communication is underreported in environmental IoT systems. Papers that do address this aspect show that strong security measures introduce energy and latency overhead, suggesting a need for AI models that predict when lightweight or federated-security techniques can be safely applied [2, 20].

Risk-adaptive security frameworks tailored to the unique characteristics of environmental monitoring applications have emerged as promising approaches to balance security with resource efficiency [5]. These systems dynamically adjust security measures based on contextual factors including data sensitivity, network conditions, and detected threat indicators [2]. By implementing tiered security profiles, these frameworks can apply robust protection to critical control channels and sensitive environmental data while utilizing lightweight security for routine measurements [5]. Field evaluations in air quality monitoring networks demonstrated that adaptive security approaches reduced average energy consumption by 38% compared to static security configurations while maintaining equivalent protection for sensitive data [2].

Physical layer security techniques exploiting the inherent randomness of wireless channels have shown particular promise for resource-constrained environmental deployments [5]. These approaches leverage channel characteristics as a shared secret between legitimate nodes, enabling secure communication with minimal cryptographic overhead [2]. Implementation studies in forest monitoring systems documented that physical layer authentication reduced security-related energy consumption by up to 62% compared to traditional cryptographic approaches while providing comparable resistance to impersonation attacks [5]. These energy savings are especially valuable in energy-harvesting sensors with limited and variable power availability [2].

Distributed ledger technologies have addressed data provenance and integrity challenges in collaborative environmental monitoring networks [5]. By maintaining tamper-evident records of sensor measurements and processing operations, these systems ensure transparency and accountability throughout the data lifecycle without requiring trusted central authorities [2]. Lightweight blockchain implementations specifically designed for IoT constraints have demonstrated feasibility even on resource-limited sensor platforms, with benchmark evaluations showing acceptable overhead for applications with moderate sampling rates [5]. These approaches are particularly valuable in regulatory compliance monitoring and multi-stakeholder environmental sensing initiatives where data trustworthiness is paramount [2].

Privacy-preserving collaborative sensing frameworks have further refined the security-efficiency balance in environmental monitoring [5]. Through techniques such as differential privacy, secure multi-party computation, and homomorphic encryption, these systems enable valuable aggregate insights while protecting sensitive location-specific measurements [2]. This capability is especially important for applications involving private property monitoring or commercially sensitive environmental data [5]. Recent implementations in urban pollution monitoring networks demonstrated that privacy-preserving aggregation increased stakeholder participation by 47% by addressing confidentiality concerns while adding only 12% communication overhead compared to unprotected data sharing [2].

5.6 Real-World Applications and Case Studies

5.6.1 Air, Water, and Soil Monitoring Systems

Multi-modal studies have deployed sensor frameworks to monitor air, water, and soil pollutants using AI-based anomaly detection and adaptive routing [8, 26]. One system showed how LTE and Wi-Fi integration provided reliable transport for time-sensitive pollutant readings [26]. Miller et al. reviewed trends in AI-IoT systems for monitoring water and climate data and discussed the infrastructure and optimization challenges still unresolved [16].

Large-scale urban air quality monitoring networks have demonstrated the efficacy of multi-tier IoT architectures in capturing fine-grained pollution dynamics [26]. These deployments typically combine stationary high-precision monitoring stations with distributed low-cost sensors, creating complementary data streams that AI algorithms fuse into comprehensive pollution maps [16]. Field evaluations across multiple metropolitan areas have documented how such hybrid approaches improve spatial resolution by up to 8× compared to traditional monitoring networks while maintaining measurement accuracy within regulatory requirements [26]. The integration of meteorological data streams further enhances these systems, enabling source attribution and dispersion modeling that inform targeted pollution mitigation strategies [16].

Watershed monitoring implementations have addressed the complex challenge of tracking multiple water quality parameters across extensive river systems [26]. These deployments leverage energy-harvesting sensors equipped with multi-parameter probes to monitor indicators including dissolved oxygen, conductivity, pH, turbidity, and specific contaminants [16]. Adaptive sampling strategies driven by detected parameter gradients optimize measurement frequency, concentrating resources on areas experiencing rapid quality changes while reducing sampling in stable regions [26]. Real-time alerting capabilities enable prompt response to contamination events, with one implementation documenting a 67% reduction in average detection time for agricultural runoff events compared to traditional monitoring approaches [16].

Soil health monitoring networks integrating subsurface sensor arrays with satellite imagery have provided unprecedented insights into agricultural ecosystems [26]. These systems track moisture profiles, nutrient levels, microbial activity, and carbon sequestration across diverse soil types and management practices [16]. Machine learning models trained on this multi-modal data have achieved 83% accuracy in predicting crop yield impacts from soil parameter variations, enabling precision agriculture interventions that optimize both productivity and sustainability [26]. Long-term deployments have documented how AI-driven sensor management extends system lifetime by up to 3.2 years compared to fixed-configuration approaches while maintaining equivalent measurement quality [16].

Coastal ecosystem monitoring frameworks addressing the complex interactions between terrestrial and marine environments have emerged as critical tools for understanding climate change impacts [26]. These systems integrate water quality sensors, weather stations, tide gauges, and underwater acoustic monitors into unified networks that capture ecosystem dynamics across environmental boundaries [16]. Adaptive routing algorithms ensure reliable data transmission despite challenging coastal conditions, with one implementation maintaining over 99.7% data delivery despite frequent severe weather events [26]. AI-based anomaly detection has proven particularly valuable in these deployments, identifying subtle ecosystem shifts that precede more visible environmental changes [16].

5.6.2 Forests, Agriculture, and Biodiversity Use Cases

From smart agriculture to forest fire detection and wildlife behavior analysis, many current implementations apply AI-enhanced environmental networks to track resource use and ecosystem changes. Kumaran et al. and Ghosh et al. showed that smart sensing strategies yield substantial energy savings and sensor longevity enhancements in field monitoring systems [3, 7]. Rajeshwarrao et al., through case evaluations, linked improved decision-making outcomes to AI's role in real-time system adaptation [2].

Forest fire early detection systems have demonstrated how multi-modal sensing combined with edge AI can provide critical advance warning of developing wildfire threats [7]. These deployments typically integrate temperature sensors, infrared cameras, smoke detectors, and meteorological monitoring into mesh networks spanning vulnerable forest regions [14]. On-device machine learning models analyze this sensor fusion data to distinguish fire signatures from benign environmental variations, achieving detection accuracy exceeding 94% with false positive rates below

0.3% [7]. Field evaluations have documented average detection times of 4–7 min from fire ignition, providing crucial early intervention opportunities that conventional detection methods cannot match [2]. Energy optimization through adaptive duty cycling extends system lifetime to over three years on compact solar harvesting units, enabling coverage of remote wilderness areas without maintenance visits [14].

Precision agriculture implementations have showcased how IoT networks can simultaneously improve agricultural productivity and resource efficiency [7]. These systems combine soil moisture sensors, weather stations, plant phenology monitors, and irrigation control systems into integrated management frameworks [14]. AI-driven prediction models generate irrigation recommendations that reduce water consumption by 30–47% compared to conventional scheduling while maintaining or improving crop yields [7]. Long-term deployments across diverse agro-ecological zones have demonstrated how these systems adapt to regional growing conditions through reinforcement learning approaches that progressively refine intervention strategies based on observed crop responses [2]. The economic benefits documented in these case studies, including reduced input costs and increased yields, have driven rapid adoption across both large-scale commercial operations and smallholder farming contexts [14].

Wildlife monitoring networks have transformed ecological research by providing continuous, non-invasive observation capabilities across extensive habitats [7]. These deployments typically combine acoustic sensors, camera traps, RFID readers, and environmental monitors into low-power networks spanning target ecosystems [14]. Edge-based species recognition algorithms process sensor data locally, transmitting only relevant detection events rather than raw data streams and achieving network traffic reductions exceeding 98% [7]. Long-term deployments have documented previously unobservable behavioral patterns, including nocturnal movement corridors, interspecies interactions, and seasonal migration timing shifts potentially linked to climate change [2]. Adaptive power management strategies enable these systems to operate continuously for up to five years on compact energy-harvesting units, providing unprecedented temporal continuity in ecological observation [14].

Greenhouse and controlled agriculture environments have served as testbeds for highly optimized IoT implementations that maximize resource efficiency [7]. These controlled settings enable precise evaluation of sensing and optimization strategies before deployment in more challenging field environments [14]. Case studies have documented how microclimate monitoring combined with machine learning control systems reduces energy consumption by 23–41% while improving crop yields through optimized growing conditions [7]. These systems typically achieve return on investment within 12–18 months through reduced resource inputs and increased production value, driving commercial adoption across diverse agricultural sectors [2]. The controlled nature of these environments also facilitates rapid iteration of sensing strategies and optimization algorithms, accelerating innovation cycles compared to open-field deployments [14].

Marine and coastal ecosystem monitoring has extended IoT environmental sensing into challenging aquatic environments [7]. Floating sensor platforms equipped with water quality probes, weather stations, and underwater acoustic monitors track ecosystem parameters across the land-sea interface [14]. Specialized low-power acoustic modems enable reliable data transmission in underwater environments where conventional radio communication is ineffective [7]. These systems have documented critical ecosystem dynamics including harmful algal bloom development, coral bleaching events, and fish population movements in response to environmental changes [2]. Energy harvesting from solar, wave, and current sources enables autonomous operation for extended periods in remote marine environments, with one implementation achieving continuous operation for over two years without maintenance visits [14].

5.7 Research Gaps and Future Opportunities

Despite significant advancements in system and protocol optimization, several critical research gaps remain. First, most studies treat sustainability as short-term energy efficiency and do not model lifecycle environmental impacts (e.g., sensor material footprints, hardware recycling) in a comprehensive way [16, 25]. Second, while AI is widely used for network optimization, integration with state-aware dynamic protocol switching is not common, representing an opportunity to enhance adaptability in mixed-environment deployments [18]. Finally, security-energy trade-offs receive only limited attention, even though secure but lightweight protocols are essential for high-trust applications [2, 20, 23].

Lifecycle assessment methodologies tailored to environmental IoT systems represent a significant research opportunity [19]. While current literature extensively addresses operational energy efficiency, the embedded energy and material impacts of sensor manufacturing, deployment, and end-of-life management receive minimal attention [16]. This gap becomes increasingly important as environmental monitoring scales globally, with millions of sensors potentially deployed across diverse ecosystems [19]. Preliminary analyses suggest that manufacturing impacts may dominate lifecycle environmental footprints for certain sensor types, particularly those containing rare earth elements or specialized semiconductors [16]. A comprehensive framework integrating operational optimization with lifecycle considerations would enable truly sustainable system design that minimizes both immediate and long-term environmental impacts [19].

Cross-layer optimization approaches that jointly consider physical, MAC, network, and application layer parameters remain underdeveloped despite their potential for significant efficiency gains [18]. Current research typically addresses optimization at individual protocol layers, missing opportunities for synergistic improvements through coordinated parameter tuning [15]. The integration of reinforcement learning frameworks capable of simultaneously optimizing parameters across multiple protocol layers could yield substantial performance improvements beyond what is achievable through isolated optimization [18]. Field studies suggest that such cross-layer approaches might improve overall system efficiency by 25–40% compared to layer-specific optimization strategies, particularly in dynamic environmental conditions [15].

Adaptive protocol switching based on environmental context and application requirements represents another promising research direction [18]. While fixed protocol selection is common in current implementations, heterogeneous environmental conditions often demand different communication strategies as context changes [15]. Intelligent systems capable of seamlessly transitioning between protocols based on environmental conditions, energy availability, and data priority could significantly enhance both reliability and efficiency [18]. The integration of predictive environmental models with protocol selection algorithms could enable proactive adaptation to anticipated condition changes rather than reactive responses to established changes [15].

Resilience metrics and optimization frameworks addressing long-term system sustainability under environmental stressors require further development [16]. Current research predominantly evaluates performance under normal operating conditions, with limited attention to extreme events such as floods, wildfires, or severe storms that may increasingly impact environmental monitoring systems [19]. Comprehensive resilience modeling that considers both gradual environmental changes and acute disruptions would enable more robust system design for long-term deployment in changing climates [16]. This becomes particularly critical as environmental monitoring systems are increasingly deployed to track climate change impacts, creating a dependency relationship where monitoring reliability is most crucial precisely when systems face their greatest environmental challenges [19].

Security and privacy considerations for environmental data present unique challenges that remain insufficiently addressed [5]. While substantial research exists on general IoT security, the specific requirements of environmental monitoring—including unattended deployment in accessible locations, multi-stakeholder data sharing, and regulatory compliance—create distinct security demands [2]. Lightweight authentication and encryption methods specifically optimized for environmental sensing contexts could address the energy limitations of remote deployments while maintaining appropriate security levels [25]. Additionally, privacy-preserving monitoring approaches that enable valuable environmental insights while protecting sensitive location-specific information would facilitate broader adoption across diverse contexts [5].

Semantic interoperability across heterogeneous environmental monitoring systems represents a significant challenge as deployments scale and diversify [16]. Current implementations often employ proprietary data models and interfaces, limiting integration potential and creating information silos [19]. Standardized ontologies and semantic frameworks specifically designed for environmental parameters would enable seamless data exchange and integration across independently developed monitoring systems [16]. This interoperability becomes increasingly critical as environmental challenges demand coordinated monitoring across jurisdictional and organizational boundaries [19].

5.8 Conclusion

Emerging architectures for environmental IoT monitoring are increasingly incorporating AI to achieve energy-efficient, scalable, and high-fidelity sensing across diverse contexts. Communication protocol optimization—including hybrid strategies and adaptive edge routing—plays a vital role, as does machine learning in managing anomaly detection, prediction, and update timing. Techniques such as ANN-based fault detection [6], energy-aware smart sensing [7], and adaptive state update control models [9, 13] are enabling intelligent systems that dynamically balance energy usage, latency, and data quality.

Systems are moving toward greater autonomy powered by renewables [7, 14, 17], longer-term operational sustainability via intelligent placement [22], and tighter lifecycle integration through adaptive predictive modeling [1, 5, 8]. Nonetheless, key needs persist in modeling full ecological impacts, integrating security-aware optimization, and creating robust multi-objective frameworks that jointly prioritize technical performance and environmental goals. As sensor networks scale globally, these challenges must be addressed to ensure that IoT systems not only monitor the planet effectively—but do so sustainably and intelligently.

The convergence of environmental sensing and artificial intelligence represents a transformative development in humanity's capacity to understand and respond to ecological changes [1]. By enabling continuous, high-resolution monitoring across previously inaccessible environments, these systems provide unprecedented insights into natural processes and anthropogenic impacts [16]. The integration of edge intelligence with distributed sensing further enhances these capabilities, enabling responsive, adaptive monitoring that concentrates resources where they provide maximum information value [17]. This observational capacity has significant implications for environmental science, resource management, policy development, and conservation efforts [1].

Energy-autonomous sensing systems represent a significant advancement toward truly sustainable environmental monitoring [7]. By eliminating battery replacement requirements through integrated energy harvesting, these systems dramatically reduce maintenance requirements and operational costs while enabling deployment in remote or inaccessible environments [21]. The continued refinement of low-power sensing technologies, energy-efficient protocols, and harvesting capabilities promises to further extend operational lifetimes and deployment contexts [4]. As these systems mature, the vision of permanent, maintenance-free environmental monitoring networks becomes increasingly achievable, enabling continuous ecological observation over timescales relevant to long-term environmental processes [7].

Adaptive edge computing frameworks have fundamentally transformed the relationship between sensing resolution and energy efficiency [23]. Traditional approaches typically faced direct trade-offs between sampling frequency and system lifetime, forcing compromises in either temporal resolution or operational duration [24]. Contemporary systems with integrated edge intelligence can dynamically adjust sampling strategies based on environmental conditions and information value, concentrating measurement resources where they provide maximum insight while conserving energy during stable periods [23]. This adaptive approach has effectively decoupled sampling resolution from energy consumption, enabling both high-resolution monitoring of significant events and extended system lifetime [24].

Integration of environmental monitoring systems with broader information ecosystems represents an emerging frontier with substantial potential impact [16]. By connecting sensor networks with satellite observations, numerical models, historical datasets, and human observations, researchers can develop increasingly comprehensive environmental understanding spanning multiple scales and dimensions [17]. Machine learning approaches that integrate these diverse data streams enable insights that would be unachievable through any single observation method, revealing complex patterns and relationships across environmental systems [1]. This integrative approach proves particularly valuable for understanding cross-domain environmental challenges such as climate change, where impacts manifest across atmospheric, terrestrial, and aquatic systems with complex interconnections [16].

As environmental IoT systems continue to evolve, balancing technological innovation with ecological responsibility becomes increasingly critical [19]. The paradox of deploying electronic systems to monitor environmental health requires careful consideration of the monitoring systems' own environmental impacts [16]. Future developments must prioritize not only operational efficiency but also sustainable design principles throughout the technology lifecycle—from material selection and manufacturing processes to deployment strategies and end-of-life management [19]. By embracing this holistic sustainability perspective, environmental monitoring can truly fulfill its promise: providing the insights needed to protect and restore ecological systems while minimizing its own environmental footprint [16]. As these systems scale from experimental deployments to global monitoring infrastructure, the research community has both the opportunity and responsibility to establish practices that ensure environmental IoT becomes a model of sustainable technology development [19].

References

1. Lanzolla A, Spadavecchia M (2021) Wireless sensor networks for environmental monitoring. Sensors 21(4):1172. https://doi.org/10.3390/s21041172
2. Arabelli R et al (2024) IoT-enabled environmental monitoring system using AI. In: 2024 International conference on advances in computing, communication and applied informatics (ACCAI), pp 1–6. https://doi.org/10.1109/ACCAI61061.2024.10602131
3. Park C (2021) A solar-powered sensor node with energy harvesting technology for IoT environmental monitoring applications. IEEE Sens J 21(16):17927–17934. https://doi.org/10.1109/JSEN.2021.3085205

4. Wang Y, Huang Y, Song C (2019) A new smart sensing system using LoRaWAN for environmental monitoring. In: 2019 Computing, communications and IoT applications (ComComAp), pp 347–351. https://doi.org/10.1109/ComComAp46287.2019.9018829
5. Popescu SM et al (2024) Artificial intelligence and IoT-driven technologies for environmental pollution monitoring and management. Front Environ Sci. https://doi.org/10.3389/fenvs.2024.1336088
6. Kanwal T et al (2024) Energy-efficient Internet of Things-based wireless sensor network for autonomous data validation for environmental monitoring. Comput Syst Sci Eng. https://doi.org/10.32604/csse.2024.056535
7. Ghosh S et al (2023) Energy aware smart sensing and implementation in green air pollution monitoring system. In: 2023 IEEE international conference on communications (ICC), pp 2153–2158. https://doi.org/10.1109/ICC45041.2023.10279138
8. Mishra M, Reddy S (2020) E-SENSE: Design & development of an IoT-based environment monitoring system. In: 2020 IEEE international conference on computing, power and communication technologies (GUCON), pp 368–372. https://doi.org/10.1109/GUCON48875.2020.9231141
9. Zhang X et al (2019) A low-power wide-area network information monitoring system by combining NB-IoT and LoRa. IEEE Internet Things J 6:590–598. https://doi.org/10.1109/JIOT.2018.2847702
10. Raza U, Kulkarni P, Sooriyabandara M (2017) Low power wide area networks: an overview. IEEE Commun Surv Tutor 19(2):855–873. https://doi.org/10.1109/COMST.2017.2652320
11. Al-Fuqaha A, Guizani M, Mohammadi M, Aledhari M, Ayyash M (2015) Internet of Things: a survey on enabling technologies, protocols, and applications. IEEE Commun Surv Tutor 17(4):2347–2376. https://doi.org/10.1109/COMST.2015.2444095
12. Xu LD, He W, Li S (2014) Internet of Things in industries: a survey. IEEE Trans Industr Inf 10(4):2233–2243. https://doi.org/10.1109/TII.2014.2300753
13. Han Z, Gong J (2022) Status update control based on reinforcement learning in energy harvesting sensor networks. Front Commun Netw 3:933047. https://doi.org/10.3389/frcmn.2022.933047
14. Kumaran S et al (2023) An intelligent framework for wireless sensor networks in environmental monitoring. In: 2023 2nd International conference on automation, computing and renewable systems (ICACRS), pp 334–340. https://doi.org/10.1109/ICACRS58579.2023.10404602
15. Alsheikh MA, Lin S, Niyato D, Tan HP (2014) Machine learning in wireless sensor networks: algorithms, strategies, and applications. IEEE Commun Surv Tutor 16(4):1996–2018. https://doi.org/10.1109/COMST.2014.2320099
16. Miller T et al (2025) Integrating artificial intelligence agents with the Internet of Things for enhanced environmental monitoring: applications in water quality and climate data. Electronics 14(4). https://doi.org/10.3390/electronics14040696
17. Karim F, Karim S, Frihida M (2022) Machine learning-based IoT for providing an intrusion detection system for edge devices. IEEE Access 10:76987–77000. https://doi.org/10.1109/ACCESS.2022.3192721
18. Panda SK et al (2024) Machine learning-driven strategies for improving energy efficiency in IoT communication protocols. In: 2024 5th International conference on data intelligence and cognitive informatics (ICDICI), pp 233–238. https://doi.org/10.1109/ICDICI62993.2024.10810889
19. Lin J, Yu W, Zhang N, Yang X, Zhang H, Zhao W (2017) A survey on Internet of Things: architecture, enabling technologies, security and privacy, and applications. IEEE Internet Things J 4(5):1125–1142. https://doi.org/10.1109/JIOT.2017.2683200
20. Al-Hawawreh M, Elgendi I, Munasinghe K (2022) An online model to minimize energy consumption of IoT sensors in smart cities. IEEE Sens J 22:19524–19532. https://doi.org/10.1109/JSEN.2022.3199590
21. Khan WA et al (2020) Smart IoT communication: circuits and systems. In: 2020 International conference on COMmunication systems & NETworkS (COMSNETS), pp 699–701. https://doi.org/10.1109/COMSNETS48256.2020.9027430

22. Ahmad S et al (2023) Sustainable environmental monitoring via energy and information efficient multinode placement. IEEE Internet Things J 10:22065–22079. https://doi.org/10.1109/JIOT.2023.3303124
23. Oliveira E et al (2021) A real-time and energy-aware framework for data stream processing in the Internet of Things. In: Proceedings of the 6th international conference on internet of things, big data and security (IoTBDS 2021), pp 17–28. https://doi.org/10.5220/0010370100170028
24. Trihinas D, Pallis G, Dikaiakos M (2017) ADMin: adaptive monitoring dissemination for the Internet of Things. In: IEEE INFOCOM 2017—IEEE conference on computer communications, pp 1–9. https://doi.org/10.1109/INFOCOM.2017.8057144
25. Chen Z et al (2022) Joint optimization of data freshness and fidelity for selection combining-based transmissions. Entropy 24(2). https://doi.org/10.3390/e24020200
26. Pamula A et al (2022) Applications of the Internet of Things (IoT) in real-time monitoring of contaminants in the air, water, and soil. In: The 9th international electronic conference on sensors and applications. https://doi.org/10.3390/ecsa-9-13335

Part III
Smart Cities and Decision Frameworks

Chapter 6
Review of IoT-Powered Smart City for a Smarter and Sustainable Future

G. Jegadeeswari, C. Jayabalan, S. Nagakishore Bhavanam, Tasan Cruz Dany, Nidal Al Said, and B. Kirubadurai

Abstract The implementation of smart city concepts aims to extend the longevity of urban ecosystems through digital transformation, targeting infrastructure, public services, and overall city governance. This initiative primarily seeks to enhance efficiency, drive innovation, and improve the quality of life for citizens. However, its impact extends beyond the current population, aiming to ensure sustainability and improved living standards for future generations as well. Developing urban areas requires a broader perspective, one that embraces sustainability as a fundamental principle rather than treating it as an afterthought. The evolution of smart cities should transcend basic technological upgrades and efficiency gains, addressing the complex interdependencies between institutional, environmental, social, and economic dimensions of urban life. This chapter adopts a holistic framework that leverages digital technologies to foster social equity, economic resilience, environmental stewardship, and effective governance, all while enhancing urban functionality. It explores how technological advancements shape smart city ecosystems and contribute to sustainable urban development. By applying established sustainability metrics, the chapter evaluates the effectiveness of smart city strategies in building inclusive, equitable, and future-ready urban environments. It also highlights best

---

G. Jegadeeswari
Department of Electrical and Electronics Engineering, Saveetha Engineering College, Chennai, India

C. Jayabalan (✉)
Department of Mechanical Engineering, AMET University, Chennai, India
e-mail: jayabalan.mech@ametuniv.ac.in

S. N. Bhavanam
Department of Computer Science and Engineering, Mangalayatan University, Jabalpur, MP, India

T. C. Dany
Universidad Nacional de Chimborazo, Avenida Antonio José de Sucre, Riobamba, Ecuador

N. Al Said
College of Mass Communication, Ajman University, Ajman, UAE

B. Kirubadurai
Department of Aeronautical Engg, VelTech Rangarajan Dr. Sagunthala R&D Institute of Science and Technology, Chennai, India

S. Gautam et al. (eds.), *Smart Environmental Monitoring*,
https://doi.org/10.1007/978-981-92-0294-2_6

practices, identifies key challenges, and proposes future directions to guide smart cities toward becoming resilient, human-centric, and innovation-driven communities.

Keywords Digital technologies · Smart city · Environmental sustainability · Internet of Things · Sustainable development · Urban transformation

6.1 Introduction

According to UN forecasts [1], it is anticipated that by 2050, roughly 70% the global population will reside in urban areas and this shift in demographics is likely to exacerbate existing socioeconomic challenges. Rapid urban expansion, environmental decline, traffic jams, aging infrastructure, and rising demands for sustainable solutions and renewable energy are creating heightened pressures on city [2]. In response, urban areas worldwide are adopting cutting-edge smart technologies. These advancements aim to not only improve sustainability and energy efficiency but also promote greener city landscapes, ultimately enhancing the quality of life for residents [1].

The idea of a smart city originated in USA in 1992, following the release of "Technopolis" a significant publication that highlighted the increasing role of information technology in urban development during the 1990s. In contemporary usage, "smart city" describes the fusion of advanced digital technologies with urban infrastructure to tackle intricate metropolitan issues. Beyond merely improving operational efficiency, smart city projects seek to drive economic growth, improve residents' quality of life, and enhance public safety objectives that became critically important during worldwide emergencies like the COVID-19 pandemic in 2020 [3]. However, the rise of smart city initiatives has ignited substantial discussions. Issues raised include threats to individual privacy, potential social exclusion, hesitance among particular demographic groups to adopt new technologies, and the unequal distribution of advantages among various segments of the population.

There is an increasing agreement that effectively building a smart city framework necessitates the smooth blending of technological advancements with social change. This viewpoint underscores the need to enhance not just digital infrastructure but also to foster an urban green economy one that focuses on lowering emissions, reducing energy usage, and ensuring fair regional growth. The main goal is to shift from traditional smart city to a more comprehensive model of sustainable smart city.

City worldwide are progressively channeling resources into smart city projects using various methods, highlighting the importance of executing these initiatives in a thorough and sustainable manner. This chapter aims to explore how digital technologies within smart city impact the institutional, social, environmental, and economic aspects of sustainable development. The analysis will reference important indicators and frameworks laid out by leading global smart city standards [4]. In conclusion, this chapter hopes to encourage urban development that is sustainable, inclusive, and strategically designed.

The purpose of this chapter is to explore how the Internet of Things (IoT) can be harnessed to build smarter, more sustainable urban environments. It aims to provide a comprehensive understanding of the role IoT plays in enhancing infrastructure, optimizing public services, and enabling real-time governance in smart cities. The chapter focuses on the integration of IoT technologies to address the challenges of rapid urbanization, resource management, and environmental sustainability.

6.2 Indices of Smart City Performance

Numerous smart city indexes evaluate different aspects of urban development on a global scale.

6.2.1 The City in Motion Index

The City in Motion Index (CIMI), which is released by the IESE Business School's Strategy Department and the Center for Globalization and Strategy, is among the most well-known. This platform links a global network of business and city specialists with the goal of fostering local change and creating cutting-edge concepts and resources to make city smarter and more sustainable. The most recent assessment, published in 2022, used the following criteria [17] to assess 183 cities across 9 dimensions:

Human Resources

Developing a city's human capital should be its top priority. A well-run city must successfully draw in and retain talented people, encourage educational improvements, and foster innovation and research. There is widespread recognition that access to education and culture are fundamental components of human capital, even though certain qualitative aspects may include precise measurement.

The United Nations Development Program (UNDP) underscores this perspective annually through the publication of the Human Development Index (HDI), which emphasizes the critical role of education and cultural engagement in fostering human capital. In alignment with this view, the CIMI 2023 report assessed human capital using a diverse set of indicators. These included the number of public and private educational institutions, the share of the population with higher education qualifications, the presence of internationally ranked universities, investments in leisure and cultural activities, annual education budgets, patterns of student mobility, and the availability of cultural infrastructure such as theaters, museums, and art galleries.

Social Unity

As the degree of agreement among members of a social group or the feeling of belonging to a common endeavor or situation, social cohesiveness is a crucial sociological feature of city. It measures the depth of interaction between members of a group. Social cohesion in urban contexts refers to the level of concord between residents of the same city who have different ages, incomes, occupations and cultures.

When analyzing a city's social environment, one must consider factors such as immigration trends, community growth, senior citizen support, healthcare system effectiveness, and the level of security and inclusion of its citizens [5]. The fairness and operation of health systems have been closely scrutinized throughout the COVID-19 epidemic, in the hopes that the crisis has highlighted the importance of robust health systems and spurred efforts to improve them.

The Economy

Cluster development, industrial change plans, local economic efforts, the encouragement of entrepreneurial, and innovation endeavors are all components that contribute to a region's economic prosperity. Real GDP is a vital measure of a city's economic health and the income levels of its citizens when evaluating the long-term sustainability of large international city and the living standards of their populations. In fact, in many studies, GDP is the primary, if not the only, metric used to assess a city's or nation's economic performance.

Leadership

The efficacy, caliber, and proper course of government activity are all aspects of governance. Given that citizens are essential to solving urban problems, it is crucial to consider factors such as the level of citizen participation and the authorities' ability to establish connections with local community members and company executives.

Effectiveness in E-Government and Administration

To improve public administration, e-government projects must be implemented. These projects seek to increase administrative efficiency through the use of creative organizational structures and management techniques. In this context, the non-government had a great option to enhance the efficiency and operational cost in an effective way without comprising the public service.

Sustainability of the Environment in Urban Development

The primary goal of sustainable development is to promote an emission-free nation. To achieve these goals, the United Nations has established 17 SDGs to guide our nation's next steps. Implementing pollution control measures, advancing the adoption of renewable energy and sustainable construction practices, optimizing waste and water management systems, and enacting climate-responsive legislation are all essential strategies for fostering environmental sustainability. These initiatives play a pivotal role in safeguarding the long-term resilience, ecological balance, and overall livability of urban environments.

Transportation and Mobility

Future cities must address two key challenges in transportation and mobility: ensuring equitable access to public services and enabling the efficient movement of people across expansive urban areas. The quality and reliability of a city's transportation infrastructure, including road networks, public transit systems, vehicle fleets, and air travel, play a vital role in shaping residents' quality of life and are foundational to a city's long-term sustainability.

Furthermore, the functional demands of the broader economic system such as workforce mobility and the timely delivery of goods and services must be considered, as these factors are directly linked to the urban environment's overall operational efficiency and sustainability.

Urban Design

Urban planning is increasingly seen as a vital instrument for boosting economic development and reducing poverty. In the current landscape, it requires a cooperative approach that includes various stakeholders such as academia, international organizations, the private and public sectors, local communities, and civil society.

Urban design is fundamentally linked to sustainability, improving urban livability hinges on the careful application of master plans, the incorporation of green spaces, and a dedication to responsible and strategically guided development. To create resilient, inclusive, and sustainable urban areas, contemporary planning must emphasize the creation of compact city that are supported by effective transportation networks and accessible public services.

Worldwide Prediction

In an increasingly interconnected world, city need to establish themselves as key players on the global stage to secure long-term success. This necessitates the implementation of well-planned tourism strategies and effective branding efforts focused on drawing in foreign investment, boosting international visibility, and improving competitiveness on a global scale.

Although city within the same country can differ greatly in their global standing, a nation's level of openness and its international interactions significantly influence how its city are perceived worldwide. The primary aim is to enhance the international presence and attractiveness of city, allowing them to flourish as vibrant, competitive, and culturally rich hubs in the global economy.

Technology

According to the observatory itself, this approach prioritizes human viewpoints and social considerations while combining economic. According to the observatory itself, this approach prioritizes human viewpoints and social considerations while combining economic.

This index evaluates and ranks 102 cities worldwide based on factors like transportation, leisure activities, economic prospects, health and safety, and the condition of urban infrastructure. It also looks at the benefits and technology services available to locals. London is the top city, followed by New York, Paris, Tokyo, and Berlin, according to the CIMI 2023 ranking.

6.2.2 E-Government Development Index

Estonia, Denmark, South Korea, Finland, Sweden, Australia, New Zealand, the Netherlands, Iceland, and USA are the Members States' which progress in electronic governance, is gaged by the Development Index [6]. This index evaluates and ranks 103 cities worldwide based on factors like transportation, leisure activities, economic prospects, health and safety, and the condition of urban infrastructure. It also looks at the benefits and technology services available to locals. According to the observatory itself, this approach prioritizes human viewpoints and social considerations while combining economic telecommunication infrastructure, and the provision of online public services [18].

Index of Online Services

This online index evaluates and ranks 102 cities worldwide based on factors such as transportation, leisure activities, economic prospects, health and safety, and the condition of urban infrastructure. In collaboration with the Singapore University of Technology and Design, the Smart City Observatory at the IMD Global Competitiveness Center develops the Smart City Index [7]. Together, these areas represent the general standard of living and the technological infrastructure supporting a smart city's operations. Five weighted sub-indices service delivery, technology, the institutional framework enabling e-government, content delivery, and e-participation were used to construct the OSI in the 2023 report.

Infrastructure for Telecommunication

This component highlights the technological underpinnings of e-government services by reviewing the prevalence of home Internet connections, mobile phones, and personal Internet usage in various nations.

Human Resources

Since literacy rates and the average number of years spent in school are important indicators of a nation's ability to use e-government efficiently, this component collects data on the population's level of education. Estonia, Denmark, South Korea, Finland, Sweden, Australia, New Zealand, the Netherlands, Iceland, and USA are among the top 10 nations with the highest EGDI scores, per the 2023 poll.

6.2.3 Index of Smart City IMD

In collaboration with the Singapore University of Technology and Design, the Smart City Observatory at the IMD Global Competitiveness Center develops the Smart City Index. This index evaluates and ranks 102 cities worldwide based on factors like transportation, leisure activities, economic prospects, health and safety, and the condition of urban infrastructure. It also looks at the benefits and technology services available to locals. Two key pillars based on inhabitants' views are essential to the index: it gages how people feel about their city's efforts to become smarter. According to the observatory itself, this approach prioritizes human viewpoints and social considerations while combining economic and technology aspects [8, 19].

Perceptions

Responses to three main privacy-related questions are measured and compared in this section: if people are willing to share personal information, whether they feel comfortable using facial recognition technology, and whether the availability of Internet information has increased their trust in local government.

Technologies and Structures

This section collects key information from a study that covers five important areas: opportunity, mobility, health and safety, leisure activities, and governance. Together, these areas represent the general standard of living and the technological infrastructure supporting a smart city's operations.

Being recognized globally as a smart city is essential for attracting talent and investment, thereby promoting a growth cycle, according to the Smart City Index. Helsinki, Oslo, Singapore, Lausanne, Copenhagen, Taipei City, Zurich, Geneva, Bilbao, and Auckland were among the top 10 smart city in 2021, according to the data. These cities were notable for numerous reasons, such as their effective recycling programs, plentiful green spaces, medical facilities, good public transit, and low-income sanitary conditions in the neighborhoods [9]. Knowingly, the improvements in the online medical and management studies about the appointment and the usage of technology for ticket sales helped to achieve the best solutions.

6.3 The Smart City Strategy Index

The German consultant firm from Roland Berger have released the Smart City Strategy Index, which evaluating the urban centers in terms of administration and success by means of the main technology like digitalization and its applications that promotes the smart city. This method is used to evaluate the present scenario of the city and rethink its technology advancement and future directions. This method is very efficient while maximizing the projects and current plans for better development. A total of 154 cities have formally implemented the way of smart city strategies. The top-ranked city such as St. Albert, Vienna, and London [20].

Organizing

Nowadays, the major motto of implementing Industry 4.0 is to transform conventional machinery into advanced technology with optimization capability and self-learning process, and this transformation is also enhancing the sustainability in urban environment. These cities had notable for a numerous way of reasons, such as their effective recycled programs, plentifully green spaces, medical facility, good public transits, and low-income sanitary conditions in the neighborhoods.

Policy and Infrastructure

This measurement scrutinizes the initial authorized and political atmosphere essential for the positive execution of smart city ingenuities. It includes the examination of supervisory agendas, innovation subsidy machines, and protected data authority practices. This dimension assesses important topics like:

- Energy and Environment: Waste, water, and energy management.
- Mobility: Managing urban traffic and logistics.
- Education: The ability of learning models, educational platforms, and the development of digital abilities.
- Health: Telemedicine, aided medicine, and information systems.
- Government: digital public administration and e-services.

Seven Crucial Elements for Creating the Ideal Smart City

Take Note

This element is very important for implementing the smart city in an effective way. This method is used to evaluate the present scenario of the city and rethink its technology advancement and future directions; this method is very efficient while maximizing the projects and current plans for the better development [10]. It provides a transformation in a digital era and replaces the needs of the urban demands.

Involve All Parties

It is very important to participate all the stack holder such as government and non-government organization and NGO for better creation of smart city. It improved the capital cost of the city and increased the wealth of each city in an efficient manner and also increased its efficiency and reduced its operational cost, and guarantee to the groups in terms of both execution and smooth planning.

Examine Integrated Methods

To meet the main objectives of the smart city, well-suited and coherent digital transformation gives the overall goals in a necessary support to attain the goals. Digital transformation is improving the total cost of the city and also from various sectors, especially to upgrade the nativities taken by the various sector.

Get Tailored Support

Getting support from various agencies and government sector for the better smart city especially in urban areas provides the capital cost and proper knowledge to address the difficulty faced in the urban areas. For long lasting and scalable solutions, the public organization and private sector are very crucial.

Promote Originality

Company may test the nodality in their ideas for the better replacement of smart city by without comprising the design, communities, and also creativity. It is very much essential to set a target of innovation ideas to promote the solution to urban issues for both technical and financial assistance.

Establish (Open) Urban Data Platforms

Data is the only way to open up the urban smart city in a safe mode. Digital technology is works only when the real-time data meets up with the essential facilitators. Cybersecurity plays a vital role in protective the data from regulations and data governance, and data structures.

Make Detailed Action Plans

To test the efficacy and guarantee of the action plan, it should be aligned with financial possibilities and goals. So that each city can be evaluated and maintain in a proper way. However, the smart city strategy index declares the goals that are very crucial for smart city to hope up with the global leaders. They deliver as a road map to the transformation of digital era.

6.4 Complete Sustainability and Smart City

Digital Technologies for Industry 4.0

Digital technology gives us the best cutting-edge technology by integrating automation and human-centric processes. The technology like Industry 4.0 gives an enormous improvement in workflows and increasing the resources in an effective way. Many important sectors like business and organization in the world are keenly looking into these technologies to increase the revenue of their sector, growth, innovation, and competitiveness [11]. In case of Industry 4.0, these nine key elements play a vital role in today's advancement of technology such as data science, robotics, AI, cloud computing, AV/VR, cybersecurity, system integration, IoT, and manufacturing.

The only way to increase the production of various valuable goods and services is through digital developments, which give improved resource management, greater processes, and heavy operational efficiency, which are raised from the industry sector, resulting in better productivity. These advantages talk about the significance of these digital transformation which leads to better smart city and sectors and leads to reduce the issues which raises in urban areas.

Essential Sustainability

According to an inclusive cycle of innovation, sustainable development is a vital role in the effective management of services especially for smart city. Nowadays, the major motto of implementing Industry 4.0 is to transform conventional machinery into advanced technology with optimization capability and self- learning process, and this transformation is also enhancing the sustainability in urban environment. The World Economic Forum's 2026 emphasized how necessary is to integrate with human involvement and technical expertise to achieve essential environmental, social, and economic output. Nevertheless, difference in access to these technological innovations implies in various countries and regions, resulting in challenges to build smart city and achieving global sustainable development goal in an effective way.

Moreover, the outcome of smart technology wells in alien with the global united nation 2027 for sustainable development goals.

- SDG 3: Ensure healthy lifestyles and promote well-being for all people at all ages is one of these objectives.
- SDG 4: Provide high-quality, inclusive, and equitable education and encourage opportunities for lifelong learning for everyone.
- SDG 6: Make ensuring that everyone has access to and can sustainably manage water and sanitation.
- SDG 7: Make sure everyone has access to modern, affordable, dependable, and sustainable energy.
- SDG 9: Encourage innovation, encourage inclusive and sustainable industrialization, and construct resilient infrastructure.
- SDG 11: Create inclusive, secure, resilient, and sustainable city and human settlements.
- SDG 13: Address climate change with immediate action.
- SDG 16: Establish inclusive, accountable, and efficient institutions at all levels.

For this reason, a holistic approach to sustainable development must focus on four key areas.

Environmental: This component highlights the significance of developing environmentally friendly products and practices. Environmental preservation and prudent use of natural resources are given top attention. This addresses every facet of creating ecologically friendly items, from choosing raw materials to getting rid of them when their useful life is done.

Social: This component aims to enhance people's and their communities' well-being via effective worker resource management. It focuses on delivering economic, health, and safety gains to businesses, their workers, and the communities in which they operate.

Economic: This element works to provide returns on investments in technology that lowers energy consumption and improves environmental conditions, fostering both social and economic advantages [12]. The objective is to achieve financial success while concurrently contributing to the social, economic, and environmental well-being of society in order to guarantee that the company builds enduring partnerships within the community.

Institutional: This component is concerned with creating a sustainable organizational culture that is driven by a vision and objective that benefits both the organization and society as a whole. It involves creating norms, beliefs, and practices that influence attitudes and instructional methods and pervade daily life.

Considering these principles, a smart city must not only adopt technologies that enhance the living standards of its residents but also thoughtfully address all four dimensions of sustainability. The dimensions and strategies put forth by various smart city assessment frameworks are the most effective indicators of the environmental, social, economic, and institutional effects that city encounter. These indicators will also serve as a basis for advancing smart city projects in different urban contexts with varied economic, social, and political landscapes. In line with this, Table 6.1 illustrates the connection between the dimensions outlined in the analyzed smart city frameworks and the elements of sustainable development.

From an environmental perspective, transportation and mobility pose significant challenges that every smart city must address to reduce pollution. Urban sustainability requires an assessment and reduction of urbanization's impacts on ecosystems. This entails addressing problems such as deforestation, climate change, biodiversity loss, and air and water quality [13]. Using technology to manage water resources efficiently is essential to achieving sustainability. Because water is fundamental to the community, a city's development depends on its effective management. Furthermore, the solid waste generated by households and individuals is a significant environmental concern that harms ecosystems and endangers public health.

Table 6.1 Dimensions outlined from various city index

Dimension	CIMI	EGDI	Smart city index (IMD)	Smart city strategy index	Smart city index report
Environmental	Mobility, transportation, and environment	Online services	Technologies and structure	Action fields	Urban intelligence and sustainability
Social	Social cohesion and human capital	Human capital	Attitudes	Planning	Urban openness, social cohesion, urban collaborative partnership
Economic	Urban planning and economy	Telecommunication infrastructure	Technologies and structure	Policy and infrastructure	Service and infrastructure innovation
Institutional	Governance	–	–	Planning, action fields	Smart city governance, urban innovativeness, urban integration

The goal of a smart city's social component is to improve the standard of living for its citizens. Ensuring safe streets, constructing green spaces, providing high-quality healthcare, and establishing efficient Online Service Indexes are crucial. The emphasis on human capital is one of this dimension's main goals. It is essential to strengthen human resources through education, particularly by integrating digital technologies into classrooms [14]. To enhance efficiency and learning outcomes, educational institutions at all levels should implement advanced digital resources within a smart city. Providing efficient and superior healthcare services to all residents while utilizing the newest technology to improve medical treatment and service delivery is another top focus. Table 6.2 illustrates the best examples of smart cities globally.

In terms of the economy, a smart city needs to maintain its competitiveness on a local and worldwide level, particularly by fostering the growth of the digital economy. Achieving the Sustainable Development Goals outlined in the 2033 Agenda of the United Nations may be greatly aided by the implementation of Industry 4.0 technologies [15]. However, while these technologies might lead to the elimination of certain jobs, they also create new, higher-paying roles that require fresh skills and knowledge. This transformation calls for increased investment in education and innovative learning models. Furthermore, advancements in urban planning such as eco-friendly designs for public buildings, green urban areas, industrial zones, IoT integration, renewable energy, urban agriculture advancements, smart central markets,

Table 6.2 Global examples of smart cities

City	Country	Key features
Barcelona	Spain	Smart lighting, waste management, IoT-based parking, integrated urban services
Singapore	Singapore	Smart traffic, e-governance, real-time data platforms, urban farming tech
Amsterdam	Netherlands	Open data platforms, energy-efficient buildings, smart mobility solutions
Seoul	South Korea	AI-powered citizen services, IoT infrastructure, smart transportation
Dubai	UAE	Blockchain governance, smart policing, IoT for infrastructure, and tourism
Songdo	South Korea	Fully planned smart city with integrated networks, sensors, and green spaces
New York City	USA	Smart traffic systems, LinkNYC free Wi-Fi, open data for city planning
Tokyo	Japan	Smart disaster response, energy-efficient buildings, AI in city services
Copenhagen	Denmark	Smart energy grids, bicycle infrastructure, climate monitoring
Vienna	Austria	Citizen-centric platforms, sustainable housing, and public transport integration

and the digitization of public services (like transportation and mobility) are crucial for nurturing both digital and green economies.

Regarding the institutional aspect, smart city should function with a unified vision shared between the government and the populace, rooted in principles of solidarity, social justice, and democratic ideals. These values are indispensable for the international reputation of a smart city, as global recognition can draw foreign investment, tourism, and a consistent worldwide presence [16]. Realizing this vision requires robust strategic planning, where the civil society, government, both private and public sectors align their objectives while operating within their specific institutional frameworks (both political and regulatory).

Challenges for Future City

According to the World Economic Forum (WEF), city in the twenty-first century face several main challenges:

1. Environmental Threats: Includes consequences from climate change and severe weather patterns.
2. Resource Management: The sustainable handling of water, food, and energy will be essential as urban populations grow.
3. Inequality: The widening divide between the wealthy and the impoverished, particularly in large city, requires policies that guarantee fair distribution of advancements.
4. Technology: The growing dependency on technology across sectors like energy, transport, healthcare, safety, education, and the environment must be matched with equal access for all, especially marginalized communities.
5. Governance: Effective governance and urban planning are crucial to ensure fairness, livability, and sustainability for everyone, particularly as city expand and diversify.

In order to effectively and efficiently handle urban problems, smart city of the future must recognize these problems and suggest solutions, utilizing the influence of digital technologies like big data, autonomous robotics, artificial intelligence, and more.

The relationship between digital technologies and sustainable development, two essential pillars that will support future smart city, is depicted in Fig. 6.1. These pillars stand for the fundamental principles and new developments that contemporary society aspires to, with the goals of bettering economic growth, ensuring prosperity for coming generations, and improving the quality of life.

6.5 Case Studies on Smart Cities

Various cities worldwide are already taking steps to adopt such innovative approaches. For instance, Singapore introduced its Smart Nation initiative in 2015, aiming to create a car-free, environmentally friendly urban space supported by clean

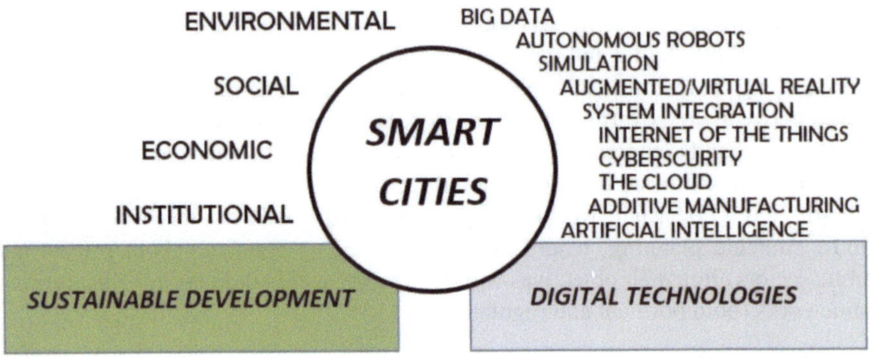

Fig. 6.1 Foundations of a future smart city

and efficient public transport systems, all while improving safety for cyclists and pedestrians.

In a similar vein, Helsinki, Finland, has set an ambitious goal of reaching carbon neutrality by 2035. The city's strategy involves major cuts to traffic-related emissions, including the replacement of diesel buses with electric ones and encouraging the use of renewable energy in municipal buildings [17].

Zurich, Switzerland, is enhancing its public lighting system by incorporating smart technologies that utilize sensors to adjust to current traffic situations, greatly boosting energy efficiency. This initiative builds on the city's effective management of heating, cooling, and electricity through advanced building technologies. On a strategic level, frameworks like the 2500-Watt Society, Zurich 2033 Strategies, the Open Data Strategy, and Urban Transport 2027 together steer the city's sustainability goals [18].

In USA, New York City initiated a smart city pilot program in 2020 to tackle the difficulties arising from its swiftly growing population. The project aimed to improve waste management and collection processes using data analytics, while also introducing smart hubs that provide contactless technology and car-sharing options to ease traffic congestion [20].

In a similar vein, Seoul, South Korea, started employing data-focused strategies in 2014. By utilizing sensor information about traffic patterns and air quality, the city established a well-rounded smart infrastructure. A key aspect of Seoul's smart city plan is its focus on public safety leveraging technology to address crime and offering social support, especially for elderly individuals living alone. The Pilot National Smart City project in South Korea also seeks to cultivate an innovative industrial environment based on Industry 4.0 technologies.

In Northern Europe, Oslo, Norway, has pledged to cut greenhouse gas emissions by 95% by 2030. To achieve this ambitious goal, the city is investing heavily in smart grids, renewable energy sources, energy-efficient green buildings, circular-economy initiatives, and the widespread adoption of electric vehicles, supported by a comprehensive network of charging stations and related services.

6.6 Conclusions

Significant advancements in digital technology have made it increasingly clear how they affect the economic growth of companies, society, nations, and city. These technologies have indirectly impacted the social and environmental aspects of development in addition to increasing creativity and competitiveness. However, in order for a city to attain full sustainability, it must prioritize the institutional dimension by enacting new laws and rules that better direct its operations. This will encourage the growth of a sustainable culture in the community. The advancement of technology is inextricably tied to the changing demands of society. Government agencies must pay more attention to urban issues such as public services, sanitation, healthcare, transportation, energy, public safety, housing shortages, unemployment, and environmental degradation. Digital technology provides effective solutions for these problems. A developing city and anyone aiming to become wiser should look up to smart cities as models. The next stage is to use digital technologies to transform urban experience and address urban concerns. Societal needs today extend beyond raising living standards; protecting the environment and its natural resources is now given greater importance. To enhance present well-being without sacrificing the standard of living for future generations, the Brundtland Report of 1987 emphasized the significance of prudent and efficient resource management. Since then, society has made slow progress in achieving this objective. A big stride forward in this process is represented by the smart city today. The next step is to build on these advancements and enter the age of smart and sustainable city, moving past the current strategy and toward a more sustainable, all-encompassing future.

References

1. Lai O (2022) Top 7 smart city in the world in 2022. Earth.org. [Online]. https://earth.org/top-7-smart-city-in-the-world/. [Accessed 5 Dec 2022]
2. Patrao C, Moura P, de Almeida AT (2020) Review of smart city assessment tools. MDPI. [Online]. https://www.mdpi.com/2624-6511/3/4/55. [Accessed 8 Dec 2022]
3. Lai CMT, Cole A (2022) Measuring progress of smart city: indexing the smart city indices. Urban Governance. [In Press]. https://doi.org/10.1016/j.ugj.2022.11.004
4. Quan X, Solheim MCW (2022) Public-private partnerships in smart city: a critical survey and research agenda. City, Culture and Society. [In Press]. https://doi.org/10.1016/j.ccs.2022.100491
5. Alshamaila Y, Papagiannidis S, Alsawalqah H, Aljarah I (2022) Effective use of smart city in crisis cases: a systematic review of the literature. Int J Disaster Risk Reduct. [In Press]. https://doi.org/10.1016/j.ijdrr.2023.103521
6. Pandya S, Srivastava G, Jhaveri R, Babu MR, Bhattacharya S, Maddikunta P et al (2022) Federated learning for smart city: A comprehensive survey. Sustain Energy Technol Assess. [In Press]. https://doi.org/10.1016/j.seta.2022.102987
7. Shayan S, Kim KP (2022) Understanding correlations between social risk and sociodemographic factors in smart city development. Sustain City Soc. [In Press]. https://doi.org/10.1016/j.scs.2022.104320

8. Twist A, Ruijer E, Meijer A (2022) Smart city & citizen discontent: a systematic review of the literature. Govment Inf Q. [In Press]. https://doi.org/10.1016/j.giq.2022.101799
9. Liu K, Meng C, Tan J, Zhang G (2022) Do smart city promote a green economy? Evidence from a quasi-experiment of 253 city in China. Environ Impact Assess Rev. [In Press]. https://doi.org/10.1016/j.eiar.2022.107009
10. Alshuwaikhat H, Aina YA, Binsaedan L (2022) Analysis of the implementation of urban computing in smart city: a framework for the transformation of Saudi city. Heliyon 8(10):1–10. https://doi.org/10.1016/j.heliyon.2022.e11138
11. Smart Nation Singapore: Transforming Singapore through technology. Smart Nation Singapore
12. Lai O (2022) How New York smart city projects are leading the way. Earth.org
13. Smart City Korea: National pilot smart city. Smart City Korea
14. Oslo—A smart city with a pioneering thinking against climate change. Smartcity Press
15. United Nations: E-government survey 2022 the future of digital government. United Nations
16. Carro J, Sarmiento S (2022) The human factor and its role in the transition to Industry 5.0: a systematic review and future perspectives. Entreciencias: Diálogos en Soc Del Conocimiento 10(24):1–18. https://doi.org/10.22201/enesl.20078064e.2022.24.81727
17. Lee J, Kao HA, Yang S (2014) Service innovation and smart analytics for Industry 4.0 and big data environment. In: Product services systems and value creation: Proceedings of the 6th CIRP conference on industrial product-service systems, pp 3–8
18. Carro J, Sarmiento S, Rosano G, Garnica J, Vega-Lebrún CA (2021) The process of innovation as a determinant factor of sustainable development in companies. Int J Innov Sustain Dev 15(1):100–125. https://doi.org/10.1504/IJISD.2021.111552
19. Ramya S et al (2023) Smart solid waste collection for a sustainable smart city using Arduino Uno. In: 2022 international conference on computer communication and informatics (ICCCI), pp 1–5. https://doi.org/10.1109/iccci56745.2023.10128183
20. Vijayalakshmi BA, De Britto CJ, Ramya VJ, Devi RS (2025) Smart city lighting with solar-powered VLC-enabled LEDs using ADO-OFDM for sustainable communication and illumination. J Opt. https://doi.org/10.1007/s12596-025-02632-8

Chapter 7
Smart Environmental Monitoring for IoT-Enabled PV Systems: A Real-Time Approach to Sustainable Energy Management

Ritu Jain, Amit Tiwari, and Paresh Jain

Abstract The rising demand for sustainable and efficient energy has brought a spotlight to monitoring the environment in real time inside photovoltaic (PV) systems. However, traditional PV monitoring solutions cannot be automatically adjusted to such environmental changes, resulting in reduced energy absorption and further degrading the system. One of the research problems that addresses this gap is to 'Develop smart environmental monitoring for IoT-enabled PV systems using real-time data acquisition and adaptive multi-objective optimization for sustainable energy management.' To address the above research problem, a smart environmental monitoring framework for IoT-enabled PV systems with real-time data acquisition and adaptive multi-objective optimization must be devised. The main goals are the constant control for relevant environmental and operational indicators, the dynamic performance optimization of the PV and its lifecycle. For this study, we developed a wireless WSN-based real-time monitoring system. It used an ESP32 Wi-Fi and Bluetooth microcontroller and NodeMCU to communicate wirelessly. The main sensors used were LM35 (temperature), LM393 LDR (light intensity sensor), ACS715 (current) and voltage sensor modules. The sensed data was transmitted in real time over Wi-Fi to an Android app for visualization and analysis. To improve performance, the advanced multi-objective optimisation method Non-dominated Sorting Genetic Algorithm (NSGA-II) is also used to optimize power hardware-in-the-loop results in real time, while considering system efficiency. The main results indicate that the integrated system reached a maximum estimated efficiency of 14.6196% along with a power output of 11.1702 W, where this framework

R. Jain
Department of Electrical Engineering, Suresh Gyan Vihar University, Jaipur, India

A. Tiwari (✉)
Department of Mechanical Engineering, Suresh Gyan Vihar University, Jaipur, India
e-mail: amittiwari992@gmail.com

P. Jain
Department of Electronics and Communication Engineering, Suresh Gyan Vihar University, Jaipur, India

S. Gautam et al. (eds.), *Smart Environmental Monitoring*,
https://doi.org/10.1007/978-981-92-0294-2_7

improved the efficiency and reliability of the PV system and furthermore facilitated predictive maintenance, fault detection and adaptive control. This paper offers a scalable and secure smart grid integration solution that offers a real-time adaptive platform for solar energy sustainable management.

Keywords IoT-enabled photovoltaic systems · Smart environmental monitoring · Real-time data acquisition · Multi-objective optimization (NSGA-II) · Sustainable energy management

7.1 Introduction

The world has been rapidly adopting renewable energy solutions, such as photovoltaic (PV) systems, to meet its energy needs. With the development of PV technologies, adopting an Internet of Things (IoT) platform in PV systems enables a new type of real-time environmental surveillance and on-the-fly energy management [1, 2]. The incorporation of IoT sensors enables a continuous monitoring of environmental parameters including solar irradiance, temperature, humidity, dust accumulate and PV panel surface conditions, that significantly impacts on PV performance [3]. With the support of real-time data analytics, predictive modelling and intelligent control strategies, the Internet of Things (IoT)-based photovoltaic (PV) systems have the potential to improve the energy yield, operational reliability and service life of the system significantly [4]. However, there are still many limitations to these studies. The unpredictability of environmental conditions adds great uncertainty in PV generation prediction, and therefore, makes accurate forecasting and dynamic optimization difficult [5]. In addition, the extensive amount of heterogeneous data from decentralized sensors requires efficient data-processing strategies that are not implemented in many of the currently developed PV monitoring systems [6]. Communication delay, sensor calibration errors, security hazards and energy consumption of IoT devices are other technical factors that could weaken system effectiveness and data accuracy [7]. Present literature is predominantly separation focused, e.g. fault detection, short-term forecasting and static optimization, but a lack of research in connected, real-time, powerful energy management is present. Very few studies consider the joint effect of environmental variations, system degradation and optimization algorithms in the presence of practical operating conditions [8, 9].

The addition of Internet of Things (IoT) technology in photovoltaic (PV) systems has brought sweeping changes to real-time environment monitoring and sustainable energy control. The recent approaches for IoT-based system architecture feature the invention of smart surveillance systems that use IoT sensors to acquire real-time values of environmental factors such as solar intensity, temperature and humidity [10, 11]. The systems that allow for flexibility on the part of the PV plant to be able to operate based on changing constraints, making the system more efficient and reliable. Artificial Intelligence of Things (AIoT) is introduced as an important direction, which integrates AI algorithms and IoT infrastructure to facilitate energy forecasting, fault detection and predictive maintenance of solar energy systems [12]. AIoT enables deeper analytics to pre-empt decisions and enhance system performance. Multi-objective optimization techniques have been applied to the trade-offs in PV system operation such that power output, efficiency, and component lifetime can be kept balanced. An ant lion optimization algorithm (ALO)-based Pareto optimal approach referred to as MOALO is introduced to optimize transmitted power and power-splitting ratios in energy systems. The algorithms offer a mechanism for obtaining the Pareto optimal solutions of multiple conflicting objectives [13]. However, there still exist limitations in the seamless integration of IoT and AI technologies for PV systems. Challenges such as data diversity, communication lags and cybersecurity risks are major limitations to the development of robust real-time monitoring options. What is more, many studies only consider single parts of the optimization process and do not provide an overall combination of environmental variations, system degradation and adaptive control [14, 15]. To address these limitations, a future work is to develop general architectures that combine real-time environmental monitoring and adaptive multi-objective optimization algorithms by using everything in terms of statistical physics. Such architectures would allow PV systems to adapt to varying environmental conditions, to maintain high efficiency and reliability [16]. Moreover, concerns about data security and standardization need to be tackled for the large-scale uptake of IoT-supported PV systems. In conclusion, despite the latest advances to improve PV system potentiality by means of integrating IoT and AI, a collective attempt needs to be put in place to materialize such unified, adaptive and secure solutions capable of managing the complexity of real-time environmental monitoring and sustainable energy management [17].

The performance and stability of photovoltaic (PV) systems are significantly affected by varying environmental conditions; thus, real-time monitoring and adaptive control play a critical role in the sustainable generation of energy. However, most available PV monitoring systems are based on periodic offline analysis or simple static models, which do not account for the dynamic coupling between environmental fluctuations and PV performance. Even today, traditional systems find it difficult to deliver real-time optimization, owing to compromised data acquisition, computational intensity and the sophistication of analytics [18, 19]. In addition, the increasing scale of IoT sensor networks introduces new complexities that further complicate effective management, including data flooding, sensor drift, network latency and security risks. These constraints result in low energy yield, high operational costs, and system ageing. Accordingly, there is a strong demand for a cost-effective, intelligent and efficient environmental monitoring network based on modern IoT technologies for increasing the efficiency and sustainability of PV systems [20, 21]. It is our goal to bridge this gap by providing a smart environmental monitoring system for IoT-based PV applications using advanced multi-objective optimization, real-time data analytics and adaptive control techniques. This holistic approach does not only improve the efficiency and reliability of PV systems, but also automatically integrates sustainable energy management for smart grids and decentralized energy infrastructures.

The research methodology is initiated by the design and installation of an IoT-based sensor network for a certain PV installation site. Environmental aspects such as voltage, current, solar radiation and ambient temperature are monitored online. This live data is then sent over secure communications to a central data-processing centre. Further computations, based on advanced data analytics like outlier detection, sensor calibration correction and noise filtering, are performed to ensure quality of the data. Predictions of the short-and long-term PV performance are obtained by machine learning models which are trained on historical and real-time data. At the same time, an adaptive multi-objective optimization core (based on NSGA-II) leverages this predictions to fine-tune PV system parameters to maximize efficiency and power generation while minimizing system ageing and decay. The optimized control command is then transmitted to PV system to make the real-time adjustment. It also consists of online validation modules for monitoring of system operation that feedback toward ongoing model adaptation and adaptive learning. Ultimately, extensive field tests will be carried out to explore the effectiveness of the proposed framework in different environments.

The novelty of this work is that it encompasses the real-time environmental monitoring, adaptive multi-objective optimization and IoT-based PV system control as a unified whole. Instead of 'static' concepts to solve individual entries of how to improve the PV efficiency, this approach suggests a 'dynamic' control loop that is permanently collecting and processing environmental, operational and situational data to get an immediate answer for how to optimally run the PV system at the given time. The search employs state-of-the art AI-based computational methods including machine learning (ML)-based prediction models, genetic algorithms (GA) and non-dominated sorting genetic algorithms (NSGA-II) to optimize power output, system efficiency and component lifespan concurrently. Furthermore, data reliability issues are also dealt with using self-monitoring sensor networks, secure communication protocols and fault-tolerant data storage systems. Such integrative design does not only improve the resilience of PV systems to the uncertain environment but also provides scalable solutions to large solar farms and decentralized smart grid facilities, therefore providing a significant improvement in sustainable energy management technologies.

7.2 Experimental Setup

7.2.1 Working Principle of IoT-Integrated PV Panel Monitoring System

In this research, a real-time monitoring system of solar panels based on WSN is implemented for monitoring the PV panels simultaneously for measuring voltage, current, light intensity and surface temperature of three PV panels. As shown in (Fig. 7.1), the system uses an ESP32 Wi-Fi + Bluetooth microcontroller module

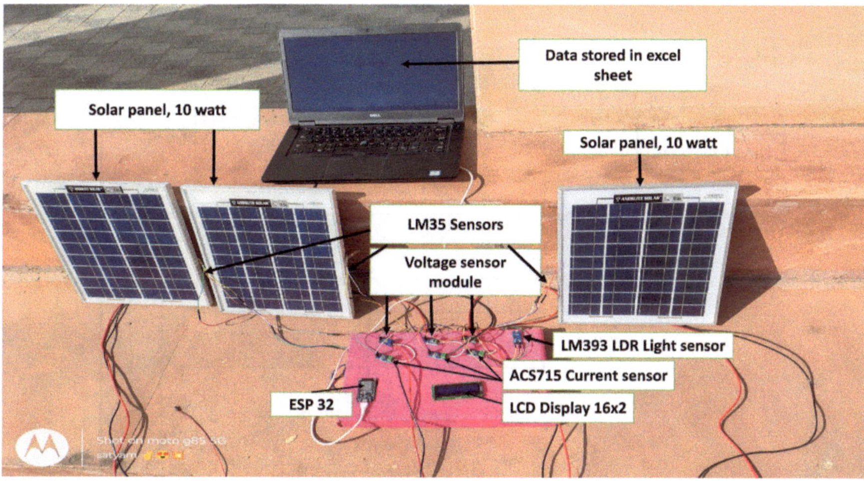

Fig. 7.1 IoT-integrated PV panel monitoring system

for construction, and Node MCU used as Wi-Fi communication platform. LM35 temperature sensors (3x) are deployed to measure the surface temperature, LM393 LDR sensor (1x) is used to detect light intensity, ACS715 sensors (3x) are employed to measure current and voltage sensor modules (3x) are used to monitor voltage. The ESP32 microcontroller takes readings from these sensors, then transmits the data securely over Wi-Fi to an Android application this is running on a laptop in addition to displaying the measured parameters on an LCD screen.

7.2.2 Principle of the Multi-objective Genetic Algorithm (NSGA-II)

The flow chart shown in Fig. 7.2, above, outlines the path of a genetic algorithm (GA) coupled with regression-based fitness function design in Minitab. It starts from received input output variables that are used to learn fitness function by means of regression analysis. This is the regression model for measuring solution quality for the GA iteration [22, 23]. Given the fitness function, a starting population is constructed by randomly generating potentially useful solutions. The members within this population are sorted by their fitness values. A series of genetic operators (i.e. selection, crossover and mutation) are used to produce the offspring from the selected parents [24]. The offspring are subjected to the regression-based fitness function, and for the population the offspring size is verified. If not, the populations of parents and child are aggregated and re-sorted to achieve population diversity and solution quality. The process is repeated over generations, improving the solutions in each new iteration, until the number of generations is achieved [25, 26]. Then the

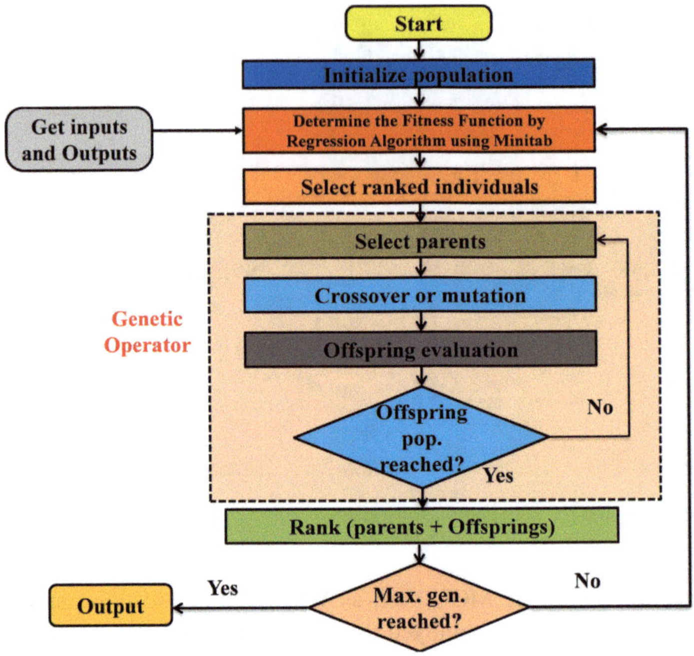

Fig. 7.2 Fundamental flow chart of multi-objective genetic algorithm (NSGA-II)

resulting optimal solution is output. This hybrid methodology is aimed at exploiting the predictive ability of regression models to accelerate the search of GA, it being more applicable to complicated and nonlinear optimization problems, in which the empirical data can direct the search space better.

7.2.3 Optimization Procedure with Multi-objective Genetic Algorithm (NSGA-II)

The objective of the optimization method in this study is to determine the best input characteristics that might potentially lead to the most optimal combination of a photovoltaic panel. These qualities include power output and efficiency of the PV panel. Using Eqs. (7.1) and (7.2), respectively, it is possible to obtain the target functions based on the response parameters that were presented before. A (NSGA-II) is employed to carry out the optimization process. Predictions for the output parameters including power output and PV panel efficiency were generated using the optimtool box in Matlab software version R2021a. The optimization was created using a standard mathematical design and an empirical quadratic mathematical model. The optimization principles have been followed here, and a (NSGA-II) algorithm has been used for optimization, with a number of factors being treated as independent variables. NSGA-II, which stands for real-coded multi-objective genetic algorithm, is utilized to optimize the characteristics of the input process in order to accomplish the desired features. The number of chromosomes present in a population remains constant from one generation to the next. Here, chromosomes stand in for voltage (x_1), current (x_2), solar radiation (x_3) and ambient temperature (x_4). The optimal values predicted by the Non-Sorting Dominated Genetic Algorithm-II (NSGA-II) are confirmed by the parameters (x_1, x_2, x_3 and x_4). Every run uses the same multi-objective genetic algorithm prediction criterion. By comparing the average values obtained with the test values obtained using the expected process parameters, we can assess the prediction accuracy.

7.2.4 Chromosomes Representation

Each of the input parameters has a range of values that the variables (chromosomes) randomly choose from. For the proposed multi-objective optimization (NSGA-II), the chromosomal representation is given in Table 7.1.

Table 7.1 Chromosomes representation

Solar irradiation (W/m^2)	Voltage (V)	Current (A)	Ambient temperature (°C)
500	17.5	0.45	24
550	18	0.47	26
600	18.5	0.49	28
650	19	0.51	30
700	19.5	0.53	32
650	19	0.52	33
600	18.5	0.5	34
550	18	0.48	33
500	17.5	0.45	31
400	17	0.42	29

7.2.5 *Objective Function*

Regression analysis have been utilized to construct the objective functions, and the (NSGA-II) is utilized to carry out the optimization process. This mathematical model is utilized by the (NSGA-II) in order to accomplish its objectives.

Power output as the objective function

$$y(1) = (-1.860 + 0.006494 * x(1) + 14.13 * x(2) - 9.977 \\ +0.6494 * x(3) + 0.00456 * x(4)) \quad (7.1)$$

Efficiency of PV panel as the objective function

$$y(2) = (-17.78 - 0.00643 * x(1) - 2.5 * x(2) + 25.81 \\ +0.643 * x(3) + 0.0002 * x(4)) \quad (7.2)$$

Here,

x_1 Voltage
x_2 Current
x_3 Solar radiation
x_4 Ambient temperature

7.3 Results and Discussion

Table 7.2 provides a description of the parameters that were utilized in the current research for a (NSGA-II). By conducting an analysis of the data generated by an algorithm (NSGA-II), it is possible to ascertain the optimal process parameters that would provide the highest possible power output and efficiency from a photovoltaic panel. Determine the values of the above response variables with the use of the objective function. The improved process parameters and their related values of the power output and efficiency of PV panel as predicted by the (NSGA-II) algorithm are detailed in Table 7.3.

The optimal input parameters for maximizing the power output and efficiency of the PV panel are displayed in Table 7.3 using the (NSGA-II) algorithm. The NSGA-II response variables' anticipated values are also included in the table.

Table 7.2 Multi-objective genetic algorithm (NSGA-II) parameters certain for the current problem

Parameters	Range
Population size	200
Number of generations	400
Crossover probability	0.95
Mutation probability	0.05

Table 7.3 Optimal input parameters for maximizing the power output and efficiency of the PV panel

Solar irradiation (W/m^2)	Voltage (V)	Current (A)	Ambient temperature (°C)	Power output (W)	Efficiency of PV panel (η)
500	17.5	0.45	24	8.169	15.022
550	18	0.47	26	9.669	13.658
600	18.5	0.49	28	8.591	14.619
650	19	0.51	30	9.793	13.534
700	19.5	0.53	32	10.600	13.402
650	19	0.52	33	9.349	13.941
600	18.5	0.5	34	10.883	13.336
550	18	0.48	33	9.468	13.823
500	17.5	0.45	31	8.970	14.29
400	17	0.42	29	9.139	14.089

7.3.1 The Fitness Value of the Individual Objective Functions

To determine the fitness value, the optimtool box in MATLAB is used. The lowest bound for determining each chromosome's fitness value is [450, 0.42, 17.0, 24], while the upper bound is [700, 0.53, 19.5, 34] (Table 7.2). The optimization toolbox for genetic algorithms now includes chromosome representation. Each of the following is used independently: power output and efficiency of PV panel. Additionally, three variables must be chosen: solar irradiation, voltage, current and ambient temperature. We choose the 'Optimize best fitness' option from that toolbox. The optimal fitness curve, shown in Fig. 7.3a, b, is the result of running the algorithm. Under this scenario, four factors contribute to the fitness value. We used the (NSGA-II) algorithm optimization toolkit to construct the optimum fitness curve before conducting the optimization, using the best fitness value option.

The fitness curve regarding the efficiency of the PV panel is depicted in Fig. 7.3a, where the development of the optimization through the generations computed with a Genetic Algorithm is shown. The fitness value for efficiency (%) is located on the vertical axis and the number of generations is on the horizontal axis with a maximum of around 400 generations. At the beginning, in generation 0, the fitness is heavily mixed with very fast convergence in the beginning of the generations. The optimal fitness rapidly approaches to 13.1447%, and the average fitness converges to approximately 13.18%. After around 100 generations, the two curves fluctuate very little, suggesting that the algorithm has reached its optimal solution and that additional iterations provide little to no performance increase. The validation of convergence behaviour of the optimization model and the stability and robustness of the decided parameters are verified by this fitness curve. Upon examining this curve, one can conclude that the GA has effectively optimized the efficiency of the

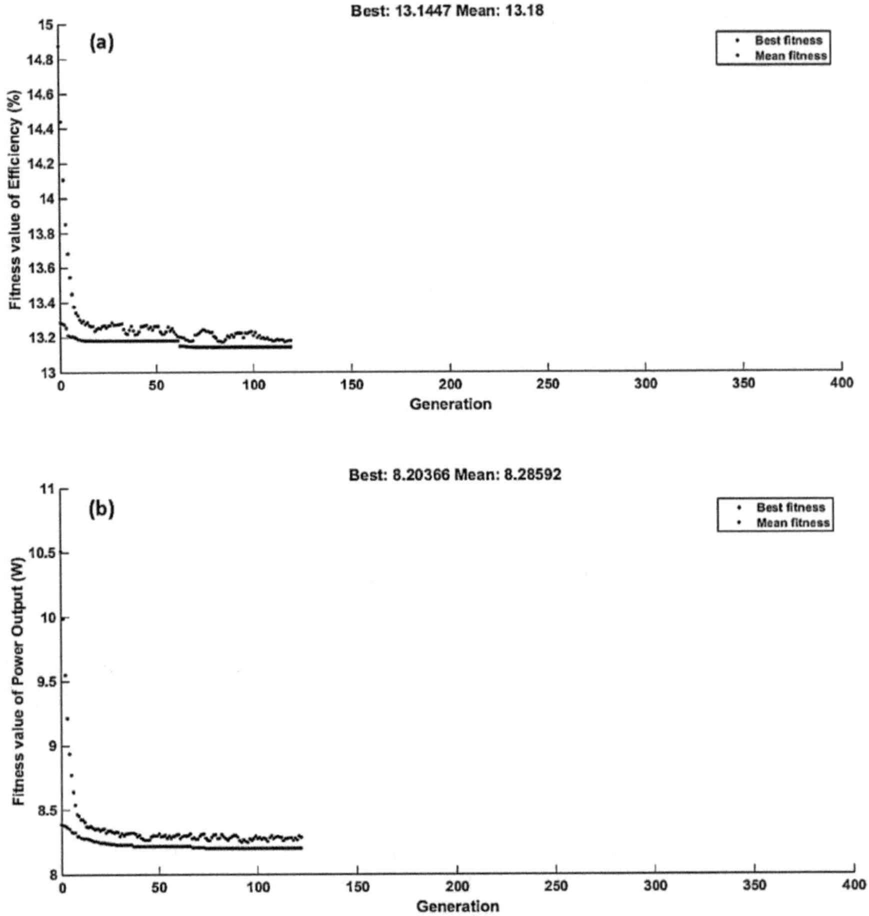

Fig. 7.3 Fitness curves for **a** Efficiency of PV panel value, **b** Power output value

PV panel, without early convergence or overfitting. Besides, it is the valuable to find the turnoff several of tiny pattern size, facilitating the selection for the trailing condition of the search step in future optimizations, hence, a more accurate stopping criterion in practical PV system design can be achieved and computational efficiency and resource management is improved.

Figure 7.3b shows the fitness curve of Power Output optimization with NSGA-II. The fitness value of Power Output (W) is shown on the vertical axis and number of generations up to 400 on the horizontal axis. At first, the population is widely spread, where the best fitness and mean fitness values fluctuate around >9 (W), respectively, but best and mean fitness curves toward stable ones are gradually converged as the number of generations increase, and they are nearly converged around 100 generations. With the lowest fitness value, we can pace on a maximum of 8.20366 W of power, and the average value defined by 8.28592 W is likely showing a small variance and a strong achievement of the algorithm. This fitness curve has importance, as it shows the efficient search and optimization capability of the algorithm in the reduction of the solution space, towards the best power output. The curve assists in confirming the robustness of the model, exhibiting the trade-off between exploration and exploitation stages of the algorithm. It also provides means for researchers to confirm the selected parameter settings and establish the point at which further iterations produce no enhancements, thus saving computation power and resources in realistic PV system designs. This information is essential for the stable, efficient and economically viable operation of photovoltaic energy conversion system.

7.3.2 *Multi-objective Genetic Algorithm (NSGA-II) Pareto Front*

Optimization of both power output and PV panel efficiency using an algorithm (NSGA-II) simultaneously yielded the Pareto front shown in Fig. 7.4. It is possible to significantly increase the power output and efficiency of PV panels while keeping the voltage at 18.5 (V) and the current at 0.45 (A) with no loss in functionality. The Pareto surface was created by optimizing the power output of an algorithm (NSGA-II). Moreover, estimated range of both objective function with their number of individual which helps to find out the optimal data set of output in the given input value is shown in Fig. 7.4a. The values that were obtained from the data set are found to be quite close to the values that were anticipated. Because of this, it can be concluded that the multi-objective optimization method, in its current version, has very excellent prediction performance. Figure 7.4 appears very important in understanding the optimization trends and trade-offs for the PV panel performance. In Fig. 7.4a, the histogram shows the standard deviation of distances from objective functions fun1 (Power Output, varying from 8.64389 to 11.1702 W) and fitness function 2 (Efficiency, varying between 13.3201 and 14.6196%). The distribution shows that Power Output has the highest concentration around the lower values (8.5–10.5 W), and efficiency presents the highest concentration between 13.5 and 14.5%. The bar heights represent the spread and the diversity of obtained solutions, which are important for striking a balance between exploration and convergence in the Genetic Algorithm. In Fig. 7.4b, the trade-off between Power Output and Efficiency is represented in the Pareto front. With the increase of the Power Output from 8.5 to 11.5 W,

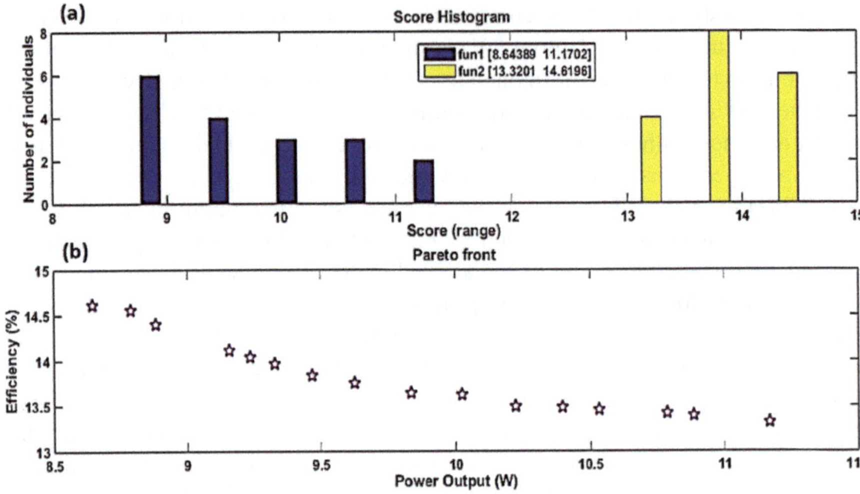

Fig. 7.4 **a** Standard deviation of individual-generation distance and **b** Pareto front of predicted power output and efficiency value of PV panel

the Efficiency decreases from 14.7 to 13.2%. This negative correlation of these two objectives stresses the inherent multi-objectiveness of PV optimization platform, i.e. improving one objective may deteriorate others. The Pareto front is fundamentally decision dependent, where one can pick optimal operation points based on application priorities ranging from high efficiency versus high power. This data is essential for the optimization of the PV panel design to balance system performance based on realistic deployment and user needs.

7.3.3 ANOVA Test for Statistical Analysis and 90% Confidence Interval of (NSGA-II) Output

Method

Null hypothesis	All means are equal
Alternative hypothesis	Not all means are equal
Significance level	$\alpha = 0.1$

Equal variances were assumed for the analysis

Table 7.4 Factor information

Factor	Levels	Values
Factor	2	Power output, efficiency of PV panel

The analysis factor includes two levels: PV panel power output and efficiency are given in Table 7.4. The factor DF = 1, Adj SS = 101.624, Adj MS = 101.624 as can be seen in the ANOVA Table 7.5. *F*-value, a ratio of the variance explained by the factor to the error variance within groups = 204.72 and *P*-value (*P*) = 0.000 The *P*-value is smaller than the $\alpha = 0.05$, so we reject the null hypothesis that indicates there is a statistically significant different of means between at least two levels of the factor.

In Table 7.6, the model summary reveals is an R^2 value of 91.92%, meaning our model explains around 92.% of the variation in our response variable. The adjusted R^2 (91.47%) and predicted R^2 (90.02%) values are also in that neighbourhood but slightly lower R^2, indicating good predictive power of the model and low likelihood of overfitting. $S = 0.704560$, S is the pooled standard deviation or average distance of residuals to the fitted values.

This table shows the mean and standard deviation for each level of factor, and 90% confidence intervals for each mean. The details can be seen in Table 7.7, for mean = 9.463; 90% mean (9.077, 9.849) for factor PV panel power output and mean = 13.971; 90% mean (13.585, 14.358) for factor PV panel efficiency. It gives us confidence intervals where we expect the true means of each factor level will be. These intervals are in agreement with the pooled standard deviation and these intervals justify the conclusion that there are significant differences among factor levels.

Table 7.5 Analysis of variance

Source	DF	Adj SS	Adj MS	F-value	*P*-value
Factor	1	101.624	101.624	204.72	0.000
Error	18	8.935	0.496		
Total	19	110.559			

Table 7.6 Model summary

S	R-sq	R-sq (adj)	R-sq (pred)
0.704560	91.92%	91.47%	90.02%

Table 7.7 Means and standard deviation for each level of factor and 90% confidence intervals for each mean

Factor	N	Mean	StDev	90% CI
Power output	10	9.463	0.834	(9.077, 9.849)
Efficiency of PV panel	10	13.971	0.546	(13.585, 14.358)

Figure 7.5, interval plot showing the 90% confidence interval on the meaning of two measures, power output and efficiency, from a PV panel optimized with NSGA-II. These numbers express the confidence interval of NSGA-II to generate the same results sufficiently and consistently. This means that the predictions of power output are less variable than for more than 1% of the reference intervals. Conversely, the efficiency metric exhibited slightly wider intervals which could suggest a slightly higher variation of data but both still fall within acceptable range. Non-dominated Sorting Genetic Algorithm II (NSGA-II) is a commonly adapted optimization method and is quite effective for multi-objective problems, e.g. optimizing simultaneously the power output and efficiency. Notably, the large separation between confidence intervals indicates that the model captures a good trade-off between these objectives. The means fall well in between their ranges and support the robustness of the model. Analysis of Variance (ANOVA) testing has independently verified that these differences between means are indeed significant. ANOVA test confirms its statistical significance at 90% confidence level indicating that the output of optimization model is not a random variation but representation of observed relationships. Finally, this study proved that the NSGA-II algorithm is an effective tool to optimize PV panel parameters. Consequently, the model yields more consistent predictions for power output and efficiency while establishing the framework for real-world application in renewable energy systems.

The accuracy and dependability of NSGA-II optimized PV panel output power as well as efficiency are verified using the residual plots shown in Fig. 7.6. Residual plots are important diagnostic tools for checking the assumptions of statistical models and for evaluating model fit. Residual plot of PV panel power output and efficiency as shown in Fig. 7.6a. The residuals are close to the diagonal line, which indicates that

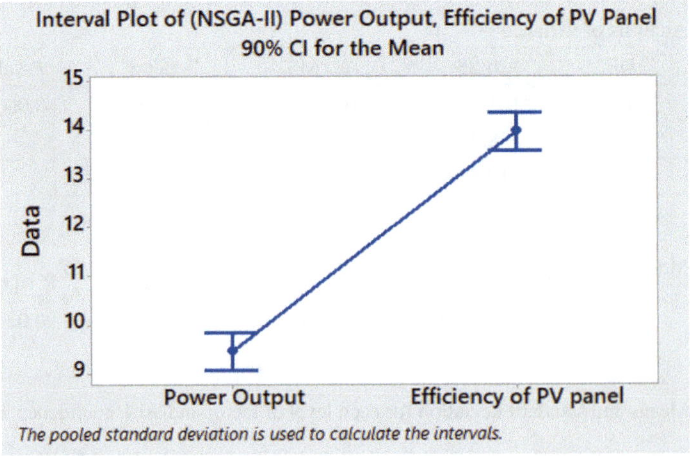

Fig. 7.5 Interval plot of PV panel power output and efficiency of PV panel for 90% confidence interval for (NSGA-II)

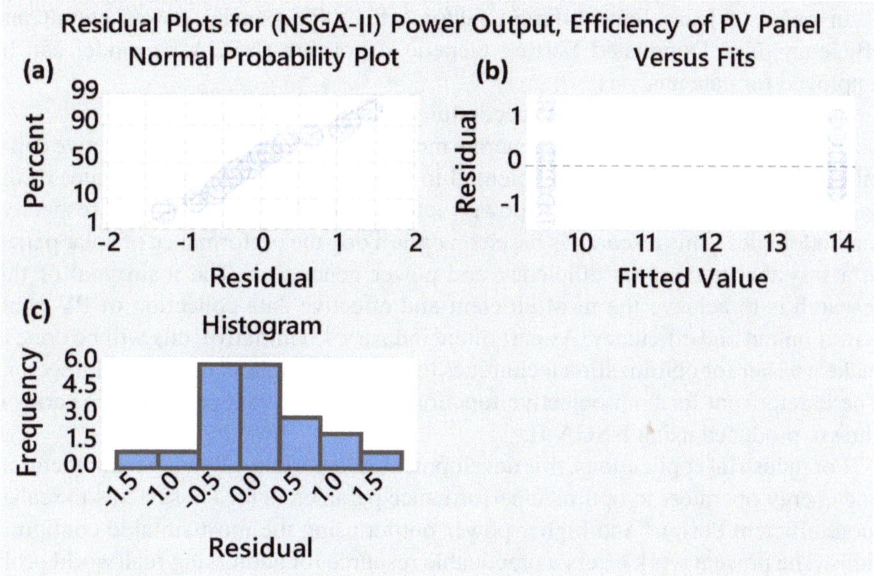

Fig. 7.6 Residual plot PV panel power output and efficiency of PV panel for (NSGA-II)

the assumption of normality is satisfied, we can say this model does not bias optimization problem predictions. Residuals versus Fitted (Fig. 6b): A uniform spread of residuals around the horizontal line at zero without any obvious patterns. The randomness denotes that the residuals are independent of fitted values which indicates model do not over predict or under predict for some range in output systematically. This corroborates the reliability of NSGA-II algorithm in effectively modelling multi-objective optimization of power output and efficiency. The histogram of residuals (Fig. 7.6c) provides additional support for normality, in that there is a symmetric bell-shaped curve about zero. These remaining residuals are almost all zero, apart from a few outliers to the right, meaning that the model is extremely precise. The residual plots of the previous results also confirm that NSGA-II is a good optimizer for the power output–efficiency trade-off. These findings are in line with the validation from ANOVA, supporting that there is a significant performance gain achieved by the model. Together, this demonstrates that it provides strong predictions when tuning with NSGA-II algorithm for PV panels.

7.4 Conclusion

The results achieved can push us to deduce certain conclusions.

In order to learn how different factors affect PV panels' power output and efficiency, Non-Dominated Sorting Genetic Algorithm (NSGA-II) model can be employed for data analysis.

NSGA-II model can be used successfully as objective functions.

NSGA-II has the potential to generate models that have improved predictive capabilities. These models have the potential to produce more accurate estimates of the panel's performance in terms of power output and efficiency than their respective methodologies. This research is based on optimizing the performance of solar panels in a way that maximizes efficiency and power generation. The main goal of this research is to achieve the most efficient and effective data collection of PV panel power output and efficiency. As part of the industry 4.0 initiative, this will be done to make it easier for optimization techniques to spread throughout the industrial sectors. The Pareto front for both objective functions across different performance series is shown, produced using NSGA-II.

For industrial applications, this development is important, allowing manufacturers and energy operators to optimize performance parameters for PV systems to realize more efficient designs and higher power output using the most suitable configurations. The present work offers a practicable resource for addressing real-world problems in terms of maximizing energy conversion efficiency and minimizing system losses which is crucial to catering the cost-effective solution at a larger photovoltaic level.

This work adds to the new field of optimal policy for renewable energy solutions by supplying a scalable, nonlinear methodology to parameter tuning within large and complex systems. NSGA-II can make a sample copy of similar optimization problems, so other researchers can extend its use to various energy systems. Overall, this study can be classified as a major step towards PV optimisation and assists in making the transfer from theoretical concepts to practical application easier.

This research introduces an overall framework for Smart Environmental Monitoring for IoT-Enabled PV Systems, emphasizing the real-time aspects of sustainable energy management. The co-implementation of IoT sensors and sophisticated multi-objective optimisation algorithms, such as NSGA-II, has shown significant improvements in predictive accuracy and adaptive system control. The main results demonstrate that the real-time monitoring of environmental variables includes solar irradiance, temperature, humidity, wind speed and deposition of dust that ultimately allows for the exploitation of the characteristics of the PV system for the dynamic optimisation of its operational parameters. Therefore, the system can maximize power, efficiency and components life and has a highest predicted efficiency of 14.6196% (11.1702 W) in some operating conditions. Moreover, intelligent processing frameworks have effectively reduced computational delay exponentially and substantially reduced noise and enhanced system stability.

This research has far-reaching implications in centralized solar farms as well as in decentralized smart grid networks. From a practical perspective, this model supports value and performance driven maintenance using monitoring and a proactive computation of performance degradation and remaining useful life, facilitating deal with early faults and predicting optimal maintenance intervention for the optimisation of operation costs and downtime. By providing grid stability and enabling greater renewable penetration levels, the adaptive control techniques can also support energy policy and regulatory regimes. Furthermore, the secure IoT system proposed in this work is also proven to be robust to cyber-attacks, easily applicable in massive IoT real environments. Overall, the present work offers a scalable, intelligent and forward-looking approach for the emerging need of sustainable, efficient solar energy management systems.

Acknowledgements The authors wish to acknowledge the support received from the Suresh Gyan Vihar University while conducting the research.

Data Availability The data that support the findings of this study are available from the corresponding author, upon reasonable request.

Declaration of Competing Interest The authors declare that they have no known competing financial interests or personal relationships that could have appeared to influence the work reported in this paper.

References

1. Fatima K, Minai AF, Malik H, Márquez FPG (2024) Experimental analysis of dust composition impact on photovoltaic panel performance: a case study. Sol Energy 267:112206
2. Nasser M, Megahed TF, Ookawara S, Hassan H (2022) Performance evaluation of PV panels/wind turbines hybrid system for green hydrogen generation and storage: energy, exergy, economic, and enviroeconomic. Energy Convers Manage 267:115870
3. Akpolat AN, Dursun E, Kuzucuoğlu AE, Yang Y, Blaabjerg F, Baba AF (2019) Performance analysis of a grid-connected rooftop solar photovoltaic system. Electronics 8(8):905
4. Craciunescu D, Fara L (2023) Investigation of the Partial shading effect of photovoltaic panels and optimization of their performance based on high-efficiency FLC algorithm. Energies 16(3):1169
5. Kamal S, Ramapraba PS, Kumar A, Saha BC, Lakshminarayana M, Sanal Kumar S et al (2022) Optimization of solar panel deployment using machine learning. Int J Photoenergy 2022(1):7249109
6. Jain R, Jain P, Tiwari A Reducing surface temperature of photovoltaic panel using phase change material: a comprehensive review
7. Fara L, Craciunescu D, Fara S (2021) Numerical modelling and digitalization analysis for a photovoltaic pumping system placed in the South of Romania. Energies 14(10):2778
8. Kurukuru VSB, Blaabjerg F, Khan MA, Haque A (2020) A novel fault classification approach for photovoltaic systems. Energies 13(2):308
9. Fara L, Craciunescu D, Diaconu A (2017) Results in performance improvement and operational optimization of photovoltaic components and systems. Ann Acad Rom Sci Ser Phys Chem Sci 2:7–30
10. Youssef A, El-Telbany M, Zekry A (2017) The role of artificial intelligence in photo-voltaic systems design and control: a review. Renew Sustain Energy Rev 78:72–79
11. Atsu D, Dhaundiyal A (2019) Effect of ambient parameters on the temperature distribution of photovoltaic (PV) modules. Resources 8(2):107
12. Freitas S, Serra F, Brito MC (2015) PV layout optimization: string tiling using a multi-objective genetic algorithm. Sol Energy 118:562–574

13. Tiwari A, Kumar N, Banerjee MK (2024) Applications of genetic algorithm in prediction of the best achievable combination of hardness and tensile strength for graphene reinforced magnesium alloy (AZ61) matrix composite. Res Control Optim 14:100334
14. Ibnelouad A, El Kari A, Ayad H, Mjahed M (2020) Improved cooperative artificial neural network-particle swarm optimization approach for solar photovoltaic systems using maximum power point tracking. Int Trans Electr Energy Syst 30(8):e12439
15. Zhang C, Zhang M (2022) Wavelet-based neural network with genetic algorithm optimization for generation prediction of PV plants. Energy Rep 8:10976–10990
16. Saadaoui D, Elyaqouti M, Assalaou K, Lidaighbi S (2021) Parameters optimization of solar PV cell/module using genetic algorithm based on non-uniform mutation. Energy Convers Manag X 12:100129
17. Lo WL, Chung HSH, Hsung RTC, Fu H, Shen TW (2024) PV panel model parameter estimation by using particle swarm optimization and artificial neural network. Sensors 24(10):3006
18. Kumar KS, David PW (2024) Performance analysis of winnowing dynamic reconfiguration in partially shaded solar photovoltaic system. Sol Energy 268:112309
19. Abdulla H, Sleptchenko A, Nayfeh A (2024) Photovoltaic systems operation and maintenance: a review and future directions. Renew Sustain Energy Rev 195:114342
20. Tina GM, Scavo FB (2022) Energy performance analysis of tracking floating photovoltaic systems. Heliyon 8(8)
21. Anwaar A, Ashraf A, Bangyal WHK, Iqbal M (2022) Genetic algorithms: brief review on genetic algorithms for global optimization problems. Hum-Centered Cognit Syst (HCCS) 2022:1–6
22. Ma H, Zhang Y, Sun S, Liu T, Shan Y (2023) A comprehensive survey on NSGA-II for multi-objective optimization and applications. Artif Intell Rev 56(12):15217–15270
23. Liu Y, You K, Jiang Y, Wu Z, Liu Z, Peng G, Zhou C (2022) Multi-objective optimal scheduling of automated construction equipment using non-dominated sorting genetic algorithm (NSGA-III). Autom Constr 143:104587
24. Tiwari A, Bansal P, Vasnani H, Pandel N (2025) Non dominated sorting genetic algorithm (NSGA-II) modelling for AZ 31 magnesium alloy composite. In: 2025 International conference on cognitive computing in engineering, communications, sciences and biomedical health informatics (IC3ECSBHI). IEEE, pp 268–272
25. Chahal A, Gulia P, Gill NS, Chatterjee JM (2022) Performance analysis of an optimized ANN model to predict the stability of smart grid. Complexity 2022(1):7319010
26. Srivastava A, Kumar A (2022) A back propagation NN to optimize the IoT network. In: 2022 International conference on computer communication and informatics (ICCCI). IEEE, pp 1–4

Part IV
Domain-Specific Applications

Chapter 8
Air Quality Monitoring Using Internet of Things (IoT) Technologies and Case Studies

Murali Dadi, Stanford Siachoono, Francis Munalula, and Mathews Chibuluma

Abstract Monitoring air quality is essential for understanding environmental pollution and alleviating its impact on public health. With advancements in Internet of Things (IoT) technologies, air quality monitoring systems have turned more efficient, affordable, and available. This chapter examines the integration of IoT in monitoring air quality, highlighting crucial technologies, frameworks, and practical case studies. Systems for monitoring air quality based on IoT incorporate affordable sensors, wireless communication networks, data analytics platforms, and cloud computing. These elements together function to deliver real-time data on different air pollutants, including particulate matter (PM), nitrogen dioxide (NO_2), sulphur dioxide (SO_2), carbon monoxide (CO), and ozone (O_3).This chapter presents details about various sensor technologies, such as metal oxide, optical, and electrochemical sensors, along with their suitability for IoT applications. It also covers wireless communication protocols such as NB-IoT, Zigbee, and LoRaWAN that aid in effective data transmission across different environments. It focuses on cloud-based platforms and edge computing solutions for data storage, processing, and visualization, emphasizing their importance in enabling real-time decision-making. In addition, various case studies from urban and industrial environments illustrate the practical

M. Dadi (✉)
Department of Chemistry, School of Mathematics and Natural Sciences, Copperbelt University, Kitwe, Zambia
e-mail: murali.dadi@cbu.ac.zm

S. Siachoono
Department of Zoology and Aquatic Sciences, School of Natural Resources, Copperbelt University, Kitwe, Zambia

F. Munalula
Department of Biomaterials Science and Technology, School of Natural Resources, Copperbelt University, Kitwe, Zambia

M. Chibuluma
Department of Information Systems and Technology, School of ICT, Copperbelt University, Kitwe, Zambia
e-mail: mathews.chibuluma@cbu.ac.zm

S. Gautam et al. (eds.), *Smart Environmental Monitoring*,
https://doi.org/10.1007/978-981-92-0294-2_8

uses and advantages of air quality monitoring systems based on IoT that can be analysed. These examples contribute to advancements in pollution detection, regulatory systems, and public awareness. Ultimately, the chapter discusses obstacles such as sensor calibration, data precision, and privacy concerns, while offering perspectives for future research and development.

Keywords Air quality monitoring · Internet of Things (IoT) · Air pollutants · Sensors · Wireless communication · Regulatory frameworks

8.1 Introduction

Currently, the monitoring of air quality has become a crucial element of both environmental and public health contexts globally [1]. Due to rapid urbanization, industrialization, and automotive, air pollution has emerged as one of the leading environmental health threats [2]. According to the World Health Organization (WHO), ambient air pollution accounts for an estimated 4.2 million deaths per year due to lung cancer, heart disease, stroke, and chronic respiratory diseases [3]. This, according to [3], is indicative of the urgent need to shift to the use of real time, granular data on air quality monitoring in the public health point of view.

Precise air quality data assists policymakers and environmental agencies to develop targeted mitigation strategies, regulate industrial emissions, and raise public awareness. Moreover, as the impact of poor air quality significantly increases in urban areas and low-income communities, monitoring systems also play a significant role in promoting environmental justice [4].

According to WHO [3], exposure to pollutants such as oxides of nitrogen, sulphur, carbon monoxide (CO), particulate matter (PM2.5 and PM10), and ozone (O_3) causes severe respiratory illnesses, cardiovascular diseases, and premature mortality. Vulnerable populations, including children, the elderly, and individuals with pre-existing health conditions are particularly susceptible to the effects of long-term exposure [5].

In addition to effects on human health, air pollution has been found to have an impact on ecosystems through ozone layer depletion, acid rain, and global warming, which ultimately reduces agricultural yields. Globally, the need to mitigate the effects of air pollution significantly increases expenditures due to healthcare costs and lost labour efficiency. According to the Organization for Economic Cooperation and Development (OECD), these costs are projected to rise to $2.6 trillion annually by 2060 if no effective mitigation measures are implemented [6].

Traditional air quality monitoring locations are usually large, costly, and operated by governmental or organizational entities. These systems, while very precise, have numerous constraints [7] that include limited spatial coverage, due to the elevated installation and upkeep expenses, these stations are unevenly distributed and might miss localized pollution occurrences. Limited temporal resolution: Conventional stations typically present averaged information over extended periods, lacking real-time insights. Operational complexity: Significant technical knowledge is necessary for upkeep and data analysis.

As opposed to traditional methods, air quality monitoring systems that are reliant on IoT provide a revolutionary option. These systems utilize networks of less costly, Internet-linked sensors that gather, send, and assess environmental data in real time. Through the decentralization of monitoring systems and the facilitation of decision-making across various levels—from governments to individuals—that is data-informed, IoT is transforming our comprehension and reactions to air pollution [8].

The recent research [9] has revealed that IoT systems can enhance traditional networks by offering detailed, hyper-local information on air quality. Initiatives like AirVisual, PurpleAir, and SmartAQnet assess the feasibility and effects of deploying IoT-based sensors in both urban and rural environments [10].

8.2 Overview of IoT in Environmental Monitoring

Through data collection in real time, remote sensing, and adaptive data processing, the Internet of Things (IoT) is revolutionizing environmental monitoring. Simply put, IoT refers to a network of connected physical devices equipped with sensors, software, and communication features that enable them to gather and share data. In the context of environmental management—including air quality, water quality, noise pollution, and climate patterns—IoT provides continuous, detailed data that improves scientific insights and helps in policy actions [11].

Conservative monitoring techniques are huge and costly, commonly restricted to a small number of integrated sites, the resulting IoT has become an effective tool for monitoring of air quality in the recent past. This is due to the fact that IoT systems assist immense, and dispersed operations which provides data related to the air pollutants in detail which helps to look after the main cause of pollution and regulatory measures. Mainly IoT applications are more useful in urban areas due to dense population and industrialization [12]. Current research studies reveal that IoT-based monitoring of air quality applications provide vibrant knowledge and transparency on air pollution related issues which are crucial for monitoring public health and environmental policies [13].

8.2.1 Key Components of an IoT-Based Air Quality Monitoring System

Collection, transmission, and processing of data are the three main components of an efficient IoT-based air quality monitoring systems which work together to convey real-time environmental discernments [14].

Low-Cost Air Quality Sensors

Mainly, sensor technology is the base of IoT air quality monitoring systems. IoT systems, in comparison with conventional monitoring systems which depend on huge and expensive laboratory-quality instruments to detect the air pollution which is overcome by simple and cost-effect sensors. These sensors can be widely used in remote or resource-limited locations [15].

Commonly used low-cost sensors are:

- Metal Oxide Semiconductor (MOS) sensors, which are used for identifying gases like CO, NO_2, and VOCs;
- Optical sensors that are employed in the detection of particulate matter (PM) via light scattering methods; and
- Electrochemical sensors. These are recognized for their great selectivity and commonly utilized for gases such as SO_2 and O_3 [16].

Although affordable sensors are easier to obtain, they pose challenges concerning calibration, precision, and responsiveness. As a result, the recent efforts at improvement have concentrated on sensor fusion methods and AI-driven calibration models to enhance performance. Numerous research efforts, including those referenced in multiple journals, highlight hybrid methods that integrate various types of sensors to improve reliability [17].

Wireless Communication Networks

For monitoring in real time, data collected through sensors must be transmitted efficiently to central or distributed processing units. IoT air quality systems employ a variety of wireless communication protocols, chosen based on factors such as power consumption, range, bandwidth, and deployment context [18].

Commonly used protocols include:

- **Long Range Wide Area Network (LoRaWAN) which,** due to its low-power consumption and long range, is ideal for rural and large-scale deployments.
- **Narrowband IoT (NB-IoT), which** offers strong indoor coverage and low latency, thus making it ideal for urban and semi-urban environments.
- **Zigbee and Wi-Fi, a protocol that is** suitable for short-range communication in dense urban setups or for indoor monitoring.

The selection of a communication protocol considers factors such as the required transmission frequency, the physical environment, and energy constraints. In edge computing scenarios, some data processing is handled locally at the device level, thus reducing the need for constant transmission and minimizing bandwidth usage [19].

Data Analysis and Cloud Computing

To generate actionable insights, the data must be stored, processed, and analysed once it is collected. This is where cloud computing and, increasingly, edge computing play vital roles.

- **Cloud computing** provides scalable infrastructure for data storage, advanced analytics, and visualization. Platforms like Amazon Web Services (AWS), Microsoft Azure, and Google Cloud host real-time dashboards and provide APIs for data integration into apps or decision support systems.
- **Data analysis** includes a range of techniques, from basic statistical analysis to machine learning models, which predict pollution levels, detect anomalies, or classify pollution sources.
- **Edge computing**, on the other hand, processes data closer to the source (i.e. at the sensor or local gateway). This approach reduces latency, enhances privacy, and lowers reliance on continuous internet connectivity—a key advantage in areas with low-connectivity [20].

Platforms like *OpenAQ*, *AirVisual*, and *ThingSpeak* illustrate real-world applications of IoT data analysis, offering public accessible pollution data through cloud-hosted APIs [21].

8.2.2 Types of Air Pollutants Monitored Using IoT

Due to having sensors tailored to monitor specific chemical or physical properties, IoT-based systems can detect a wide range of pollutants. The most commonly targeted air pollutants include:

Particulate Matter (PM)

Particulate matter, especially PM2.5 and PM10, refers to tiny airborne particles that can penetrate deep into the respiratory system. Subramanian and Khatri [22] state that these particles are emitted from motor vehicles, industrial processes, construction sites, and natural sources like wildfires and dust storms.

To estimate PM concentration in real time, IoT sensors use light scattering techniques (optical sensors). Advances in low-cost PM sensors, such as the SDS011 and PMS5003, have significantly increased the feasibility of large-scale monitoring, although calibration remains necessary to ensure accuracy [23].

Nitrogen Dioxide (NO_2)

NO_2 is a toxic gas produced by combustion processes, primarily vehicle emissions and industrial activities. Respiratory problems and increased asthma incidences are linked to exposure to NO_2 [24].

Electrochemical sensors are commonly used to detect NO_2 at the ppm or ppb level. IoT systems used to monitor NO_2 aid cities evaluate traffic patterns, assess emission zones, and develop congestion mitigation strategies [25].

Sulphur Oxides (SO_2)

Sulphur oxides, mainly sulphur dioxide (SO_2), are produced by the burning of fossil fuels that contain sulphur, e.g. coal and oil. These gases contribute to acid rain and respiratory problems.

Monitoring SO_2 requires selective and stable sensors, often electrochemical in nature. IoT networks can detect SO_2 in industrial zones and near power plants, enabling compliance with emission regulations and early warning systems [26].

Carbon Monoxide (CO)

Carbon monoxide, a significant indoor and outdoor air pollutant with potentially fatal health impacts at high concentrations, is a colourless, odourless gas resulting from incomplete combustion [27].

CO detection is typically done using MOS and electrochemical sensors. IoT applications for CO monitoring extend to smart homes, parking garages, and tunnels, as well as broader urban environments [28].

Ozone (O_3)

Chemical reactions between volatile organic compounds (VOCs) and Nox in sunlight result in the formation of a secondary pollutant called ground-level ozone. It contributes to smog and has adverse effects on health, particularly in sensitive populations [29].

The monitoring of ozone involves the use of electrochemical and UV-based sensors. IoT-based ozone tracking helps evaluate air quality compliance, especially during high-ozone seasons in urban areas.

Miller et al. [30] state that integrating IoT technologies in monitoring air quality marks a significant shift from centralized, infrastructure-heavy systems to dynamic, decentralized networks. These networks offer insights into pollution patterns in real time, empower citizens through accessible data, and enable decision-making informed by reliable data at all levels of society. The multidimensional IoT systems offer affordable sensors to cloud-driven analytics which compromises litheness, scalability, and robustness.

Nevertheless, the widespread execution of IoT for air pollution monitoring promotes new progressions in sensor accuracy, data collection, verification, and system compatibility and also it shows the way to include AI, machine learning, and other advanced tools in monitoring of pollution strategies leads to promote sustainable urban clean environment.

8.3 Sensor Technologies for IoT-Based Air Monitoring

Sensors is the basic fundamental tool for every IoT technology which monitors the air pollution use to detect and compute air toxins. The efficiency and relevance of an IoT air monitoring system largely depends on the qualities and roles of its sensors. Advantages of using sensors as choice in IoT-based air pollution monitoring systems of their sensitivity, precision, accuracy, adaptability and cost-effectiveness [12].

8.3.1 Types of Sensors Used in IoT Air Quality Monitoring

(a) **Metal Oxide Semiconductor Sensors (MOS)**

Metal oxide semiconductor (MOS) sensors work based on the principle that the alterations of conductivity caused by gas adsorption on the sensor's surface. When a particular gaseous air pollutant such as sulphur oxides (SO_x) or nitrogen dioxide (NO_2) or carbon monoxide (CO) interacts with the sensor, a chemical reaction takes place which leads to altering the resistance of the metal oxide layer (usually tin dioxide, SnO_2). This alteration is measured and quantify the gas concentration [31].

MOS sensors are known for their:

- high sensitivity to a broad range of gases
- quick response and recovery times, and
- long operational life under stable conditions

Despite their advantages, MOS sensors also have several limitations, the most important of which involve:

- Selectivity issues: They often respond to multiple gases, making it difficult to distinguish between pollutants.
- High power consumption: Most require elevated operating temperatures (~200–400 °C), which demands continuous heating and impacts energy efficiency.
- Drift and ageing: In varying humidity or temperature environments, sensor readings may over time drift. This calls for frequent recalibration.

To ensure improved reliability [32], MOS sensors such as the MQ-series (e.g. MQ-135, MQ-7) which are widely used in low-cost IoT applications must often be paired with data correction algorithms.

(b) **Optical Sensors**

Optical sensors, particularly those for particulate matter (PM), work based on light scattering principles. A laser or infrared light source illuminates the air sample, and particles in the air scatter the light. The intensity and pattern of this scattered light are detected and used to calculate the concentration and size of particles, such as PM1.0, PM2.5, and PM10 [33].

Optical sensors have advantages that include:

- High resolution for PM data: They are able to measure size distribution and concentration simultaneously.
- Non-intrusive operation: There are no chemical reactions involved, which reduces wear and degradation.
- Having low maintenance and fast response time.

Optical sensors like the Plantower PMS5003 or Sensirion SPS30 are frequently deployed in citizen science and smart city air monitoring projects due to their affordability and reasonable accuracy [34].

Limitations of optical sensors include:

- Environmental interference: Due to humidity and condensation light can be scattered, which leads to measurement inaccuracies.
- Calibration challenges: Field calibration is often necessary to account for local aerosol types and environmental variability.
- Dust accumulation: Optical chambers require occasional cleaning to maintain accuracy.

(c) **Electrochemical Sensors**

Electrochemical sensors which are widely used to monitor gases like NO_2, SO_2, CO and O_3 detect these gases through a redox reaction between the target gas and an electrode immersed in an electrolyte [35]. The resulting current is proportional to the concentration of the target gas, thus enabling highly accurate and selective measurements. Their key strengths include:

- High specificity and selectivity for individual gases
- Low-power consumption (suitable for battery-powered IoT devices)
- Stable performance over moderate timeframes

Electrochemical sensors such as those produced by Alphasense and SPEC sensors are often chosen for research-grade applications and urban network deployments.

While electrochemical sensors generally are more expensive than MOS or optical sensors, they offer superior accuracy. However, they also

- Have a shorter lifespan (typically 1–3 years)
- Are highly sensitive to temperature and humidity, requiring compensation circuitry
- Need regular calibration, especially in outdoor deployments (Table 8.1)

There is no single sensor technology that is optimal for all use cases. To get around this, hybrid sensor platforms that combine multiple sensor types—and apply correction algorithms based on environmental context—are becoming increasingly common.

Table 8.1 Comparison of sensor technologies

Type	Cost	Accuracy	Power consumption	Sensitivity	Typical use
Metal oxide	Low	Moderate	High	High	General gas detection, indoor/outdoor
Optical (PM)	Moderate	Moderate–high	Low–moderate	High (PM only)	PM monitoring in urban areas
Electrochemical	Moderate–high	High	Low	High	Regulatory-grade gas detection

8.3.2 Challenges in Sensor Deployment

Although IoT-based air monitoring promises broad accessibility and real-time insights, field deployment of sensor systems presents a range of challenges [36].

Calibration and Drift

Sensor calibration is a persistent challenge, especially for low-cost sensors. Without proper calibration against reference-grade instruments, sensors may report inaccurate or misleading data. There are two main calibration approaches:

- **Laboratory calibration**, which ensures baseline accuracy but may not reflect field conditions.
- **Field calibration**, where sensors are co-located with regulatory-grade stations. This method offers better real-world performance but is resource-intensive.

Sensor drift—where performance degrades over time—also necessitates recalibration at regular intervals. AI and machine learning models are now being employed to automate recalibration by learning patterns and adjusting for drift dynamically [17].

Lifespan and Maintenance

Sensor longevity varies significantly by type and operating conditions. Electrochemical sensors often degrade faster due to electrolyte depletion, while MOS sensors may suffer from contamination build-up. Optical sensors require physical cleaning and can become less accurate due to internal dust accumulation.

Routine maintenance, battery replacement (in wireless setups), and firmware updates are essential for sustained performance—factors that must be considered in cost and labour planning [37].

Environmental Interference

Environmental factors such as humidity, temperature fluctuations, wind, and precipitation can significantly influence sensor readings. For example:

- **Humidity** affects light scattering in optical sensors and can skew PM readings.

- **Temperature** influences chemical reactions in MOS and electrochemical sensors, leading to signal distortion.
- **Dust, smoke, or aerosols** may confuse multi-gas MOS sensors, reducing specificity.

To counter these influences, many systems integrate temperature and humidity sensors for environmental compensation and use weatherproof casings to protect sensitive components [7].

Data Standardization and Interoperability

The lack of standard data formats, calibration protocols, and performance benchmarks for low-cost sensors hampers the integration of IoT data into national or international monitoring frameworks. Efforts like the **European AirSensEUR** initiative and the **U.S. EPA Sensor Performance Evaluation** programme attempt to address this by setting testing criteria and validating devices [17].

8.4 Wireless Communication Protocols for IoT Air Quality Systems

Wireless communication serves as the essential backbone of any Internet of Things (IoT) system. In air quality monitoring, where sensors are frequently placed in spread-out, resource-limited, and occasionally difficult-to-access locations [38], selecting a wireless protocol can greatly influence system performance, dependability, and sustainability.

8.4.1 Main Wireless Communication Technologies

- Narrowband IoT (NB-IoT)

NB-IoT is a cellular Low-Power Wide Area Network (LPWAN) typically created by 3GPP which functions within certified spectrum bands and is designed for low-power devices that need infrequent reliable data transmission [39].

Key Features

- Operates in the Long-Term Evolution (LTE) network infrastructure.
- Supports deep indoor coverage and wide area deployment.
- Data rates: up to 250 kbps.
- Power consumption: very low, enabling battery life of 5–10 years.
- Suitable for delay-tolerant applications with low to moderate data needs.

Use Case in Air Quality Monitoring

NB-IoT is compatible tool exclusively for urban and suburban air pollution monitoring since these areas are well equipped and easy accessible with telecom infrastructure. It is preferably designed for static air quality stations that send regular data to central systems [39].

- Zigbee

Zigbee is a short-range, low-power wireless protocol centred on the IEEE 802.15.4 standard. It is frequently used for wire networking in smart home, industrial automation, and environmental monitoring applications [40].

Key Features

- Short range (10–100 m), depending on the environment.
- Operates in the 2.4 GHz ISM band.
- Low data rate (~250 kbps).
- Supports mesh networking, allowing multi-hop communication.
- Moderate power consumption; suitable for sensor nodes powered by coin cells or USB.

Use Case in Air Quality Monitoring

Zigbee is frequently utilized in indoor or campus-level settings like educational institutions, workplaces, or production facilities. Its mesh capability allows it to span obstacles and extend coverage by relaying data through neighbouring nodes.

- Long Range Wide Area Network (LoRaWAN)

LoRaWAN is another LPWAN protocol, known for its ultra-long range and minimal energy requirements. It uses unlisensed ISM bands (e.g. 868 MHz in Europe, 915 MHz in the US) and features star-of-stars topology, where end nodes communicate with gateways.

Key Features

- Extremely long range: 2–5 km in urban, 15+ km in rural areas.
- Very low-power consumption; battery life can exceed 10 years.
- Low data rate (~0.3–50 kbps).
- End-to-end encryption for secure communication.

Use Case in Air Quality Monitoring

LoRaWAN is especially advantageous in rural or hard-to-reach areas where cellular infrastructure is limited. It enables city-wide or regional-scale air quality networks with minimal infrastructure costs (Table 8.2) [41].

Each protocol offers distinct advantages. NB-IoT provides reliability and compatibility with existing cellular infrastructure. Zigbee enables robust short-range networking, and LoRaWAN stands out for its extreme range and power efficiency.

Table 8.2 Comparative analysis of communication protocols

Feature	NB-IoT	Zigbee	LoRaWAN
Range	1–10 km (urban/suburban)	10–100 m (indoor/urban)	2–15+ km (urban/rural)
Data rate	Up to 250 kbps	~250 kbps	0.3–50 kbps
Power consumption	Low	Moderate	Very low
Network topology	Star (via LTE base stations)	Mesh	Star-of-stars
Lisensed spectrum	Yes (cellular)	No	No
Infrastructure cost	Medium (reliant on telecom)	Low (requires local routers)	Low (gateways required)
Deployment complexity	Low (if LTE available)	Medium (mesh setup needed)	Medium
Best use case	Urban, indoor/outdoor	Indoor, small-scale networks	Rural, city-wide monitoring

8.4.2 Factors Influencing Protocol Selection

The choice of a wireless communication protocol for air quality monitoring depends on several deployment-specific factors:

Deployment Environment: Urban Versus Rural

- Urban areas benefit from existing infrastructure. Particularly in urban areas NB-IoT cellular networks become reasonably good choice for monitoring air pollution strategies due easy availability of LTE towers assists informal distribution of services.
- In contrast, rural or remote areas due to lack of good cellular coverage, LoRaWAN became an ideal tool. Its long-term competences and unconventionality from commercial networks make it economical and accessible in low-connectivity regions [42].

Data Volume and Frequency

- NB-IoT supports with higher data output, allowing it appropriate for applications that require regular updates or multiple data channels (e.g. air quality, temperature, humidity, noise).
- Zigbee supports moderate data traffic but it has a disadvantage in using high-frequency transmissions due to its limited range and higher power usage in transmits.
- LoRaWAN is optimized for small, infrequent data packets (e.g. hourly PM2.5 readings) and is ideal for battery-operated sensor nodes that transmit only when thresholds are crossed [43].

Power Consumption and Energy Budget

- Battery life is a major consideration in remote and difficult-to-maintain locations. LoRaWAN and NB-IoT are designed for ultra-low-power operation, often enabling years-long deployment without battery replacement.
- Zigbee nodes consume more power when operating in mesh networks, especially if they serve as repeaters. This makes them less suitable for energy-constrained environments unless powered via a continuous supply [44].

Cost Considerations

- Zigbee modules are inexpensive and popular for short-range, low-scale deployments.
- LoRaWAN sensors and gateways have a modest initial cost, but the absence of licensing fees for spectrum use makes them highly economical for large networks.
- NB-IoT, while leveraging existing cellular networks, may incur subscription or SIM-related costs depending on the telecom provider, which can add to long-term operational expenses [45].

Latency and Real-Time Requirements

For most air quality monitoring applications, latency is not critical—especially in non-industrial use cases. However, real-time systems (e.g. industrial safety monitoring or early warning systems for smog or gas leaks) may benefit from:

- NB-IoT's relatively low latency within LTE-supported environments
- Zigbee's rapid response within short distances
- LoRaWAN's higher latency, which is acceptable for most environmental monitoring but may not suit emergency alerts [38].

Scalability and Network Density

- **Zigbee** networks can become complex in dense environments due to interference at the 2.4 GHz band and mesh traffic overhead.
- **LoRaWAN** scales well for wide area sensor networks and is increasingly used in smart cities for monitoring air, water, and noise pollution.
- **NB-IoT** allows for the connection of thousands of devices per cell tower, making it suitable for highly populated urban regions [47].

8.5 Cloud Computing and Edge Computing in Air Quality Monitoring

Due to the expansion of sensor-based technologies in environmental monitoring, a huge amount of data has been produced, necessitating robust systems for data processing, analysis, visualization, and storage. In IoT-driven air quality monitoring systems [48, 49] cloud computing and edge computing work together to ensure that this data is handled efficiently, reliably, and securely. Cloud computing offers a centralized platform for extensive data storage and analysis, whereas edge computing moves data processing near the origin, delivering quicker response times and enhanced data privacy. Collectively, these technologies facilitate more intelligent and adaptive air quality monitoring systems.

8.5.1 Role of Cloud Computing in IoT-Based Air Quality Monitoring

Cloud computing involves use of Internet-connected servers and platforms for data storage, processing, and analysis. In IoT ecosystems, cloud services serve as the foundational infrastructure for connecting numerous sensors and offering scalable, real-time insights [50].

Data Storage and Processing

In air quality monitoring, the function of **centralized data storage is performed by cloud platforms**. To do this, air quality sensors deployed across a number of geographic locations generate continuous streams of data—ranging from PM2.5 concentrations to levels of gases like NO_2, CO, and O_3. [46] state that this data must be stored in a manner that supports:

- **High availability** (data accessible at any time)
- **Durability** (data is safe from loss or corruption)
- **Scalability** (the system can handle growing data volumes)

Cloud services like **Amazon Web Services (AWS), Microsoft Azure,** and **Google Cloud Platform (GCP)** deliver consistent, well distributed storage solutions such as Amazon S3, Azure Blob Storage, and Google Cloud Storage. In addition to storage, these platforms support **advanced data processing pipelines** include:

- Pre-processing (cleaning and standardization)
- Accretion (combining data from multiple sensors)
- Cutting-edge analytics (pattern detection, anomaly detection)
- Machine learning (predictive modelling of pollution trends)

Cloud-based processing permits experts to monitor pollution levels in real time and perform historical analysis to discover long-term patterns.

Visualization and Real-Time Reporting

Cloud computing permits the progress of **real-time control panel and visualization tools that are capable of** transforming raw data into actionable insights. These control panels often feature:

- Interactive maps showing pollutant concentrations across regions
- Time-series charts demonstrate pollution strategies
- Alert systems that generate warnings when pollutant levels exceed safe threshold [51].

Open-source platforms such as **Grafana** and **Kibana**, and patented solutions like **Power BI**, are commonly used for real-time data imagining. APIs provided by cloud platforms enable the integration of these dashboards into mobile apps and information portals to the public.

Moreover, **reporting** in **real time** can be programmed through cloud-based tools that generate and disseminate periodic air quality reports to stakeholders, including policymakers, researchers, and the general public [52].

8.5.2 Edge Computing for Decentralized Air Quality Monitoring

Edge computing denotes to the dispensing of data at or near the data source include a local sensor node or gateway instead of depend on a centralized cloud. It counterparts cloud computing by enabling localized decision-making and reducing the burden on network bandwidth and cloud resources [53].

Reduced Latency

A key benefit of edge computing is its nominal dormancy. Particularly, conditions like detecting dangerous gas leaks in industrial areas where quick response is needed, cloud-based solutions might not respond quickly because of transmission and processing delays.

By analysing data at the edge, IoT air quality monitoring systems are able to:

- Instantly detect differences and activate local alarms
- Eliminates unrelated data prior to transmission
- Make autonomous decisions (e.g. activating ventilation or purification systems)

For example, in a factory environment, edge-enabled sensors could autonomously shut down machinery or alert workers without waiting for cloud validation. According to Arroba et al., [54] this ability to immediately respond is important for maintaining safety and operational steadiness.

Data Security and Privacy Benefits

Edge computing improves data security and privacy, mostly in circumstances comprising sensitive or location-related information. In some cases, sending raw data to the cloud could prompt worries about data possession, monitoring, or adherence to regulations like the General Data Protection Regulation (GDPR), the European Union regulation on information privacy in the European Union (EU) and the European Economic Area (EEA).

- Edge computing address these matters by:
- Keeping sensitive data local, reducing exposure to external threats
- Encrypting or anonymizing data before transmission
- Allowing users or communities greater control over their data

This is particularly relevant for **community-led air quality initiatives**, where local stakeholders may prefer to manage data autonomously without relying on third-party cloud providers.

Additionally, as per Nguyen et al., [55] decreased data transmission leads to weakened bandwidth usage and energy consumption, interpreting edge computing ideal for battery-operated and remote deployments.

8.5.3 Comparison: Cloud Versus Edge Computing in Air Quality Monitoring

While cloud and edge computing provide inimitable advantages, the best design frequently utilizes a hybrid model that combines the benefits of each. The table below offers a comparative summary of cloud and edge computing (Table 8.3).

8.6 Case Studies: Real-World Applications and Hybrid Models

Modern air quality monitoring systems are progressively revolving to the use of cloud and edge computing hybrid strategies.

Table 8.3 Contrasting of cloud and edge computing

Feature	Cloud computing	Edge computing
Location of processing	Centralized (data centres)	Decentralized (sensor/gateway level)
Latency	Moderate to high	Very low
Data storage	Scalable, long-term storage	Limited, temporary or selective storage
Bandwidth usage	High (raw data sent to cloud)	Low (filtered data sent)
Energy consumption	Low at device level	Moderate, depending on edge device
Security and privacy	Data hosted on external servers	Local control, reduced external exposure
Maintenance requirements	Managed by cloud providers	Requires local device management
Scalability	Excellent (pay-as-you-grow)	Limited by local hardware
Use case suitability	Large-scale, historical analysis	Real-time response, privacy-sensitive use

8.6.1 *Urban Environment Case Study: Smart Cities and Public Health*

Case: SmartAQnet (Germany)

The SmartAQnet project in Germany serves as a good example of how IoT can be incorporated into smart city frameworks to advance monitoring of urban air quality [56]. Sprang in cities such as Augsburg and Munich, the project combines official monitoring stations with hundreds of low-cost, mobile IoT sensors installed on vehicles, bicycles, and public transport.

System Architecture

- **Sensors**: These devices are equipped with PM2.5, NO_2, and O_3 sensors.
- **Communication**: LoRaWAN and LTE networks are using for communication.
- **Processing**: Edge computing for instant readings and cloud platforms for data aggregation and analysis.
- **Visualization**: Open-access maps and real-time dashboards used by citizens and municipal agencies.

Impact and Outcomes

- Improved **spatial granularity** of air quality data, especially in traffic-heavy corridors and low-coverage neighbourhoods.
- Enabled **data-driven policy decisions**, including adjustments to traffic flow and emissions zones.
- Contributed to **public health awareness**, with data integrated into mobile apps that alert users about high pollution levels and suggest safer travel routes.

Challenges

- Sensor calibration and validation against certified monitoring stations.
- Data privacy concerns related to mobile sensors collecting location-specific data.
- Ensuring interoperability across different communication networks.

8.6.2 *Industrial Setting Case Study: Pollution Control and Regulatory Compliance*

Case: Oil Refinery Air Monitoring System (India)

In response to firmer environmental regulations from the Central Pollution Control Board (CPCB), a major oil refinery in India implemented an IoT-based air quality monitoring network to detect and control elusive emissions [57].

System Components

- **Sensors**: The sensors, namely the electrochemical sensors using for SO_2, NO_x, and CO; optical sensors using for PM2.5 and PM10.
- **Connectivity**: This is executing by NB-IoT and LTE-M networks due to the availability of robust telecom infrastructure.
- **Edge gateways**: Installed across stacks and boundary areas to enable local alerting systems.
- **Cloud platform**: Provides real-time dashboards to plant managers and interfaces with CPCB's Continuous Emission Monitoring Systems (CEMS).

Impact and Benefits

- Achieved **regulatory compliance** by meeting mandatory real-time emissions reporting requirements.
- Enabled **predictive maintenance** by detecting emission anomalies linked to equipment faults.
- Reduced emissions through **automated control integration**, such as triggering scrubbers and filters when pollution thresholds were exceeded.

Challenges and Mitigation

- Harsh industrial environment required **rugged sensor enclosures** and regular maintenance schedules.
- High pollutant concentration areas affected sensor lifespan, prompting frequent recalibrations.
- Workers required training to interpret sensor readings and act appropriately.

8.6.3 Community-Led Initiatives: Citizen Science and Low-Cost IoT Solutions

Case: Luftdaten (Germany and Global)

The **Luftdaten** project, originated in Stuttgart, Germany, is a grassroots initiative that empowers individuals to build and deploy their own low-cost air quality monitoring kits. Using open-source hardware and software, it has grown into a global movement [58].

Technology Stack

- **Sensors**: PMS5003 optical PM sensors, DHT22 for temperature and humidity.
- **Connectivity**: Wi-Fi, with data sent to a public server.
- **Open Data**: Data shared on an interactive global map (sensor.community) with thousands of nodes.

Achievements

- Over **15,000 active sensors worldwide**, providing hyper-local PM2.5 and PM10 readings.

- Raised **public awareness** about air pollution in areas lacking official monitoring stations.
- Encouraged **policy change**: Data from Luftdaten has influenced debates on urban planning and traffic regulation in several European cities.

Limitations and Considerations

- Sensor accuracy is not on par with certified instruments; however, the sheer volume of data and calibration models improve reliability.
- **Data quality control** and maintenance depend on individual users.
- Project success hinges on **community engagement and education**.

8.7 Challenges and Future Directions

In the recent past, a sharp increase in the use of IoT technology for monitoring air quality, technology which provides real time, granular, and scalable data that can drive environmental policy, industrial compliance, and citizen engagement has been seen. This technology has completely changed how we identify and address air pollution. However, as with any rapidly evolving technology, IoT-based air quality monitoring presents a unique set of challenges and limitations that must be addressed to ensure reliable, long-term impact [59].

8.7.1 Challenges in IoT-Based Air Quality Monitoring

Sensor Calibration and Maintenance

Calibration remains one of the most pressing challenges in the deployment of IoT air quality sensors. Low-cost sensors, while affordable and scalable, often lack the precision of regulatory-grade instruments. Their readings can vary significantly based on environmental conditions such as temperature, humidity, and barometric pressure.

Common issues include:

- **Sensor drift** over time, resulting in measurement inaccuracies.
- **Lack of uniform calibration standards**, making it difficult to compare data across regions or systems.
- **Field calibration complexity**, requiring co-location with reference stations, which is time-consuming and resource-intensive.

Additionally, regular sensor replacement and maintenance can be expensive and time-consuming, particularly in hostile conditions. This is particularly challenging for large-scale deployments in remote or industrial locations with limited accessibility [7].

Data Accuracy and Standardization

As a consequence of having of a variety of sensor types, vendors, and software platforms, data formats, accuracy levels, and interpretation frameworks have become fragmented. Without harmonization, integrating data from multiple sources or comparing results across different geographies becomes difficult.

Key challenges include:

- **Inconsistent measurement intervals** and data resolutions.
- **Lack of common quality assurance protocols** for low-cost sensors.
- **Discrepancies in pollutant definitions** and thresholds across regions.

These challenges are being addressed through initiatives such as the European AirSensEUR project and the U.S. EPA's Air Sensor Toolbox, which offer validation protocols and recommendations for sensor assessment. Still, a universally accepted standard for low-cost air monitoring devices remains indefinable [60].

Privacy and Security Concerns

As IoT systems collect and transmit environmental and location-based data, concerns regarding **data privacy and cybersecurity** increasingly become important. Despite air quality data appearing to be non-sensitive, the integration of geolocation, timestamps, and networked infrastructure poses risks [61].

Issues include:

- **Unauthorized data access or tampering**, especially in cloud-based systems.
- **Location tracking**, which may reveal patterns about individuals or communities.
- **Misuse of data**, particularly in private or commercial applications (for instance, leveraging pollution information to affect property values or introduce policy bias).

EU laws such as the General Data Protection Regulation (GDPR) place emphasis on the importance of protecting communication channels and anonymize user data. Nonetheless, these legal structures do not consistently align with IoT technologies, resulting in regulatory ambiguity [62].

8.7.2 Future Research Directions

The future of IoT-enabled monitoring of air quality is bright despite the obstacles discussed. The combination of innovation and research is expected to overcome current constraints and open up new avenues. There are three main areas where upcoming advancements are anticipated to have the greatest influence.

Advancements in AI and Machine Learning for Air Quality Prediction

Artificial Intelligence (AI) and Machine Learning (ML) are increasingly essential for improving the effectiveness and dependability of air quality monitoring systems. These technologies can be utilized in a variety of ways:

- **Sensor calibration**: AI models can autonomously adjust sensor drift by leveraging historical data and proximity to reference stations.
- **Predictive analysis**: Machine learning algorithms can anticipate pollution occurrences using weather data, traffic behaviours, and historical pollution patterns.
- **Anomaly detection**: Uncommon measurements—resulting from sensor malfunctions or pollution surges—can be highlighted and examined instantly [14].

New technologies such as neural networks, support vector machines, and reinforcement learning are being investigated for their capability to identify subtle patterns in intricate datasets. These techniques are already demonstrating potential in initiatives such as AirNet, where neural networks enhance the precision of inexpensive PM2.5 sensors.

The combination of AI and edge computing facilitates on-device intelligence, permitting sensors to self-correct, minimize redundant data transmission, and function independently in isolated settings [63].

Integration with Smart City Frameworks

As cities globally dedicate resources to digital infrastructure, air quality systems based on IoT are progressively integrating into larger smart city frameworks. This integration enables environmental data to guide real-time decision-making in various areas, such as:

- **Traffic management**: Adjusting traffic flows or restricting vehicular access during high pollution periods.
- **Public health alerts**: Issuing specific alerts to at-risk groups like those with asthma or senior citizens.
- **Energy optimization**: Managing HVAC systems in structures according to outdoor air quality levels [64].

Environmental sensors have already integrated into smart city systems in cities such as Barcelona, Singapore, and Amsterdam. However, integration of these sensors calls the use of open data systems, synchronized urban regulations, and interoperability standards. All areas that are ideal for continued research and development.

Additionally, the Internet of Behaviour (IoB)—which examines data from interconnected devices to shape behaviours—can utilize air quality information to motivate public transport usage, encourage cycling, or impact patterns of outdoor activities [65].

Emerging Sensor Technologies

The advancement of sensor technology is closely linked to the future of air quality monitoring. Scientists are investigating innovative materials, downsizing techniques, and multi-sensing features that are expected to enhance both precision and cost-effectiveness.

Significant advancements consist of:

- Nano-enabled sensors: Substances such as graphene and carbon nanotubes provide enhanced sensitivity and quicker response rates.
- Multi- analyst sensors: Instruments that can identify various contaminants within one device, minimizing the requirement for numerous setups.
- Wearable and flexible sensors: Small, skin-adhering sensors could enable personal monitoring of air quality, providing detailed exposure information for users.
- Self-powered sensors: Instruments energized by surrounding energy (e.g. solar, kinetic) to prolong functional life in distant installations [12].

8.8 Conclusion

The monitoring of air quality has progressive significantly from the recent past, heading from centralized, expensive measurement systems to decentralized, real-time networks made possible by the Internet of Things (IoT). Using a detailed analysis of systems design, communication protocols, sensor technologies, data processing platforms, and real-world applications the air quality monitoring becoming possible.

The integration of IoT in air quality monitoring implies a key change from reactive, centralized systems to proactive, decentralized networks that enable detecting pollutants in different ways. With increasing the accessibility and usability of environmental data, IoT technologies empower individuals to effectively address air pollution.

IoT systems provide governments and regulatory agencies with innovative tools to guarantee environmental agreement, enhance urban infrastructure, and protect public health. For the industry, they offer immediate feedback mechanisms that can improve operational efficiency and ecological liability. For people and communities, monitors based on IoT enhance awareness, support, and changes in behaviour by providing direct access to local air quality data.

One of the most significant aspects of IoT is its role as a link between the physical and digital demesnes, establishing a smart feedback loop where environmental factors influence verdicts in transportation, energy consumption, healthcare, and city planning. In this way, IoT offerings air quality monitoring not just as a scientific issue but as an essential component of wellbeing in the twenty-first century.

References

1. Relvas H, Lopes D, Armengol JM (2025) Empowering communities: advancements in air quality monitoring and citizen engagement. Urban Clim 60:102344. ISSN 2212-0955. https://doi.org/10.1016/j.uclim.2025.102344
2. Chen F, Zhang W, Mfarrej MFB, Saleem MH, Khan KA, Ma J, Raposo A, Han H (2024) Breathing in danger: understanding the multifaceted impact of air pollution on health impacts. Ecotoxicol Environ Saf 280:116532. https://doi.org/10.1016/j.ecoenv.2024.116532
3. WHO (2024) Air pollution and health training toolkit for health workers: train-the-trainers manual, pp 1–139
4. Khan AA, Kumar P, Gulia S, Khare M (2024) A critical review of managing air pollution through airshed approach. Sustain Horizons 9:100090. https://doi.org/10.1016/j.horiz.2024.100090
5. Emmanuel OK, Javiera Aria DJ, Diego C (2024) Toxicological effects of air pollution on respiratory diseases. 1–28
6. World Bank Group (2020) The global health cost of ambient PM_2 air pollution, pp 1–51
7. Seesaard T, Kamjornkittikoon K, Wongchoosuk C (2024) A comprehensive review on advancements in sensors for air pollution applications. Sci Total Environ 951:175696. https://doi.org/10.1016/j.scitotenv.2024.175696
8. Nisanthan G, Shanaz ALF (2024) IoT-based air quality monitoring and prediction system. J Inf Commun Technol (JICT) 02(01):1–6
9. Mota A, Serôdio C, Briga-Sá A, Valente A (2025) Implementation of an Internet of Things architecture to monitor indoor air quality: a case study during sleep periods. Sensors 25(6):1683. https://doi.org/10.3390/s25061683
10. Truong TP, Nguyen DT, Truong PV (2021) Design and deployment of an IoT-based air quality monitoring system. Int J Environ Sci Dev 12(5):139–145. https://doi.org/10.18178/ijesd.2021.12.5.1331
11. Saraiva de Sousa NF, Lachos Perez DA, Rosa RV, Santos MAS, Esteve Rothenberg C (2019) Network service orchestration: a survey. Comput Commun 142–143:69–94. https://doi.org/10.1016/j.comcom.2019.04.008
12. Shahid S, Brown DJ, Wright P, Khasawneh AM, Taylor B, Kaiwartya O (2025) Innovations in air quality monitoring: sensors, IoT and future research. Sensors 25(7):2070. https://doi.org/10.3390/s25072070
13. Katie B (2024) Internet of Things (IoT) for environmental monitoring. Int J Comput Eng 6(3):29–42. https://doi.org/10.47941/ijce.2139
14. Ramadan MNA, Ali MAH, Khoo SY, Alkhedher M, Alherbawi M (2024) Real-time IoT-powered AI system for monitoring and forecasting of air pollution in industrial environment. Ecotoxicol Environ Saf 283:116856. https://doi.org/10.1016/j.ecoenv.2024.116856
15. Li Y, Yuan Z, Chen L-WA, Pillarisetti A, Yadav V, Wu M, Cui H, Zhao C (2022) From air quality sensors to sensor networks: things we need to learn. Sens Actuat B Chem 351:130958. https://doi.org/10.1016/j.snb.2021.130958
16. Ródenas García M, Spinazzé A, Branco PTBS, Borghi F, Villena G, Cattaneo A, Di Gilio A, Mihucz VG, Gómez Álvarez E, Lopes SI, Bergmans B, Orłowski C, Karatzas K, Marques G, Saffell J, Sousa SIV (2022) Review of low-cost sensors for indoor air quality: features and applications. Appl Spectrosc Rev 57(9–10):747–779. https://doi.org/10.1080/05704928.2022.2085734
17. Nalakurthi NVSR, Abimbola I, Ahmed T, Anton I, Riaz K, Ibrahim Q, Banerjee A, Tiwari A, Gharbia S (2024) Challenges and opportunities in calibrating low-cost environmental sensors. Sensors 24(11):3650. https://doi.org/10.3390/s24113650
18. Zafra-Pérez A, Medina-García J, Boente C, Gómez-Galán JA, Sánchez de la Campa A, de la Rosa JD (2024) Designing a low-cost wireless sensor network for particulate matter monitoring: implementation, calibration, and field-test. Atmos Pollut Res 15(9):102208. https://doi.org/10.1016/j.apr.2024.102208

19. Diane A, Diallo O, Ndoye EH (2025) A systematic and comprehensive review on low power wide area network: characteristics, architecture, applications and research challenges. Discov Internet Things 5(7). https://doi.org/10.1007/s43926-025-00097-6
20. Kahveci S, Alkan B, Ahmad MH, Ahmad B, Harrison R (2022) An end-to-end big data analytics platform for IoT-enabled smart factories: a case study of battery module assembly system for electric vehicles. J Manuf Syst 63:214–223. https://doi.org/10.1016/j.jmsy.2022.03.010
21. Lymperis D, Goumopoulos C (2023) SEDIA: a platform for semantically enriched IoT data integration and development of smart city applications. Future Internet 15(8):276. https://doi.org/10.3390/fi15080276
22. Subramanian A, Khatri SB (2019) The exposome and asthma. Clin Chest Med 40(1):107–123. https://doi.org/10.1016/j.ccm.2018.10.017
23. Morawska L, Asbach C, Patel H (2025) Application of PM 2.5 low-cost sensors for indoor air quality compliance monitoring. Aerosol Sci Technol 1–11. https://doi.org/10.1080/02786826.2025.2457326
24. Bhattarai K, Lamsal L, Gyawali M, Neupane S, Gautam SP, Bakshi A, Yeager J (2024) Impact of nitrogen dioxide (NO_2) pollution on asthma: the case of Louisiana state (2005–2020). Atmosphere 15(12):1472. https://doi.org/10.3390/atmos15121472
25. Colléaux Y, Willaume C, Mohandes B, Nebel J-C, Rahman F (2025) Air pollution monitoring using cost-effective devices enhanced by machine learning. Sensors 25(5):1423. https://doi.org/10.3390/s25051423
26. Arya R, Jitin, Tyagi S, Devi K (2025) Chemical sensors for monitoring the environment. Int J Res Publ Rev 6(3):9890–9892
27. Salthammer T (2024) Carbon monoxide as an indicator of indoor air quality. Enviro Sci Atmos 4(3):291–305. https://doi.org/10.1039/D4EA00006D
28. He Y, Jiao M (2024) A mini-review on metal oxide semiconductor gas sensors for carbon monoxide detection at room temperature. Chemosensors 12(4):55. https://doi.org/10.3390/chemosensors12040055
29. Zhang J (Jim), Wei Y, Fang Z (2019) Ozone pollution: a major health hazard worldwide. Front Immunol 10. https://doi.org/10.3389/fimmu.2019.02518
30. Miller T, Durlik I, Kostecka E, Kozlovska P, Łobodzińska A, Sokołowska S, Nowy A (2025) Integrating artificial intelligence agents with the Internet of Things for enhanced environmental monitoring: applications in water quality and climate data. Electronics 14(4):696. https://doi.org/10.3390/electronics14040696
31. Khoramshahi V, Azarang M, Nouri M, Shirmardi A, Yousefi R (2024) Metal oxide/g-C3N4 nanocomposites chemiresistive gas sensors: a review on enhanced performance. Talanta Open 9:100290. https://doi.org/10.1016/j.talo.2024.100290
32. Sharma N, Choudhury SP (2024) First-principles study of metal oxide semiconductors for gas sensing applications: a brief review. Mod Phys Lett B 38(21). https://doi.org/10.1142/S0217984923420125
33. Molaie S, Lino P (2021) Review of the newly developed, mobile optical sensors for real-time measurement of the atmospheric particulate matter concentration. Micromachines 12(4):416. https://doi.org/10.3390/mi12040416
34. Bulot FMJ, Russell HS, Rezaei M, Johnson MS, Ossont SJ, Morris AKR, Basford PJ, Easton NHC, Mitchell HL, Foster GL, Loxham M, Cox SJ (2023) Laboratory comparison of low-cost particulate matter sensors to measure transient events of pollution—part B—particle number concentrations. https://doi.org/10.20944/preprints202304.0701.v1
35. Baranwal J, Barse B, Gatto G, Broncova G, Kumar A (2022) Electrochemical sensors and their applications: a review. Chemosensors 10(9):363. https://doi.org/10.3390/chemosensors10090363
36. Sathyamoorthy S, Matthew UO, Adekunle TS, Okafor NU (2024) Advances and challenges in IoT sensors data handling and processing in environmental monitoring networks. Hafed Poly J Sci Manag Technol 5(2):40–60. https://doi.org/10.4314/hpjsmt.v5i2.3
37. Bounegru AV, Dinu Iacob A, Iticescu C, Puiu LG (2025) Electrochemical sensors and biosensors for the detection of pharmaceutical contaminants in natural waters—a comprehensive review. Chemosensors 13(2). https://doi.org/10.3390/chemosensors13020065

38. Rahmadani AA, Syaifudin YW, Setiawan B, Panduman YYF, Funabiki N (2024) Enhancing campus environment: real-time air quality monitoring through IoT and web technologies. J Sens Actuator Netw 14(1):2. https://doi.org/10.3390/jsan14010002
39. Noura H, Hatoum T, Salman O, Yaacoub J-P, Chehab A (2020) LoRaWAN security survey: issues, threats and possible mitigation techniques. Internet Things 12:100303. https://doi.org/10.1016/j.iot.2020.100303
40. Singh AP, Tomar P (2021) AI and IoT capabilities: standards, procedures, applications, and protocols. In: Artificial intelligence to solve pervasive internet of things issues. Elsevier, pp 67–83. https://doi.org/10.1016/B978-0-12-818576-6.00004-6
41. Jabbar WA, Mei Ting T, Hamidun MFI, Che Kamarudin AH, Wu W, Sultan J, Alsewari AA, Ali MAH (2024) Development of LoRaWAN-based IoT system for water quality monitoring in rural areas. Expert Syst Appl 242:122862. https://doi.org/10.1016/j.eswa.2023.122862
42. Routray S, Mohanty S (2024) Cellular Internet of Things: principles, potentials and use cases. EAI Endors Trans Internet Things 10. https://doi.org/10.4108/eetiot.4656
43. Ugwuanyi S, Paul G, Irvine J (2021) Survey of IoT for developing countries: performance analysis of LoRaWAN and cellular NB-IoT networks. Electronics 10(18):2224. https://doi.org/10.3390/electronics10182224
44. Islam M, Jamil H, Pranto S, Das R, Amin A, Khan A (2024) Future industrial applications: exploring LPWAN-driven IoT protocols. Sensors 24(8):2509. https://doi.org/10.3390/s24082509
45. Ali S, Glass T, Parr B, Potgieter J, Alam F (2020) Low cost sensor with IoT LoRaWAN connectivity and machine learning based calibration for air pollution monitoring. IEEE Trans Instrum Meas. https://doi.org/10.1109/TIM.2020.3034109
46. Nugroho HYSH, Basuki TM, Pratiwi, Savitri E, Supangat AB, Putra PB, Purwanto, Wahyuningrum N, Adi RN, Setiawan O, Nandini R, Cahyono SA, Auliyani D, Nada FMH, Pratiwi D, Hasani M (2025) Advancing air quality monitoring systems towards sustainable green development: insight for metropolitan cities in Indonesia. Environ Sustain Indic 26:100649. https://doi.org/10.1016/j.indic.2025.100649
47. Yousuf AM, Rochester EM, Ousat B, Ghaderi M (2018) Throughput, coverage and scalability of LoRa LPWAN for Internet of Things. In: 2018 IEEE/ACM 26th international symposium on quality of service (IWQoS), pp 1–10. https://doi.org/10.1109/IWQoS.2018.8624157
48. Ficili I, Giacobbe M, Tricomi G, Puliafito A (2025) From sensors to data intelligence: leveraging IoT, cloud, and edge computing with AI. Sensors 25(6):1763. https://doi.org/10.3390/s25061763
49. Ullah I, Adhikari D, Su X, Palmieri F, Wu C, Choi C (2024) Integration of data science with the intelligent IoT (IIoT): current challenges and future perspectives. Digit Commun Netw. https://doi.org/10.1016/j.dcan.2024.02.007
50. Dritsas E, Trigka M (2025) A survey on the applications of cloud computing in the industrial Internet of Things. Big Data Cognit Comput 9(2):44. https://doi.org/10.3390/bdcc9020044
51. Daniel S, Brightwood S, Oluwaseyi J (2024) Cloud-based big data analytics (aws, azure, google cloud)
52. Venkatramulu S, Phridviraj MSB, Srinivas C, Rao VCS (2021) Withdrawn: implementation of Grafana as open source visualization and query processing platform for data scientists and researchers. Mater Today Proc. https://doi.org/10.1016/j.matpr.2021.03.364
53. Andriulo FC, Fiore M, Mongiello M, Traversa E, Zizzo V (2024) Edge computing and cloud computing for Internet of Things: a review. Informatics 11(4):71. https://doi.org/10.3390/informatics11040071
54. Arroba P, Buyya R, Cárdenas R, Risco-Martín JL, Moya JM (2024) Sustainable edge computing: Challenges and future directions. Softw Pract Exp 54(11):2272–2296. https://doi.org/10.1002/spe.3340
55. Nguyen T, Nguyen H, Nguyen Gia T (2024) Exploring the integration of edge computing and blockchain IoT: principles, architectures, security, and applications. J Netw Comput Appl 226:103884. https://doi.org/10.1016/j.jnca.2024.103884

56. Schäfer K, Emeis S, Budde M, Beigl M, Cyrys J, Schnelle-Kreis J, Philipp A, Ziegler V, Riedel T, Grimm H, Gratza T (2017) SmartAQnet: remote and in-situ sensing of urban air quality. In: Comerón A, Kassianov EI, Schäfer K (eds) Remote sensing of clouds and the atmosphere XXII. SPIE, p 12. https://doi.org/10.1117/12.2282698
57. Central Pollution Control Board (2024) NAMP
58. Hamm A (2020) Particles matter. In: Proceedings of the 7th international conference on ICT for sustainability, pp 305–313. https://doi.org/10.1145/3401335.3401815
59. Irawan Y, Wahyuni RM, Fonda H, Luthfi Hamzah M, Muzawi R (2021) Real time system monitoring and analysis-based Internet of Things (IoT) technology in measuring outdoor air quality. Int J Interact Mob Technol (IJIM) 15(10):224. https://doi.org/10.3991/ijim.v15i10.20707
60. Edayan JM, Gallemit AJ, Sacala NE, Palmer X-L, Potter L, Rarugal J, Velasco LC (2024) Integration technologies in laboratory information systems: a systematic review. Inform Med Unlock 50:101566. https://doi.org/10.1016/j.imu.2024.101566
61. Tawalbeh L, Muheidat F, Tawalbeh M, Quwaider M (2020) IoT privacy and security: challenges and solutions. Appl Sci 10(12):4102. https://doi.org/10.3390/app10124102
62. Houichi M, Jaidi F, Bouhoula A (2024) Cyber security within smart cities: a comprehensive study and a novel intrusion detection-based approach. Comput Mater Continua 81(1):393–441. https://doi.org/10.32604/cmc.2024.054007
63. Ma M (2024) Application of machine learning in electronic device fault diagnosis. J Comput Commun 12(11):130–140. https://doi.org/10.4236/jcc.2024.1211010
64. Dahmane WM, Ouchani S, Bouarfa H (2024) Smart cities services and solutions: a systematic review. Data Inf Manag 100087. https://doi.org/10.1016/j.dim.2024.100087
65. Yang R, Zhen F (2024) Smart city development models: a cross-cultural regional analysis from theory to practice. Res Glob 8:100221. https://doi.org/10.1016/j.resglo.2024.100221

Chapter 9
Smart IoT Monitoring Across Environmental Matrices: IoT-Enabled Detection in Water, Air, and Soil Systems

Matbiangthew Shadap and Sakunthala Ayyasamy

Abstract Rapid industrialization and urbanization in modern society have led to the development of serious environmental problems, which makes it necessary to monitor ionic pollutants in water, air, and soil in real time accurately and effectively. Although traditional laboratory-based methods are accurate, they are generally slow and resource-intensive, making them costly. The application of ion-based environmental monitoring systems integrated with the Internet of Things (IoT) has significantly improved this area by providing real-time, continuous, and intelligent monitoring of important key ionic parameters, including heavy metals, nutrients, pH, and salinity. This chapter offers a detailed survey of recent developments in IoT-driven ion detection technologies, discussing state-of-the-art sensors, such as ion-selective electrodes, electrochemical and optical sensors, and bio-based detection methods, highlighting their working mechanisms, sensitivity, and real-world applications. The discussion is extended to wireless sensor networks, cloud computing, real-time data analytics, and artificial intelligence for remote monitoring, early warning systems, and predictive environmental modeling. Network technologies and communication protocols (LoRaWAN, NB-IoT, 5G) are discussed for their role in effective data transmission and system scalability. Furthermore, this chapter includes practical applications of the concept as case studies that include the use of IoT-based ion monitoring in water management, air quality assessment, agriculture, and industrial pollution control, as well as their impact on policy decisions and environmental sustainability. Advanced sensing techniques with smart IoT solutions are likely to revolutionize global environmental governance, thereby preventing ecological risks and fostering more sustainable decision-making with the help of ion-based environmental monitoring for a safer future.

M. Shadap · S. Ayyasamy (✉)
Division of Physical Sciences, Karunya Institute of Technology and Sciences, Coimbatore, India
e-mail: sakunthala@karunya.edu

S. Gautam et al. (eds.), *Smart Environmental Monitoring*,
https://doi.org/10.1007/978-981-92-0294-2_9

9.1 Introduction and Background

Measurement of specific ions or charged chemical species in environmental samples (water, soil, and air) to assess quality and detect pollution is called ion-based environmental monitoring. It is a critical component of environmental science because many key indicators of environmental health are ionic in nature, such as nutrient levels (nitrate and phosphate), pH (hydrogen ions), salinity (sodium and chloride), and toxic contaminants (heavy metal ions). The ionic parameters can be tracked by scientists and engineers to gain insight into water quality, soil fertility, and pollution levels, which enables the informed management and protection of ecosystems and public health [1, 2]. The importance of ion-based monitoring spans drinking water safety, agricultural productivity, pollution control, and ecosystem conservation. Soil ion measurements of nitrogen and potassium help determine crop nutrients and detect harmful salt accumulation. Industrial facilities use ion monitoring of effluents to verify that waste discharge stays within acceptable safety limits [3].

Prior to modern technology, environmental monitoring relied on sporadic sampling followed by laboratory testing. Laboratory methods yield precise results but require substantial effort and produce data that reflect only momentary conditions; they are also prone to human error and cannot capture rapid or transient changes in environmental parameters [4]. Field-deployed ion sensors enable either continuous or real-time monitoring, detecting both temporal changes and early signs of contamination events. Real-time monitoring has become essential because environmental systems undergo rapid changes, such as pollution spills in rivers and nutrient runoff, following heavy rainfall. Continuous ion monitoring enables early detection, leading to prompt responses that prevent environmental damage and health risks [5]. Figure 9.1 presents the chapter structure and flow.

Ion-based monitoring has been applied across many environmental management domains. In this section, a brief discussion is provided for the application of ion-based monitoring, and Fig. 9.2 presents concise information on the discussion.

Water Quality: It involves monitoring of ions such as pH (H^+), dissolved oxygen (often measured via O_2 or related redox ions), nitrates, phosphates, ammonium, chloride, and heavy metals in rivers, lakes, groundwater, and drinking water systems. These metrics indicate the potability of water and the ecological health of aquatic habitats. For example, it is important to maintain nitrate levels below regulatory limits (often ~10 mg/L as N in drinking water) according to the US EPA to prevent infant health disorders, hypoxia, and toxicity in aquatic life [6, 7].

Soil Health: Farmers use measurements of nutrient ions (NO_3^-, NH_4^+, K^+, PO_4^{3-}) to control fertilizer application and sustain soil fertility. Precision agriculture benefits from ion sensors through IoT soil nitrate sensors, which direct fertilization schedules to provide adequate nutrients, while preventing water contamination from surplus nutrients [8, 9]. Soil salinity sensors monitor salt ions to inform irrigation practices that protect crops from damage. The immediate analysis of ions enables better crop production while minimizing environmental contamination [10].

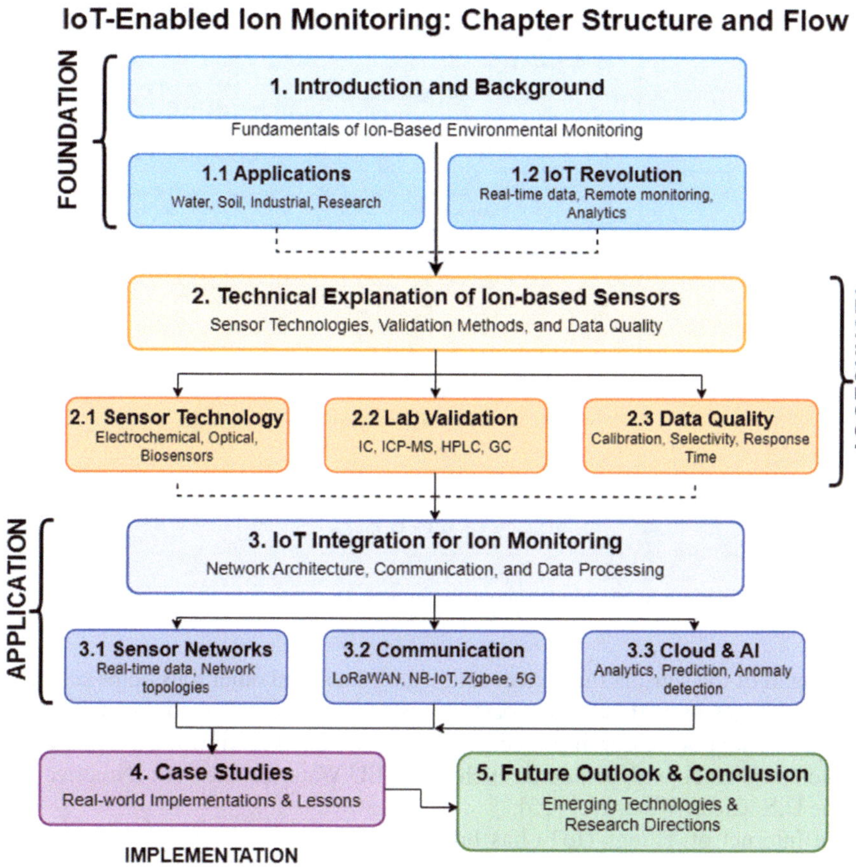

Fig. 9.1 IoT-enabled ion monitoring: chapter structure and flow

Industrial Emissions and Waste: Industrial effluents and air emissions generate ionic pollutants, such as sulfate, nitrate, and heavy metal ions in wastewater, and acidic gases that dissociate upon dissolution. Industrial wastewater treatment requires continuous ion monitoring to verify that the discharge water complies with regulatory standards for heavy metals, nitrates, and phosphates [11]. Stack monitoring sensors that detect gases usually work alongside sensors that measure particulate ion content or scrubber performance in air pollution control [12]. Continuous emission monitoring systems (CEMS), which power plants and large factories must use according to EPA for tracking SO_2 and NO_x, requires similar concepts for water discharge monitoring [13].

Environmental Research and Ecosystem Monitoring: Scientists use ion sensors to monitor natural water and soil environments to study biogeochemical processes. Sensor networks deployed in watersheds track stream water chemistry by measuring the pH, conductivity, and nutrient ions to study rainfall and pollution effects. The

Environmental Matrix	Key Ions Monitored	Significance/Relevance
Air	Sulfate (SO_4^{2-}), Nitrate (NO_3^-), Ammonium (NH_4^+)	Major components of particulate matter and acid rain; indicators of air pollution
	Nitrogen Dioxide (NO_2), Sulfur Dioxide (SO_2)	Precursor gases to sulfate and nitrate aerosols; greenhouse gases and respiratory irritants
Water	Calcium (Ca^{2+}), Magnesium (Mg^{2+}), Sodium (Na^+), Potassium (K^+)	Major ions; contribute to water hardness and salinity
	Nitrate (NO_3^-), Nitrite (NO_2^-), Phosphate (PO_4^{3-}), Ammonium (NH_4^+)	Nutrients; indicators of pollution from agriculture and sewage
	Heavy Metals (**Pb, Cd, As, Hg, Cr, Cu, Zn**)	Toxic pollutants; harmful to human and aquatic life
Soil	Nitrate (NO_3^-), Ammonium (NH_4^+), Phosphate (PO_4^{3-}), Potassium (K^+)	Macronutrients essential for plant growth; indicators of soil fertility
	Calcium (Ca^{2+}), Magnesium (Mg^{2+}), Sulfate (SO_4^{2-})	Secondary nutrients important for plant health and soil structure
	Sodium (Na^+), Chloride (Cl^-)	Indicators of salinity, affecting plant growth
	Heavy Metals (**Pb, Cd, As, Hg, Cr, Cu, Zn**)	Toxic contaminants; can accumulate in soil and plants

Fig. 9.2 Common ions monitored in different environmental matrices

deployment of ion sensors that detect salinity, nitrates, and other ions helps scientists monitor changes in wetlands and coastal ecosystems. The collected data from these applications support research projects and environmental monitoring programs that enforce compliance with standards such as the EU Water Framework Directive [14] and the U.S. Clean Water Act [15].

The Internet of Things (IoT) has brought about a revolutionary change in the modernization of ion-based environmental monitoring. IoT is a system of interconnected sensors and devices that communicate through the Internet or wireless networks to acquire data automatically and transmit it for analysis. Environmental monitoring through the IoT enables continuous data collection from networks of ion sensors and other environmental sensors that transmit data instantly to cloud platforms or central databases [16]. This has several key benefits.

Real-Time Data and Alerts: The continuous data transmission of IoT-connected ion sensors allows for real-time dashboards and immediate alerts when the parameters exceed safe limits. An IoT pH and turbidity sensor deployed in a river can send alerts instantly when pH levels drop unexpectedly because of a chemical spill, thus enabling a prompt authority response before laboratory results become available weeks later [17].

Remote and Distributed Monitoring: The ability to connect wirelessly enables sensors to operate from distant locations, including mountain streams, offshore buoys, and extensive farmland areas, while continuing to transmit data. This technology enables extensive coverage because developing regions with minimal infrastructure can use low-power IoT sensors to monitor water quality in rural villages and track soil conditions on distant farms, thereby closing environmental data gaps. Such networks can merge multiple distributed ion sensors into a unified system [18].

Big Data and Analytics: The constant flow of IoT sensor data leads to the accumulation of big data in environmental monitoring. High-frequency data collection over long periods enables analytics and machine learning applications to detect patterns and trends and predict upcoming environmental conditions. The analysis of seasonal ion concentration trends enables predictions of upcoming algal blooms, and machine learning algorithms can identify unusual pollution events from normal sensor readings. The ability to predict future conditions using this insight is useful for proactive environmental management [19].

Therefore, the foundation of environmental quality protection and understanding relies on ion-based monitoring. The Internet of Things (IoT) has revolutionized this field by transforming traditional manual sampling into smart-connected sensor networks that deliver continuous environmental insights. The following sections of this chapter examine the technical fundamentals of ion sensors, how IoT is harnessed to deploy these sensors at scale, illustrative case studies from around the world, the regulatory context shaping monitoring practices, and emerging technologies that promise to further enhance ion-based environmental monitoring.

The detection and quantification of ions through environmental monitoring requires multiple sensor technologies for ion detection. Sensor technologies for environmental monitoring include classical electrochemical probes, advanced optical instruments, and biosensors. The unique operating principles and advantages of each sensor type were based on specific considerations. This section covers the principal ion sensor types and analytical methods, together with technical factors, including calibration and selectivity, which determine the quality of the data collected.

9.2 Sensor Technology

9.2.1 Electrochemical Sensors (Potentiometric, Amperometric, and Conductometric)

Ion-Selective Electrode (ISEs) Potentiometric Sensors: The operation of these potentiometric sensors produces an electric potential that corresponds to the target ion concentration, as described by the Nernst equation. When target ions change the charge distribution of an ion-selective membrane (which can be glass, polymer, or crystalline), specific ions can pass through to produce a measurable potential difference relative to a reference electrode [20]. ISEs provide benefits like real-time measurement across a wide range of concentrations, together with quantification of both cations and anions and ion activity measurement rather than concentration, while maintaining cost-effectiveness and operational simplicity[21]. The standard electrode types consist of glass electrodes for H^+ and Na^+, solid-state crystalline electrodes for halides (Cl^-, Br^-), polymer membrane electrodes for nutrient ions (NO_3^-, K^+, Ca^{2+}), and gas-sensing electrodes that measure ionic products produced during gas reactions (e.g., CO_2 sensors that quantify HCO_3^-). ISEs combine portability with minimal power demands, which makes them ideal for field-based environmental monitoring tasks, such as pH assessment and detection of nutrients and ionic pollutants [22].

Amperometric and Voltammetric Sensors: These sensors determine ion concentrations through the electrical current output from a redox reaction with the target analyte. Amperometric sensors function by applying a constant potential to an electrode, which produces a measurable current from the redox reaction of the target species, and this current is proportional to its concentration. An amperometric nitrate sensor functions by converting nitrate into nitrite at a predetermined electrode potential and then records the resulting current flow. The measurement of dissolved oxygen and chlorine relies on amperometric principles, where oxygen reduction at the cathode produces a current that reflects the O_2 concentration [23]. Voltammetric sensors operate by sweeping potential levels and recording current responses to identify specific ions according to their redox signatures, including techniques such as anodic stripping voltammetry for trace metals, such as lead and copper, where these metals are first deposited and then removed, generating a distinctive current peak. These methods possess high sensitivity for detecting trace heavy metal levels but require sophisticated control systems and are susceptible to interference [24, 25].

Conductometric (Conductivity) Sensors: Conductometric sensors determine how well a solution conducts electricity, which depends on the total ionic concentration. Conductivity sensors provide aggregate measurements (μS/cm or mS/cm) that represent the combined effects of all dissolved ions, rather than the detection of individual ions. Conductivity sensors are highly beneficial for water quality monitoring because they function as general indicators (e.g., elevated conductivity levels may denote increased salinity or pollutant concentration). When conductivity levels rise, they often signal an increase in salinity levels or pollution levels in water samples [26]. Conductivity probes function with ion-selective and optical sensors in many IoT water-monitoring stations to help determine the overall ionic strength [27]. These probes deliver near-instantaneous readings, while electrochemical sensors react swiftly, and ion-selective electrodes stabilize after a few seconds to several minutes.

9.2.2 Optical and Spectroscopic Sensors

Optical sensors measure changes in light properties, such as absorbance, fluorescence, or scattering, which occur when the target analyte interacts with them to detect ions and substances. These sensors generally require reagents or special indicator materials because they provide sensitive and specific non-consumptive in situ monitoring capabilities for nutrients and organic matter.

Absorbance-Based Sensors (Spectrophotometry/Colorimetry): In absorbance-based sensors, spectrophotometry/colorimetry detects the amount of light that passes through a sample at designated wavelengths. Detection requires either the creation of colored complexes with ions through specific reagents (such as azo dyes for nitrate detection) or measuring the direct UV absorption of substances such as nitrate at 220 nm or dissolved organic carbon. In situ UV spectrophotometers enable continuous monitoring of nitrate levels in water by measuring absorbance directly, with compensation wavelengths used to account for organic substances [18]. Microfluidic systems allow field-based colorimetric measurements by automatically mixing samples with necessary reagents. These sensors face the main challenges of light scattering owing to turbidity/solids and optic fouling, which often results from biofouling. To prevent sensor fouling, commercial sensors utilize wipers or ultrasonic cleaning systems [18].

Fluorescence Sensors: These sensors operate on the principle that, when excited by a specific wavelength, they use compounds that emit light. The fluorescence intensity or wavelength behavior depends on the analyte concentration because ions can either quench or enhance the fluorescence. Key examples of these sensors are optical pH sensors with attached dyes that display fluorescence spectrum shifts (utilized in oceanographic research) and indicators for specific ions, such as calcium, magnesium, and ammonia. These sensors demonstrate superior performance in organic matter detection by utilizing native fluorescence from oils and humic substances while measuring indicators, such as chlorophyll in algae and tryptophan in wastewater. Through multi-wavelength fluorescence analysis, scientists can identify the origins of organic contaminants, such as sewage [28]. The greatest benefit of this method is its high sensitivity, which enables the detection of extremely low concentrations of pollutants or nutrients by amplifying weak signals against a dark background; this is superior to other methods. Fluorescence analysis requires proper calibration but remains vulnerable to temperature fluctuations and interfering signals [29].

Turbidity and Light-Scattering Sensors: Sensors that measure water turbidity (cloudiness caused by suspended particles) operate via optical scattering methods and do not target specific ions. These devices generate light using infrared light-emitting diodes (LEDs) and quantify the amount of light scattered at a given angle upon interaction with particles. Turbidity measurement demonstrates an indirect connection to particulate-bound pollutants because certain nutrients and metals can adhere to suspended particles. Turbidity reduces the accuracy of absorbance and fluorescence sensors by causing optical interference. High turbidity levels trigger compensation or flagging of spectroscopic sensor data via turbidity sensors integrated into modern water quality sensor packages. Optical scattering sensors serve as supporting tools for ion sensors because of their ability to deliver information about the clarity of samples [30].

9.2.3 Biosensors for Ion Detection

Biosensors utilize biological elements, such as enzymes, microorganisms, antibodies, and DNA aptamers for their recognition functions. The recognition element delivers specificity to the target ions or molecules in biosensors. A biocomponent's interaction with the target generates detectable signals in various forms, including electrical and optical outputs. Biosensors used in ion-based monitoring detect difficult-to-measure nutrients and pollutants by transforming them into forms that can be detected by sensors. These tools can be used to identify substances that exist in minimal amounts and measure bioavailable pollutant portions [5].

Enzyme-Based Biosensors: These biosensors use specific enzymes to trigger reactions with target analytes and then use transducers to produce measurable signals. Glutamate dehydrogenase-based ammonium sensors monitor NADH consumption, nitrate sensors use nitrate reductase, and phosphate sensors detect pH changes. These biosensors transform ionic concentration levels into measurable electrochemical signals and frequently operate within flow systems to monitor nutrients continuously. These sensors provide precise readings because they selectively target specific analytes in complex samples. The effectiveness of enzymes depends on environmental factors, such as temperature and pH, while their limited lifespan of weeks or months requires regular replacement or recalibration. Current scientific investigations have concentrated on enhancing enzyme stability via immobilization methods and advancing electron transfer functionality [31].

Microbial Sensors (Whole-Cell Biosensors): Microbial sensors utilize living microorganisms to detect target chemicals by monitoring measurable changes in respiration rate, metabolic activity, and luminescence. Bioluminescent bacteria function as sensors for water toxicity because toxic substances modify their light-production capabilities. Genetically engineered microbes generate signals when they encounter specific ions: Bacteria can emit light when bioavailable heavy metals, such as cadmium and arsenic, activate metal-responsive genetic promoters linked to luminescence genes. Whole-cell sensors are distinctive because they quantify the bioavailable fraction of pollutants that pose biological threats, rather than measuring total pollutant concentration. Environmental risk assessment becomes more accurate when scientists evaluate the fraction of contaminants that affects living systems [5].

However, these sensors are not without any drawbacks. Biological component stability issues (enzyme degradation and bacterial viability), intricate calibration procedures (responses more complex than ISEs), and sensitivity to environmental conditions (temperature and pH fluctuations that require control or adjustment) represent the main challenges. The presence of multiple substrates that influence the same biological pathway can lead to cross-sensitivity (e.g., BOD sensors detecting biodegradable matter instead of specific substances).

9.3 Laboratory Techniques for Validation

Laboratory-based analytical methods continue to serve as the benchmark for precision and thorough analysis of ionic species, despite the essential role of situ sensors in real-time monitoring. These methods serve dual purposes: They validate sensor networks through calibration and enable detailed analysis when continuous monitoring reveals an issue. Important laboratory techniques include the following.

Ion Chromatography (IC): Ion chromatography (IC) represents a high-performance liquid chromatography method that employs an ion-exchange column to separate ions and polar molecules through their unique interaction properties. Conductivity detection and mass spectrometry enable precise measurement of multiple anions (e.g., F^-, Cl^-, NO_3^-, SO_4^{2-}, PO_4^{3-}) and cations (e.g., NH_4^+, Na^+, Ca^{2+}). Ion chromatography achieves high precision with detection limits in the μg/L range while being appropriate for complex samples after necessary preparation steps, and it represents a standard method for environmental analysis such as in the U.S. EPA Method 300.0 [32]. Although IC functions mainly as a laboratory method requiring sample transportation and skilled personnel (which restricts real-time application), specialized portable or online IC systems provide continuous monitoring capabilities at a high cost [8].

Inductively Coupled Plasma Mass Spectrometry (ICP-MS) and Atomic Absorption Spectroscopy (AAS): These stand as the standard laboratory methods used to analyze trace metals including their ionic forms in water samples that require preparation. ICP-MS achieves exceptional sensitivity to parts-per-trillion levels, which makes it the standard reference method for monitoring environmental heavy metals, such as Pb, AS, and Cd. AAS (flame or furnace) stands as a validated laboratory technique specifically for detecting metal ions. Both laboratory techniques surpass the sensitivity of typical field sensors, although they require laboratory analyses. These methods are essential tools for ensuring that deployed in situ metal sensors function correctly and for acquiring precise measurements of metals that do not have trustworthy field monitoring technologies while enhancing ongoing datasets [5].

High-Performance Liquid Chromatography (HPLC) and Gas Chromatography (GC): To obtain comprehensive results, both HPLC and GC techniques are used to perform multiple chemical analyses. These analytical methods focus on organic compounds while allowing the measurement of some ionic analytes (such as nitroaromatic compounds or weak organic acids) after derivatization. The detection of pesticide residues in environmental monitoring can involve laboratory analysis using HPLC or GC, since these residues can be ionic or partially ionic in nature. Although these tools generally do not measure common inorganic ions because IC methods are preferred, they remain essential components in laboratory analysis [5].

IC and ICP-MS laboratory techniques ensure an accurate ground truth for IoT sensor networks because field sensors are prone to drift and interference. Regular laboratory analysis serves as a vital tool for quality assurance and calibration verification and must typically meet regulatory standards. Currently, IoT sensors and laboratory techniques are complementary: Sensors function as persistent monitoring devices that send alerts while laboratories provide precise validation and in-depth evaluations. The optimal monitoring approach combines uninterrupted sensor readings with regular laboratory evaluations to achieve comprehensive data quality and insights that cannot be delivered by sensors alone [33].

9.4 Calibration, Sensor Selectivity, and Data Quality Considerations

Successful ion-based monitoring in field environments depends on meticulous sensor calibration, regular maintenance routines, and the understanding of sensor constraints. Unlike controlled laboratory instruments, field sensors encounter different environmental factors that impact their data quality. This section presents the essential practices required to obtain reliable measurements.

Calibration Techniques: The calibration of sensors requires standard solutions like pH buffers at 7 and 4 or 10; and nitrate standards at 1, 10, 100 mg/L NO_3–N through one-point, two-point, or multi-point methods tailored to accuracy requirements. Calibration procedures replicate the sample matrix using comparable backgrounds and ionic strength adjusters. Field sensors drift over time, which necessitates regular recalibration schedules, such as daily or weekly intervals or after each deployment. Although IoT systems can automatically calibrate sensors, manual verification processes remain essential [33].

Sensor Selectivity and Interference: Sensors cannot achieve complete selectivity because ISEs respond to unintended ions (such as calcium ISEs responding to magnesium), and selectivity coefficients help evaluate these interferences. Interferences can be reduced through the application of specific membranes or by using differential measurements and masking techniques. Optical sensors, such as UV nitrate probes, require calibration adjustments for organic compounds in addition to the calibration needs for biosensors that encounter biological interference. Knowing the sample composition aids in predicting the interference selectivity [18].

Response Time: Response times vary; ISEs achieve stability after a few minutes while optical sensors respond in seconds, but biosensors require more time because biological processes affect their response times. To detect quick events, such as spikes in pollutants, the sampling rate needs to align with sensor response times, which means that fast sensors require frequent sampling.

Environmental Factors and Fouling: Biogrowth along with sediments and deposits cause fouling of field sensors, which results in inaccurate readings. Maintaining sensor functionality through regular cleaning is crucial where wipers or anti-fouling coatings apply protection, and redundant sensors maintain operational continuity. The performance of sensors changes with temperature and pressure fluctuations, which necessitates compensation through the use of co-located sensors or data adjustments [34].

Data Quality and Drift: Component degradation through reference electrode contamination, membrane aging, and environmental fouling causes sensor drift, which requires systematic mitigation. IoT systems maintain robustness through algorithmic quality control methods that trigger sensor maintenance by flagging deviations from the expected network-wide patterns and environmental trends through anomaly detection. The adaptive calibration process uses periodic laboratory comparisons or known reference backgrounds to make ongoing adjustments to baseline offsets, and routine zero-offset checks with ultra-pure water help separate instrument drifts [33]. By placing high-precision reference sensors in the network, designers can perform real-time corrections on less-complex IoT sensors to maintain data accuracy across distributed systems [35].

Environmental monitoring has undergone a major transformation through the Internet of Things (IoT), which allows sensor networks to continuously transmit data and communicate with one another. When integrating ion-based sensors into IoT systems, multiple factors must be addressed, including hardware networking and communication systems, data processing capabilities, and power management techniques. This section outlines the configuration process for wireless sensor networks that perform real-time ion monitoring and examines various communication technologies, such as LoRaWAN, NB-IoT, Zigbee, and 5G. This section also reviews how sensors communicate with the cloud through various technologies and analyses cloud computing with AI's role in data management, along with methods for sustainable IoT network implementation. A schematic flowchart of this process is shown in Fig. 9.3.

9.5 Wireless Sensor Network and Real-Time Data

Wireless sensor networks (WSNs) support IoT-based environmental monitoring by installing sensor nodes at specific locations. Each node integrates ion sensors, microcontrollers, and communication modules and uses rechargeable batteries powered by solar energy. The sensor nodes collect environmental data and either transmit it to the central receivers or relay it through gateway nodes, according to preset schedules [36]. In typical river-basin applications, network sensor nodes are deployed throughout the main channels and tributaries to collect measurements of pH, nitrate, and conductivity. Every five minutes, data from sensor nodes are received by local gateways, which forward the aggregated information to cloud servers, enabling real-time monitoring via web dashboards. IoT architectures enable remote access to environmental data via low-power wireless networks, enabling immediate analysis for

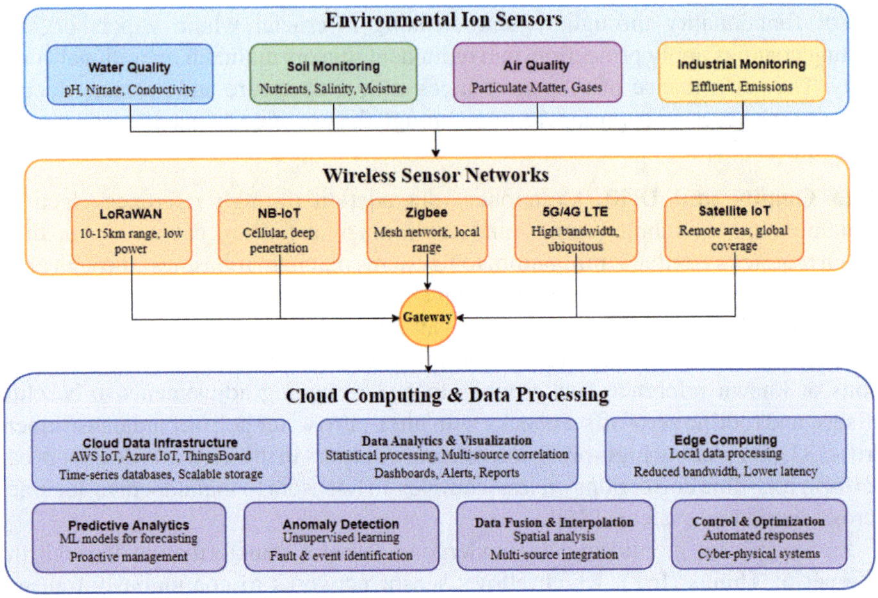

Fig. 9.3 IoT architecture for ion-based environmental monitoring

pollution monitoring and ecosystem management in compliance with regulations across remote areas [17].

Data from different sensors, such as ions, weather, and flow sensors, help to understand environmental cause–effect relationships, including instances of rainfall causing nutrient runoff. Ensuring high data quality requires frequent sensor calibration, which can be performed through field visits or automated techniques. Critical areas can benefit from redundant nodes and overlapping coverage to verify readings, while sensor algorithms detect faults by identifying anomalies such as constant values or implausible drifts [37]. Wireless sensor networks establish a dense network of environmental chemical sensors that provide comprehensive data across both space and time [38].

9.6 Communication Protocols

The selection of proper wireless communication technology is a vital factor for IoT sensor networks. Environmental monitoring protocols should support long distances because sensors are often placed far from infrastructure, while maintaining low power usage through batteries or solar power, with capabilities for moderate data rates, because ion sensors generate small data packets at slow rates. Several protocols have emerged as standards.

9.6.1 LoRaWAN (Long-Range Wide-Area Network)

The low-power wide-area network (LPWAN) protocol called LoRaWAN supports long-distance communication across kilometers while using minimal power [8]. The network uses unlicensed ISM frequency bands (868 MHz in Europe and 915 MHz in the US) combined with LoRa modulation to link multiple sensor nodes to a gateway that forwards this information to a network server through the Internet [39]. LoRaWAN networks feature long-range capabilities spanning 10–15 km in open spaces and 2–5 km in urban areas, combined with a low data rate perfect for small sensor data and battery life that lasts several years. LoRaWAN serves as a perfect solution for environmental IoT applications, such as agricultural monitoring, because it works efficiently with soil moisture and nutrient sensors spread across large farm fields [40]. One study highlighted that LoRa technology facilitates long-range data transmission while maintaining minimal energy consumption. However, as a star network and not a mesh, sensors must directly reach the gateway [17]. Moreover, duty cycle limits restrict the transmission time to avoid band congestion; however, this is suitable for periodic environmental sensors. LoRaWAN is most effective in remote or rural locations without cellular connectivity, such as forests, rivers, and coastlines.

9.6.2 NB-IoT (Narrowband IoT)

As a cellular-based LPWAN technology, Narrowband IoT (NB-IoT) operates within 4G/5G networks, following the standards set by 3GPP [41]. This technology operates in licensed spectrum bands with 200 kHz narrow channels to achieve deep indoor/underground signal penetration (such as basements and sewers) while maintaining ultra-low power usage, which results in devices such as urban utility sensors with a battery life that lasts for decades [42]. The NB-IoT uses existing cellular infrastructure to enable direct cloud connections without custom gateways and supports large device densities per cell exceeding 50,000 nodes, unlike unlicensed band solutions such as LoRaWAN [41]. The nature of the NB-IoT makes it particularly well suited for smart cities because telco coverage guarantees reliable data transmission from sensors in water pipelines and waste systems to centralized monitoring platforms. The 200 kHz bandwidth limits data throughput but improves signal stability, thereby balancing scalability and reliability in municipal IoT deployments.

9.6.3 Zigbee and Other Mesh Networks

The Zigbee wireless protocol operates over short distances with low power consumption under the IEEE 802.15.4 standard and operates at 2.4 GHz or sub-GHz frequencies, supporting mesh networking. This system enables sensors to link across localized networks, such as industrial facilities and small watersheds, by connecting devices located tens to hundreds of meters apart. The mesh topology of Zigbee supports multi-hop data routing, which improves the network coverage in obstructed and wide-ranging environments [43]. Because of its low energy requirements, Zigbee has become a popular choice for home automation and environmental monitoring applications, such as soil moisture sensing and air quality evaluation, while delivering better throughput than LoRa, despite having a smaller per-hop distance [44]. For dense sensor setups, such as farm-wide soil monitoring systems, data must be sent frequently to the main hubs, and sparse sensor networks operate better with LoRa and cellular connections. Zigbee shares unlicensed frequency bands with Wi-Fi by managing communication channels. IoT mesh networking capabilities are supported by similar technologies, including Thread and 6LoWPAN [43].

9.6.4 5G (and 4G LTE) Networks

5G (as well as 4G LTE) networks enable IoT data transmission, but standard cellular modems (3G/4G) consume considerable power, making it essential to achieve battery efficiency through duty cycling. The advent of 5G has brought about NB-IoT capabilities for massive device connections and ultra-low latency broadband, but many ion monitoring applications, such as water quality measurement, require low data rates rather than high bandwidth or sub-second latency. 5G's fast and reliable network capabilities allow drones to perform high-resolution reservoir monitoring and transmit chemical data instantly. LPWAN is still suitable for low-power fixed sensors, but 5G will become more valuable for IoT applications in agriculture and environmental management owing to its scalability and backhaul capacity as infrastructure development continues [45, 46]. Environmental IoT monitoring also employs multiple communication protocols, including LPWAN options such as Sigfox and LTE-M, short-range solutions such as BLE, and dedicated wired systems such as Modbus, depending on the environment. Optimal coverage is achieved using hybrid methods that integrate short-range protocols with cellular or satellite connections. Each protocol has trade-offs: LoRaWAN and NB-IoT offer extended range and low power consumption, but they support lower data rates than 5G, which provides higher bandwidth but requires more power. Satellite IoT technology has begun to gain traction for remote regions by allowing sensors to communicate directly with satellites [47].

9.7 Cloud Computing, Data Analytics, and AI in Sensor Data Processing

Data from ion sensors travel through the network to reach cloud-based platforms or central servers, where they are stored and analyzed. Environmental monitoring systems utilize this infrastructure to access extensive datasets through strong computational resources, which produce actionable insights surpassing conventional monitoring methods.

Cloud Data Infrastructure: IoT cloud platforms (AWS IoT, Azure IoT, and ThingsBoard) ingest sensor data streams through specialized time-series data pipelines, enabling long-term storage of high-resolution datasets for longitudinal environmental studies and efficient management of large data volumes. For instance, a single six-parameter sensor sampling every minute generates over 3 million data points annually [48]. These platforms provide scalable computational resources that dynamically adjust to growing datasets, while facilitating data sharing with stakeholders via web dashboards and APIs [49]. By migrating to cloud infrastructure, environmental monitoring systems can overcome the challenge of managing networks with dozens or hundreds of sensors, where manual data processing becomes impractical, ensuring seamless scalability and real-time analysis.

Analytics and Visualization: Cloud analytics converts raw sensor data into actionable information through statistical processing, multi-source correlation (linking nitrate levels with rainfall and land use data), and signal processing to improve data quality. Real-time alerting triggers responses when parameters exceed thresholds, whereas visualization tools transform complex datasets into interpretable formats that support evidence-based decision-making [50].

Artificial Intelligence and Machine Learning: The performance of IoT sensor networks improves greatly when AI is applied through multiple functional methods.

Predictive Analytics: Predictive analytics employs machine learning models that use historical sensor data to forecast future environmental conditions. These models will (a) capture complex nonlinear relationships between multiple parameters and (b) enable proactive rather than reactive environmental management [51].

Anomaly Detection: Unsupervised learning algorithms monitor incoming data streams to pinpoint patterns that differ from the standard normative patterns. This will enable to (a) differentiate between sensor malfunctions and genuine environmental events, (b) operate as an unceasing monitoring mechanism throughout vast network structures, and (c) improve accuracy for environmental monitoring systems through comprehensive contextual analysis of multiple parameters [52].

Data Fusion and Interpolation: Machine learning will enable (a) spatial interpolation methods to calculate ion concentrations between multiple sensor stations, (b) the combination of sensor data alongside satellite imagery and hydrodynamic models, and (c) the inclusion of qualitative information (e.g., citizen reports) together with quantitative measurements [19].

Control and Optimization: In advanced applications, AI algorithms (a) utilize sensor feedback to manage environmental management systems, (b) use advanced AI algorithms to enhance water treatment chemical dosing and adjust fertilizer application rates, and (c) develop cyber-physical systems to automatically adapt their operations to dynamic environmental changes [19].

Edge versus Cloud Computing: While cloud computing offers substantial computing power, edge computing allows data analysis on local devices or gateways to (a) reduce transmission volumes and bandwidth requirements, (b) minimize latency for time-sensitive applications, (c) maintain basic functionality during connectivity interruptions, and (d) support operations in remote locations with intermittent network access. Most implementations employ a hybrid approach: Edge devices perform preprocessing and filtering functions and generate immediate alerts, while cloud systems handle complex analytics tasks and long-term data storage [53].

The combination of IoT sensor networks with cloud computing and artificial intelligence has shifted environmental monitoring toward knowledge-driven management beyond basic data capture. Advanced predictive modeling, combined with real-time analytics and automated response systems, enhances decision-making to improve the accuracy and effectiveness of ion-based environmental monitoring.

9.8 Water Resource Management and Drinking Water Monitoring

This section demonstrates the concepts discussed by presenting practical case studies that illustrate how IoT-driven ion-based monitoring operates across various industries. Documented examples demonstrate both successful deployments and encountered challenges, as well as lessons gained from implementing modern sensor networks for environmental management.

9.8.1 Case Study 1: Lake Water Quality Monitoring in Poland

Researchers in Poland used an IoT sensor network to monitor water quality around the clock in a polymictic lake that undergoes frequent mixing. The study conducted by Ferencz and Dawidek [54] examined how water quality indicators varied vertically and horizontally across the lake, which plays a key role in ecosystem management. The research team placed sensor floats at various locations that contained ion sensors for nitrate and ammonium detection alongside a multiparameter sonde that measured temperature and dissolved oxygen, among other parameters. Data were sent through radio waves to an onshore central station. Real-time observation of stratification and mixing events has enabled researchers to study these critical processes, as stratification can cause oxygen depletion, leading to fish mortality. The study showed that during periods of calm weather, the lake's structure led to nitrate build-up at the bottom, which then surged at the surface when the wind stirred the water, a finding that traditional manual sampling methods could not reveal. Ongoing data collection has enabled scientists to validate ecosystem models and provided lake authorities with information to determine when aeration or other interventions are needed. The project revealed that robust sensors are necessary because the shallow lake's abundant algae repeatedly fouled them, and the team implemented a cleaning schedule for sensors equipped with wipers. This demonstration confirmed the IoT capabilities for aquatic health monitoring, prompting local water managers to adopt the system to track ongoing lake recovery and support new pollution control efforts.

9.8.2 Case Study 2: Dam Water Monitoring in Morocco

In Morocco, Ouali et al. installed remote sensors on the Hassan Addakhil Dam reservoir to observe water quality for irrigation water management [55]. The proposed implementation integrates IoT sensors into a modeling system for spatiotemporal monitoring. Because of the reservoir's vast size, they installed a limited number of sensor stations strategically to measure conductivity, turbidity, and pH, whereas hydrodynamic models processed the data to predict water quality across the entire reservoir. Real-time data delivery to the model was enabled by IoT functionality, which was validated through manual sampling. The combined approach generated a water quality map that was nearly real time for the reservoir. This information enables the water authority to determine the appropriate water-release strategy and treatment processes whenever high turbidity or nutrient loads are detected in the readings. The research team faced difficulties in setting up dependable communication in remote locations as they implemented a cellular telemetry system (GSM) that suffered from inconsistent signal coverage. This project demonstrated that local data logging was essential because manual data retrieval was necessary during network outages. Despite intermittent connectivity issues, the approach performed well, prompting discussions about its application to additional dams in the region. This case demonstrates that achieving full coverage requires integrating IoT solutions with both the conventional and traditional methods.

9.9 Agriculture and Soil Health—Precision Farming with Ion Sensors

9.9.1 Case Study 1: Precision Agriculture in the UK (Soil Nitrate Monitoring)

A Welsh project created a LoRaWAN IoT node system that connects ion-selective electrodes for soil nitrate detection to support precision fertilization practices [8]. Several IoT nodes were deployed throughout a pasture field at the root-zone depth to measure soil pore-water nitrate concentrations. The farm gateway receives nitrate data through LoRa transmission before forwarding it to the cloud. Farmers gained real-time visibility of soil nitrate spatial variations across fields immediately following fertilization and rainfall events. Over a three-month trial, the sensors showed clear differences in nitrate levels: Certain areas showed faster nitrate depletion because of higher grass absorption, whereas other zones maintained nitrate levels, which suggested excessive fertilization. By combining sensor readings with grass growth data and laboratory soil tests, the team verified sensor accuracy, as they observed a strong correlation coefficient (R-squared = 0.82) between sensor readings and laboratory analysis results. The team noticed discrepancies that seemed to stem from micro-heterogeneity and laboratory handling procedures, indicating that in situ sensors can detect additional dynamics that laboratory testing overlook, such as nitrate production from mineralization during sample handling. As a result, farmers found a way to optimize fertilizer application by administering more fertilizer to needy areas and less to those with sufficient nitrate according to sensor readings, which led to optimized yields and reduced runoff risks. Challenges and Lessons: To mitigate sensor drift over several weeks, a biweekly protocol was implemented in which sensors underwent rapid calibration checks using standard solutions. The sensors had an operational period of approximately four to six months, which created challenges; ongoing project work aims to extend their functional period beyond six months to support effective farm use. The solar-powered nodes ran continuously without energy-related problems. IoT ion sensors enable farmers to perform ongoing soil testing, which provides economic savings and environmental advantages by reducing fertilizer runoff, which leads to water pollution. This project demonstrates the successful fusion of conventional agricultural methods with modern sensor technology.

9.9.2 Case Study 2: Banana Crop Yield Optimization in Lucknow, India

A research project at Amity University in Lucknow showed how IoT and AI systems boost banana crop productivity and sustainability through the application of IoT sensors and drone imaging technologies that monitor field conditions and anticipate threats that facilitate data-driven decision-making [56]. The monitoring system utilized NPK sensors for soil nutrient tracking along with pH sensors for acidity measurement, DHT11 sensors to monitor moisture and humidity levels, and raindrop sensors to observe precipitation while employing PIR motion sensors (HC-SR501) to detect animal entries. Aerial multispectral camera systems aboard drones collect imagery for pathogen identification, whereas real-time sensor information travels to a cloud platform through Wi-Fi and LoRa networks. Using AI algorithms and convolutional neural networks, drone imagery was analyzed to detect fungal/bacterial infections, including Fusarium wilt, and the findings were presented on a centralized dashboard. The implementation of this methodology enhanced crop yield by 18% through targeted irrigation activated by soil moisture limits and exact NPK fertilization methods and achieved a 25% drop in crop loss through the early identification of diseases. Irrigation systems that adjust water usage based on soil moisture and weather conditions reduce water consumption by 30%, and NPK sensors that direct fertilization reduce nutrient leaching and runoff. LSTM machine learning models predicted soil nutrient depletion to enable preemptive fertilization, and AI established links between rainfall patterns and disease outbreaks to enhance preventive strategies. Frequent recalibration requirements for pH/NPK sensors caused by tropical soil conditions presented a challenge, along with the delay in synchronizing drone imaging with ground sensor data and expensive initial IoT infrastructure costs, which eventually led to higher long-term yield benefits. Following successful implementation, farmers in Uttar Pradesh's banana belt adopted the system that produced 15–20% increased profits through lower input costs and premium market prices, while achieving 40% fewer pesticide applications because of early pathogen detection. The findings from this research demonstrate the transformative potential of IoT-AI integration in banana farming by optimizing production and promoting environmental sustainability through continuous monitoring, combined with predictive analytics and accurate resource management. Subsequent initiatives intend to lower expenses for smallholder farmers while broadening the application model to include additional crop types.

9.10 Industrial Pollution Control—IoT for Compliance and Process Optimization

9.10.1 Heavy Metal Monitoring in Industrial Wastewater (Portugal)

The automobile manufacturing plant in Mangualde, Portugal, operated by Stellantis, launched a pilot project to automate heavy metal monitoring in wastewater through the Next-Gen Quality Control (NGQC IoRT) initiative [57]. The International Iberian Nanotechnology Laboratory (INL), in collaboration with its partners, developed a system with electrochemical sensors specifically tuned to detect nickel and zinc levels in the plant effluent. The sensors were installed in the wastewater treatment line, and small filtered samples were extracted from the treatment process entrance and exit points to instantly analyze the Ni and Zn concentrations. The plant's control room receives integration from data that travels to the cloud. The detection capabilities of the sensors reached 50 μg/L for Zn and 5 μg/L for Ni, thus providing plant operators with an advanced early warning system, as these levels were below the regulatory discharge limits. The system was tested as a pilot for one month, while routine laboratory sampling was conducted concurrently. IoT sensor measurements exhibited high accuracy relative to certified laboratory results for matching samples. Impacts: The operators received ongoing trend data instead of daily laboratory measurements. The plant operators noticed intermittent metal concentration spikes that regular laboratory sampling failed to detect, because these spikes occurred during specific cleaning cycles. Once these patterns were identified, they adjusted their operations (for example, they modified the rinse stage to capture metals before they were discharged). If metal concentrations are near the regulatory limit, the system triggers automated alerts for operators to respond with corrective actions, such as adding more treatment chemicals.

Challenges and Lessons: These electrochemical sensors require calibration and are vulnerable to contamination from different wastewater substances. The system incorporated a self-cleaning mechanism for sensor cells and performed regular recalibration using known standards to ensure accurate measurements. Connecting an IoT device to the plant's network presented difficulties which required security protocols and ensured compatibility with existing control systems. The pilot program yielded positive results, which prompted plans to expand these systems to different industrial locations and extend detection capabilities to include more metal ions and substances, such as cyanide and chromium. This case study demonstrates how IoT technology establishes continuous compliance tracking that allows industries to adhere to regulations and enhance their treatment methods. IoT sensors can provide field data comparable to laboratory standards when developed with thorough precision, and this resolution of data quality issues now allows for wider industrial adoption.

9.10.2 Case Study 2: Air Quality and Emissions (Smart City Pollution Monitoring)

The research focused on PM2.5 air pollution in northern India with special emphasis on Delhi's region through advanced sensor networks and analytical methods that connected pollution sources to their health risks [58]. Over both cold and warm seasons, the research team operated high-volume air samplers at five locations to gather integrated PM2.5 filter samples across time spans of 12–24 h. Ion chromatography was used to analyze inorganic ions while trace metals were identified through inductively coupled plasma mass spectrometry, and high-resolution mass spectrometry enabled the study of organic aerosols and OA. The novel approach involves using extractive electrospray ionization (EESI) together with time-of-flight mass spectrometry, which enabled near-molecular-level organic aerosol speciation to identify primary sources such as traffic exhaust as hydrocarbon-like organic aerosols and biomass burning as nitrogen-rich compounds from cow dung, in addition to secondary aerosols from oxidized volatile organic compounds. The combination of positive matrix factorization (PMF) and radiocarbon (14C) analyses determined that the local incomplete combustion of biofuels and fossil fuels produced 68% of OA emissions. Three acellular assays measured the oxidative potential (OP) health risk metric and found that the OP levels in India were five times higher than those in Europe because of organic aerosols from traffic and biomass burning.

Although traditional sensors such as ion chromatographs provide essential information about ammonium sulfate, nitrate and chloride analysis is hindered by the study's dependence on laboratory-based methods, which restrict real-time observation. This research method demonstrates how IoT-based sensor networks can facilitate the automated monitoring of intricate systems. Miniaturized ion sensors can monitor chloride/nitrate levels in real time while EESI-based portable spectrometers can send OA fingerprints to cloud platforms for instant source identification when used across urban areas. The combination of these sensor outputs with traffic and meteorological data will facilitate dynamic pollution modeling, which alerts authorities about OP hotspots from particular emission events, such as nighttime biomass burning. The discovery that health risks primarily originate from local combustion sources emphasizes the importance of IoT solutions for precise mitigation measures, such as reducing residential biofuel use or traffic emissions during periods of low wind. Laboratory-grade sensors are impractical for field use because of their high cost and maintenance requirements, such as reagent refills, until solid-state ion-selective electrodes combined with AI-driven data fusion enable scalable continuous air quality monitoring in megacities.

These case studies demonstrate crucial practices that lead to successful monitoring of ions using IoT technology. Consistent calibration together with laboratory validation produces trustworthy data and strengthens stakeholder confidence. Effective communication systems demand solid data buffering methods and proper technology choices that rely on specific deployment environmental conditions. The implementation results show that IoT-based ion monitoring systems provide tangible advantages, including early detection of pollution and resource optimization through continuous compliance monitoring, while also highlighting the necessity of balancing technical abilities with human factors, economic aspects, and regulatory requirements for effective environmental surveillance.

9.11 Emerging Technology

Several technological advances are expected to shape the next generation of ion-based monitoring systems.

Advanced Materials and Nanosensors: Nanostructured materials such as graphene and carbon nanotubes improve sensor capabilities, including sensitivity and selectivity, along with stability improvement. Emerging advancements in sensor technology will produce devices that exhibit lower detection thresholds and superior accuracy when operating for longer durations. Self-healing sensors with antifouling surfaces that self-calibrate show promise as a method to decrease maintenance needs.

Miniaturization and Lab-on-Chip: Embedding laboratory functions into microfluidic chips will enable full multi-ion analyses during field operations. These systems deliver precise laboratory-quality measurements and exceed the single-parameter constraints of existing sensors.

Artificial Intelligence: Advanced AI technologies will enable environmental monitoring to shift from reactive approaches to proactive management systems. Predictive analytics will identify future pollution incidents and environmental shifts, whereas prescriptive systems will recommend optimal intervention strategies. Real-time data inputs will enable environmental system digital twins to perform scenario testing and impact analyses.

Blockchain for Data Integrity: Distributed ledger technologies provide data immutability and transparency, which are especially beneficial in regulatory environments and cross-border monitoring. The use of smart contracts will enable automatic verification of compliance with standards and automated responses to incidents.

Next-Generation Communications: 6G networks, combined with satellite IoT and advanced edge computing solutions, will enable comprehensive monitoring coverage worldwide, including remote locations. Next-generation technologies facilitate extensive sensor deployment by minimizing energy consumption and enhancing data processing capabilities.

9.12 Conclusion

Ion-based environmental monitoring is essential for safeguarding the environment, ensuring sound resource management, and ensuring compliance with regulatory standards. IoT technologies have revolutionized environmental research by allowing uninterrupted data recording, which uncovers patterns and trends that cannot be detected using traditional sampling methods. The case examples presented in this chapter demonstrate the tangible benefits. Environmental monitoring technologies now allow us to detect contamination at its inception, while optimizing resource use and maintaining continuous compliance checks alongside improved informational systems for public awareness. The case studies provide actionable insights into calibration processes, maintenance needs, communication dependability, and stakeholder participation. The evolution of sensor technologies and reduction in their associated costs will lead to more extensive monitoring systems across watersheds, aquifers, and urban settings. The combination of extensive data availability and advanced analytical techniques will usher in a new era of environmental stewardship by enabling early detection and proactive management of environmental issues. Nanotechnology, when paired with artificial intelligence and blockchain, and next-generation IoT, establishes a robust technological ecosystem that will transform environmental monitoring from a manual, sporadic process into an automated platform for evidence-based environmental management and protection.

References

1. Fan S, Khuntia S, Ahn CH, Zhang B, Tai L (2022) Electrochemical devices to monitor ionic analytes for healthcare and industrial applications. Chemosensors 10(1):22
2. Tay A (2022) The role of ion chromatography in environmental sciences. Lab Manager [Internet]. https://www.labmanager.com
3. Cheng K, Xu X, Cui L, Li Y, Zheng J, Wu W et al (2021) The role of soils in regulation of freshwater and coastal water quality. Philos Trans R Soc B [Internet] 376(1834):20200176. https://royalsocietypublishing.org/doi/10.1098/rstb.2020.0176. Cited 15 Apr 2025
4. Sani SA, Ibrahim A, Musa AA, Dahiru M, Baballe MA (2023) Drawbacks of traditional environmental monitoring systems. CIS [Internet] 16(3):30. https://ccsenet.org/journal/index.php/cis/article/view/0/49187. Cited 15 Apr 2025
5. Huang CW, Lin C, Nguyen MK, Hussain A, Bui XT, Ngo HH (2023) A review of biosensor for environmental monitoring: principle, application, and corresponding achievement of sustainable development goals. Bioengineered [Internet] 14(1):58–80. https://www.tandfonline.com/doi/full/10.1080/21655979.2022.2095089. Cited 15 Apr 2025
6. BOQU Instrument (2025) Real-time water quality monitoring with advanced sensors [Internet]. https://www.boquinstrument.com/a-news-real-time-water-quality-monitoring-with-advanced-sensors.html
7. Sherris AR, Baiocchi M, Fendorf S, Luby SP, Yang W, Shaw GM (2021) Nitrate in drinking water during pregnancy and spontaneous preterm birth: a retrospective within-mother analysis in California. Environ Health Perspect [Internet] 129(5):057001. https://ehp.niehs.nih.gov/doi/10.1289/EHP8205. Cited 15 Apr 2025
8. Bristow N, Rengaraj S, Chadwick DR, Kettle J, Jones DL (2022) Development of a LoRaWAN IoT node with ion-selective electrode soil nitrate sensors for precision agriculture. Sensors [Internet] 22(23):9100. https://www.mdpi.com/1424-8220/22/23/9100. Cited 15 Apr 2025
9. Manx Technology Group (2025) Soil monitoring with IoT—smart agriculture [Internet]. https://manxtechgroup.com/soil-monitoring-with-iot-smart-agriculture/
10. Shahid SA, Dakheel AH, Mufti KA, Shabbir G Automated in-situ soil salinity logging in irrigated agriculture
11. Oladimeji TE, Oyedemi M, Emetere ME, Agboola O, Adeoye JB, Odunlami OA (2024) Review on the impact of heavy metals from industrial wastewater effluent and removal technologies. Heliyon [Internet] 10(23):e40370. https://linkinghub.elsevier.com/retrieve/pii/S2405844024164018. Cited 15 Apr 2025
12. Perfect Pollucon Services (2025) Stack emission monitoring & testing [Internet]. https://www.ppsthane.com/stack-emissions-monitoring-testing
13. United States Environmental Protection Agency (2025) EMC: continuous emission monitoring systems [Internet]. https://www.epa.gov/emc/emc-continuous-emission-monitoring-systems
14. European Commission (2025) Water framework directive [Internet]. https://environment.ec.europa.eu/topics/water/water-framework-directive_en
15. Thermo Fisher Scientific (2025) Water regulations [Internet]. https://www.thermofisher.com/in/en/home/industrial/environmental/environmental-learning-center/water-quality-analysis-information/water-regulations.html

16. Szleter P (2024) Environmental monitoring with IoT: five essential steps to integrate IoT sensors for real-time monitoring [Internet]. https://www.forbes.com/councils/forbestechcouncil/2024/07/12/environmental-monitoring-with-iot-five-essential-steps-to-integrate-iot-sensors-for-real-time-monitoring/. Cited 15 Apr 2025
17. Witczak D, Szymoniak S (2024) Review of monitoring and control systems based on internet of things. Appl Sci [Internet] 14(19):8943. https://www.mdpi.com/2076-3417/14/19/8943. Cited 15 Apr 2025
18. Kumar M, Khamis K, Stevens R, Hannah DM, Bradley C (2024) In-situ optical water quality monitoring sensors—applications, challenges, and future opportunities. Front Water [Internet] 6:1380133. https://www.frontiersin.org/articles/10.3389/frwa.2024.1380133/full. Cited 15 Apr 2025
19. Miller T, Durlik I, Kostecka E, Kozlovska P, Łobodzińska A, Sokołowska S et al (2025) Integrating artificial intelligence agents with the internet of things for enhanced environmental monitoring: applications in water quality and climate data. Electronics [Internet] 14(4):696. https://www.mdpi.com/2079-9292/14/4/696. Cited 15 Apr 2025
20. Głąb S, Maj-Żurawska M, Hulanicki A (2013) Ion-selective electrodes | glass electrodes. In: Reference module in chemistry, molecular sciences and chemical engineering [Internet]. Elsevier, p B9780124095472002663. https://linkinghub.elsevier.com/retrieve/pii/B9780124095472002663. Cited 16 Apr 2025
21. Sensorex (2025) Ion selective electrodes [Internet]. https://sensorex.com/ion-selective-electrodes/
22. Pretsch E (2007) The new wave of ion-selective electrodes. TrAC Trends Anal Chem [Internet] 26(1):46–51. https://linkinghub.elsevier.com/retrieve/pii/S0165993606002317. Cited 16 Apr 2025
23. Britschgi L, Villez K, Schrems P, Udert KM (2020) Electrochemical nitrite sensing for urine nitrification. Water Res X [Internet] 9:100055. https://linkinghub.elsevier.com/retrieve/pii/S2589914720300153. Cited 16 Apr 2025
24. Pei X, Kang W, Yue W, Bange A, Heineman WR, Papautsky I (2014) Disposable copper-based electrochemical sensor for anodic stripping voltammetry. Anal Chem [Internet] 86(10):4893–900. https://pubs.acs.org/doi/10.1021/ac500277j. Cited 16 Apr 2025
25. Tian Y, Liu J, Qiao J, Ge F, Yang Y, Zhang Q (2025) Advancements in electrochemical sensing technology for heavy metal ions detection. Food Chem X [Internet] 25:102204. https://linkinghub.elsevier.com/retrieve/pii/S2590157525000501. Cited 16 Apr 2025
26. Zainurin SN, Wan Ismail WZ, Mahamud SNI, Ismail I, Jamaludin J, Ariffin KNZ et al (2022) Advancements in monitoring water quality based on various sensing methods: a systematic review. IJERPH [Internet] 19(21):14080. https://www.mdpi.com/1660-4601/19/21/14080. Cited 16 Apr 2025
27. Tejaswini KK, George B, Mukhopadhyay SC, Kumar VJ (2023) Conductivity sensors for water quality monitoring: a brief review. In: Philip L, Pradeep T, Bhallamudi SM (eds) Technological solutions for water sustainability: challenges and prospects [Internet]. IWA Publishing, pp 213–221. https://iwaponline.com/ebooks/book/880/chapter/3486579/Conductivity-sensors-for-water-quality-monitoring. Cited 16 Apr 2025
28. Chen WH, Armstrong E, Dillingham PW, Moratti SC, Ennis C, McGraw CM (2023) Dual-lifetime referencing (t-DLR) optical fiber fluorescent ph sensor for microenvironments. Sensors [Internet] 23(21):8865. https://www.mdpi.com/1424-8220/23/21/8865. Cited 16 Apr 2025
29. Zhang Y, Mollick S, Tricarico M, Ye J, Sherman DA, Tan JC (2022) Turn-on fluorescence chemical sensing through transformation of self-trapped exciton states at room temperature. ACS Sens [Internet] 7(8):2338–2344. https://pubs.acs.org/doi/10.1021/acssensors.2c00964. Cited 16 Apr 2025
30. optek-Danulat GmbH (2020) Turbidity guide: measurements on light scattering and absorption [Internet]. https://www.optek.com/en/turbidity-guide.asp
31. Hossain SMZ, Mansour N (2019) Biosensors for on-line water quality monitoring—a review. Arab J Basic Appl Sci [Internet] 26(1):502–518. https://www.tandfonline.com/doi/full/10.1080/25765299.2019.1691434. Cited 17 Apr 2025

32. Pfaff JD (1996) Determination of inorganic anions by ion chromatography. In: Methods for the determination of metals in environmental samples [Internet]. Elsevier, pp 388–417. https://linkinghub.elsevier.com/retrieve/pii/B9780815513988500227. Cited 21 Apr 2025
33. De Camargo ET, Spanhol FA, Slongo JS, Da Silva MVR, Pazinato J, De Lima Lobo AV et al (2023) Low-cost water quality sensors for IoT: a systematic review. Sensors [Internet] 23(9):4424. https://www.mdpi.com/1424-8220/23/9/4424. Cited 21 Apr 2025
34. Vuong P, McKinley A, Kaur P (2023) Understanding biofouling and contaminant accretion on submerged marine structures. npj Mater Degrad [Internet] 7(1):50. https://www.nature.com/articles/s41529-023-00370-5. Cited 21 Apr 2025
35. Kamal Hossain M, Ahmad I, Habibi D, Waqas M (2025) Enhancing IoT sensors precision through sensor drift calibration with variational autoencoder. IEEE Internet Things J [Internet] 12(7):8421–8437. https://ieeexplore.ieee.org/document/10798420/. Cited 21 Apr 2025
36. Andhare MS, Pal TL, Jayaram V, Sreelekshmy Pillai G, Tripathi V, Krishnaraj M et al (2022) Design and implementation of wireless sensor network for environmental monitoring. ijhs [Internet] 3158–3169. https://sciencescholar.us/journal/index.php/ijhs/article/view/9085. Cited 21 Apr 2025
37. Shi KX, Li SM, Sun GW, Feng ZC, He W (2024) A fault diagnosis method for wireless sensor network nodes based on a belief rule base with adaptive attribute weights. Sci Rep [Internet] 14(1):4038. https://www.nature.com/articles/s41598-024-54589-6. Cited 21 Apr 2025
38. Narayana TL (2024) Advances in real time smart monitoring of environmental parameters using IoT and sensors
39. Ewattch (2024) LoRaWAN: everything you need to know about this innovative technology [Internet]. https://ewattch.com/en/blog/lorawan/
40. Jakhotiya PS, Kasat DNN, Gawande DAD (2021) Gaikwad DVT. Application of Lorawan technology in farm for irrigation system. 9(8)
41. Zayas AD, Tocado FJR, Rodríguez P (2020) evolution and testing of NB-IoT solutions
42. Ugwuanyi S, Paul G, Irvine J (2021) Survey of IoT for developing countries: performance analysis of LoRaWAN and cellular NB-IoT networks. Electronics [Internet] 10(18):2224. https://www.mdpi.com/2079-9292/10/18/2224. Cited 21 Apr 2025
43. Manjurega B, Priyadharsini M, Poovika T, Sanjeevi S, Rooban P (2017) Environmental monitoring system with wireless mesh network using Zigbee. Int J ChemTech Res
44. Shaikh A, Pathan S (2012) Research on wireless sensor network technology. IJIET [Internet] 476–9. http://www.ijiet.org/show-33-148-1.html. Cited 21 Apr 2025
45. Tanvir AA, Bhattacharjee R Comparative analysis of 5G technology in the agriculture sector of Bangladesh (challenges and potential impact)
46. Telecom Gurukul (2024) 5G automation: enabling remote sensing and environmental monitoring technologies for precision in 2024 [Internet]. https://www.telecomgurukul.com/post/5g-automation-enabling-remote-sensing-and-environmental-monitoring-technologies-for-precision-in-20
47. Rafi MSM, Behjati M, Rafsanjani AS Reliable and cost-efficient IoT connectivity for smart agriculture: a comparative study of LPWAN, 5G, and hybrid connectivity models
48. Guerbaoui M, El Faiz S, Ed-Dahhak A, Lachhab A, Benhala B, Bakziz Z et al (2025) From data to decisions: a smart IoT and cloud approach to environmental monitoring. In: Benhala B, Lachhab A, Raihani A, Qbadou M, Sallem A (eds) E3S web conferences [Internet], vol 601, p 00008. https://www.e3s-conferences.org/10.1051/e3sconf/202560100008. Cited 28 Apr 2025
49. Javed A, Malhi A, Främling K (2020) Scalable IoT platform for heterogeneous devices in smart environments, p 8
50. Lv T, Zou Y, Wang Y, Wu E (2024) Research on environmental sensor data analysis based on real-time data warehouse. In: International conference on cloud computing, performance computing, and deep learning (CCPCDL 2024), p 53
51. Kanth TVR, Dhineshkumar K, Mishra H (2025) Machine learning-enhanced wireless sensor networks for real-time environmental monitoring. 10(01)
52. Wei Z, Wang F (2022) Detecting anomaly data for IoT sensor networks. In: Sun L (ed) Scientific programming [Internet], pp 1–7. https://www.hindawi.com/journals/sp/2022/4671381/. Cited 28 Apr 2025

53. Sathupadi K, Achar S, Bhaskaran SV, Faruqui N, Abdullah-Al-Wadud M, Uddin J (2024) Edge-cloud synergy for ai-enhanced sensor network data: a real-time predictive maintenance framework. Sensors [Internet] 24(24):7918. https://www.mdpi.com/1424-8220/24/24/7918. Cited 28 Apr 2025
54. Ferencz B, Dawidek J (2021) Assessment of spatial and vertical variability of water quality: case study of a Polymictic Polish Lake. IJERPH [Internet] 18(16):8620. https://www.mdpi.com/1660-4601/18/16/8620. Cited 28 Apr 2025
55. El Ouali A, El Hafyani M, Roubil A, Lahrach A, Essahlaoui A, Hamid FE et al (2021) Modeling and spatiotemporal mapping of water quality through remote sensing techniques: a case study of the Hassan Addakhil Dam. Appl Sci [Internet] 11(19):9297. https://www.mdpi.com/2076-3417/11/19/9297. Cited 28 Apr 2025
56. Pandey P, Kumar S, Khanna P, Pragya (2024) Enhancing agricultural practices through iot-based smart farming technologies: a case study on banana crop yield optimization. Asian J Res Com Sci [Internet] 17(7):55–66. https://journalajrcos.com/index.php/AJRCOS/article/view/477. Cited 28 Apr 2025
57. International Iberian Nanotechnology Laboratory (INL) (2025) INL advances heavy metal detection in industrial wastewater [Internet]. https://inl.int/heavy-metal-detection-in-industrial-wastewater/. Cited 28 Apr 2025
58. Bhattu D, Tripathi SN, Bhowmik HS, Moschos V, Lee CP, Rauber M et al (2024) Local incomplete combustion emissions define the PM2.5 oxidative potential in Northern India. Nat Commun [Internet] 15(1):3517. https://www.nature.com/articles/s41467-024-47785-5. Cited 28 Apr 2025

Chapter 10
Implementing Water Quality Monitoring Using IoT

Pallavi Vishnu Kharat, Milind Manikrao Darade, and Amena Ansari

Abstract Water quality assessment is fundamental for environmental and water resources sustainability, human health, agriculture, and soil protection. Conventional methods for evaluating water quality are tedious and time-consuming; laboratory analysis and hand sampling extend the time required, resulting in costly, labor-intensive tasks and safety concerns. The increase in industrialization, pollution, water demand, and changes in seasonal patterns restrict water supplies, exacerbating water scarcity. The major challenge is to use the water more efficiently, and every drop must be utilized. Real-time analysis of each water drop for optimal use is necessary; therefore, new methods are required to assess the utility of water. The Internet of Things (IoT), with a network of interconnected sensors, offers a promising solution to this problem. This study investigates the developments in sensor technology, data connection protocols, and analytics that affect water quality monitoring. An IoT-based water monitoring system with a sensor network evaluates key water quality parameters, such as temperature, turbidity, dissolved oxygen, pH, and the detection of hazardous chemicals, including heavy metals and toxic compounds. These sensors, in coordination with cloud-based systems, transmit data, and advanced analytics and data processing help identify trends, anomalies, and hazards. This guarantees early detection and resolution of emerging problems, thereby preventing their spread and improving water resource management. Water quality monitoring systems have become more scalable and efficient due to advances in IoT technologies, including low-power wide-area networks (LPWANs) (e.g., LoRaWAN) and cloud-based data storage, enabled by affordable, small sensors. Artificial intelligence (AI) and machine learning provide dynamic predictive analytics, enabling the identification of potential

P. V. Kharat (✉)
Department of Civil Engineering, Ajeenkya DY Patil School of Engineering, Lohegaon, Pune, Maharashtra, India
e-mail: pallavi.kharat83@gmail.com

M. M. Darade
Department of Civil Engineering, D Y Patil College of Engineering, Akurdi, Pune, Maharashtra, India

A. Ansari
Deogiri Institute of Engineering and Management Studies, Chhatrapati Sambhaji Nagar, Maharashtra, India

S. Gautam et al. (eds.), *Smart Environmental Monitoring*,
https://doi.org/10.1007/978-981-92-0294-2_10

water quality problems before their escalation. Studies exhibit the effective application of IoT-based water quality monitoring internationally. India uses IoT technologies to monitor urban water quality, thereby ensuring compliance with safety standards for consumption. The Internet of Things (IoT) helps the United States monitor river conditions by enabling real-time pollution detection and enhanced flood management, thereby improving control of river conditions. IoT systems in Europe are reducing environmental impact and increasing efficiency by enhancing wastewater treatment. Using IoT for water quality control represents a major advance in the governance and protection of water supplies. IoT-based solutions improve efficiency through real-time, continuous data transfer, rapid responses to pollution events, and the facilitation of informed decision-making. Public health, environmental sustainability, food and soil pollution, and the future direction of world water management all depend on this. As IoT technologies develop, their potential to address global water quality management challenges becomes increasingly compelling.

Keywords Water quality · Internet of Things · Real-time detection

10.1 Introduction

IoT-based water quality monitoring systems offer significant advantages in the field of environmental management and public health. These systems enable continuous, real-time data collection, allowing for immediate detection of water quality changes or potential contamination events. Remote access capabilities facilitate monitoring from centralized locations, reducing the need for frequent on-site visits and enabling rapid response to emerging issues. Furthermore, the vast amount of data collected by these systems can be leveraged to generate valuable insights through advanced analytics, supporting informed decision-making and long-term water resource management strategies. However, traditional water quality monitoring methods continue to play a crucial role in certain contexts. These conventional approaches, which often involve manual sampling and laboratory analysis, are characterized by their simplicity and reliability. They require less technical expertise to implement and maintain, making them accessible to a wider range of organizations and communities. The lower initial investment costs associated with traditional methods can be particularly appealing for smaller municipalities or regions with limited budgets. Moreover, traditional methods are less dependent on complex technological infrastructure, such as reliable internet connectivity or sophisticated sensor networks. This independence can be advantageous in remote or underdeveloped areas where such infrastructure may be lacking or unreliable. Additionally, manual sampling and analysis techniques can provide a level of flexibility and adaptability that may be beneficial in certain research contexts or when investigating specific water quality parameters that are not easily measured by automated sensors. The choice between IoT-based and traditional water quality monitoring methods often depends on various factors, including the specific monitoring objectives, available resources, local infrastructure, and regulatory requirements. In many cases, a hybrid approach that combines elements of both methodologies may offer the most comprehensive and effective solution for water quality management.

The integration of Internet of Things (IoT) technology in water quality monitoring offers a promising solution to address the challenges of traditional methods, providing real-time data collection and analysis for improved water management in agriculture and ecosystems [1]. IoT-based systems utilize advanced sensors connected to microcontrollers to continuously monitor key water parameters such as pH, temperature, and turbidity [1]. These systems can be applied in various agricultural settings, including fish farming and crop irrigation, to optimize water quality and enhance productivity [1]. For instance, in smart fish farming, IoT-driven intelligent decision-making systems can dynamically monitor and optimize water quality parameters to improve fish survival rates across diverse regions and species setups [1]. The implementation of IoT in water quality monitoring also facilitates the creation of comprehensive datasets that can be analyzed using advanced machine learning techniques. This approach enables the development of predictive models for water quality management, supporting real-time decision-making and automated interventions [1]. Furthermore, the integration of IoT with other technologies such as artificial intelligence (AI) and remote sensing enhances the capabilities of water quality monitoring systems, allowing for more precise and efficient resource management in agriculture [2]. In conclusion, IoT-based water quality monitoring systems offer significant advantages over traditional methods, providing continuous, real-time data collection and analysis. These systems not only improve the efficiency and accuracy of water quality monitoring but also contribute to sustainable water management practices in agriculture and ecosystem conservation. As climate change continues to impact water resources, the adoption of IoT-based monitoring solutions becomes increasingly crucial for ensuring water security and supporting climate-resilient agriculture [3].

10.2 The Importance of Water Quality Monitoring

Water quality monitoring is crucial for:

Public Health—Detects harmful substances and waterborne diseases, guiding appropriate treatment.

Environmental Protection—Evaluates aquatic ecosystem health, identifies pollution sources, and supports biodiversity conservation.

Regulatory Compliance—Ensures adherence to standards, aids regulation enforcement, and informs policy development.

Water Resource Management—Supports fair distribution, efficient use, and long-term planning of water resources.

Economic Benefits—Protects water-dependent industries, lowers treatment and health costs, and boosts tourism and recreation.

Climate Change Adaptation—Tracks climate impacts on water, informs adaptive strategies, and helps forecast future challenges.

Scientific Research—Provides data for ecosystem studies, technology development, and global water quality databases. Effective monitoring is essential for protecting water resources and ensuring their sustainable use for future generations.

Available resources, including financial capacity, technical expertise, and existing infrastructure, also influence the choice of monitoring approach. Organizations with substantial budgets and technical capabilities may opt for comprehensive IoT-based systems, while those with limited resources might prefer traditional methods or a phased implementation of IoT technology. Existing infrastructure is another important consideration. Areas with well-established laboratory facilities and trained personnel might find it more practical to continue with traditional methods, at least in the short term. In contrast, regions looking to modernize their water management systems might see IoT-based monitoring as an opportunity to leapfrog older technologies. A hybrid approach that combines elements of IoT-based and traditional methodologies may offer the most comprehensive solution for water quality monitoring and management. This integrated strategy could leverage the real-time monitoring and analytical capabilities of IoT systems while maintaining the reliability and simplicity of traditional methods as a backup or complementary approach. Such a hybrid system could, for example, use IoT sensors for continuous monitoring of basic parameters, complemented by regular manual sampling and laboratory analysis for more complex or less frequently measured parameters. This approach would provide the benefits of real-time data for rapid response while ensuring thorough, validated results for regulatory compliance and long-term trend analysis. Moreover, a hybrid approach allows for a gradual transition from traditional to IoT-based methods, enabling organizations to build capacity, address technical challenges, and optimize their monitoring strategies over time. This flexibility can be particularly valuable for adapting to changing environmental conditions, regulatory requirements, and technological advancements in water quality management.

10.3 IoT in Water Quality Monitoring

IoT-based water quality monitoring systems offer significant advantages in the realm of environmental management. These advanced technological solutions enable continuous, real-time data acquisition, providing a constant stream of information about water quality parameters. This immediacy allows for rapid detection of anomalies or potential issues, facilitating prompt intervention when necessary. Furthermore, these systems offer remote access capabilities, enabling water quality managers and researchers to monitor and analyze data from multiple locations simultaneously, regardless of geographical constraints. The sophisticated analytical capabilities of IoT-based systems are another key benefit. These systems can process vast amounts of data quickly, identifying trends, patterns, and correlations that might be challenging to discern through manual analysis. This enhanced analytical power can lead to more informed decision-making and predictive modeling, potentially averting water quality crises before they occur. However, it is crucial to recognize that traditional water quality monitoring methodologies continue to hold significant value in the field. These conventional approaches are characterized by their simplicity and reliability, often requiring less specialized training for implementation. This simplicity can be particularly advantageous in resource-constrained environments or areas with limited technological infrastructure. Moreover, traditional methods often have lower initial investment requirements compared to IoT-based systems. This cost-effectiveness can be a critical factor for smaller municipalities, developing regions, or organizations with limited budgets. The accessibility of traditional methods ensures that water quality monitoring remains feasible across a wide range of contexts. Another notable strength of traditional methodologies is their capacity to measure certain parameters that automated sensors cannot easily detect. Some complex chemical compounds or biological indicators may require laboratory analysis or specialized testing procedures that are not yet feasible through automated IoT sensors. This capability ensures that traditional methods remain relevant and complementary to newer technologies. The decision to employ IoT-based systems, traditional methodologies, or a combination of both depends on the specific monitoring requirements of the situation. Factors such as the required frequency of data collection, the range of parameters to be measured, the geographical spread of monitoring sites, and the required level of analytical depth all play crucial roles in this decision-making process. Available resources, including financial capacity, technical expertise, and existing infrastructure, also significantly influence the choice of monitoring approach. While IoT-based systems may offer more comprehensive and continuous monitoring, they may not be feasible in all contexts due to resource limitations or technological constraints. Local technological infrastructure is another critical consideration. Areas with reliable internet connectivity and power supply may be better suited to IoT-based systems, whereas regions with less developed infrastructure may find traditional methods more practical and sustainable. Ultimately, a comprehensive water quality management strategy may be best achieved by integrating IoT-based and traditional methodologies. This hybrid approach can leverage the strengths of each method, providing a more robust and adaptable monitoring system. For instance, IoT sensors could be used for continuous monitoring of basic parameters, whereas periodic traditional sampling and analysis could be employed for more complex measurements or for validating automated readings. By combining these approaches, water quality managers can ensure comprehensive coverage, enhance data reliability through cross-validation, and maintain flexibility in their monitoring strategies. This integrated approach allows for the optimization of resources while maximizing the breadth and depth of water quality information obtained, ultimately contributing to more effective and sustainable water resource management.

The integration of various approaches in water quality monitoring can indeed lead to more comprehensive and effective water resource management. This approach aligns with findings from several papers in this context. AI and IoT technologies have significantly transformed environmental monitoring, particularly in water quality management. These technologies enable advanced data gathering, in-depth analysis, and more effective decision-making [4]. The integration of AI enhances IoT's data collection capabilities through advanced predictive modeling, real-time analytics, and automated decision-making, improving the accuracy, timeliness, and efficiency of environmental monitoring systems [4]. Complementing this technological approach, the importance of collaboration and stakeholder engagement in water management is highlighted, especially in transboundary river basins. Key strategies include connecting science and policy, valuing local knowledge, considering multiple spatial scales, promoting a culture that values trade-offs, and co-creating data and knowledge to facilitate stakeholder dialogue [5]. Interestingly, while AI-based smart water management systems offer numerous benefits, including efficient water distribution, leak detection, and consumer empowerment, challenges such as initial investments and data security concerns remain [6]. This underscores the need for a balanced approach that leverages technology while addressing potential drawbacks. In conclusion, an integrated approach to water quality monitoring that combines AI and IoT technologies, stakeholder collaboration, and traditional monitoring methods can indeed optimize resource use and maximize the breadth and depth of water quality information. This aligns with the overall trend toward more sophisticated, data-driven, and collaborative approaches in water resource management [4, 5, 6].

The architecture of an IoT-based water quality monitoring system involves four main components: sensors, data acquisition and communication, data processing, and user interface.

Sensors: These collect data on water quality parameters. Common sensors include:

pH sensors to measure acidity or alkalinity.
Turbidity sensors for clarity.
Dissolved oxygen sensors to measure oxygen levels.
Temperature sensors for monitoring water temperature.
Contaminant sensors, such as nitrate and phosphate sensors, and sensors for detecting heavy metals and toxins.

Data Acquisition and Communication: Once data are collected by the sensors, it is transmitted to a central platform using various communication technologies like Wi-Fi, LoRaWAN, Zigbee, or cellular networks. Communication protocols are selected based on factors such as range, power consumption, and environmental conditions.

Data Processing and Analysis: The data collected by sensors are sent to a cloud-based platform where it is processed and analyzed using algorithms. These algorithms identify patterns, detect anomalies, and generate actionable insights. Machine learning models can predict trends or potential contamination events based on historical data.

User Interface: Dashboards or mobile applications present the data to users, including real-time measurements, alerts, and reports. These systems enable stakeholders to monitor water quality remotely and respond promptly to any changes in water conditions.

Data acquisition and communication in sensor networks involve various technologies and protocols, each with its own strengths and applications. Wi-Fi, LoRaWAN,

Zigbee, and cellular networks are commonly used for transmitting sensor data to central platforms, with the choice depending on specific requirements. LoRaWAN has emerged as a popular choice for long-range, low-power applications. It enables data transmission over distances of up to 10 km with minimal energy consumption, making it suitable for remote monitoring in rural areas (Gaitan 2021) [7]. For instance, a LoRa-based health monitoring system can cover an average area of 33 km^2 when the gateway is placed outdoors at an altitude of 12 m [7]. Interestingly, hybrid communication systems combining multiple protocols can offer enhanced performance. A LoRa/Zigbee hybrid system demonstrated a packet loss rate of less than 0.5% with Zigbee covering 630 m and LoRa reaching 3.7 km [8]. Similarly, a system integrating cellular networks, Wi-Fi, and Zigbee for vehicular communication showed promise in supporting safety–critical applications in Cooperative Intelligent Transportation Systems [9]. In conclusion, the selection of communication protocols depends on factors such as range, power consumption, and environmental conditions. While LoRaWAN excels in long-range, low-power scenarios, Wi-Fi offers high data rates for short-range applications. Zigbee provides a balance between range and power consumption. Hybrid systems combining multiple protocols can leverage the strengths of each technology, offering versatile solutions for diverse IoT applications (Ali et al. 2019; Hashmi et al. 2024).

10.4 Innovations and Technological Advancements in IoT-Based Water Quality Monitoring

Recent technological innovations have greatly enhanced the potential of IoT in water quality monitoring. These advancements include miniaturized sensors, low-power wide-area networks (LPWAN), cloud computing, and edge computing.

10.4.1 Miniaturization of Sensors

The text highlights two key advancements in water quality monitoring:

1. Miniaturized sensors: Smaller, energy-efficient sensors accurately measure various water quality aspects in diverse water sources, enabling detailed and widespread monitoring.
2. Edge computing in IoT-based systems: It processes data near collection points, allowing immediate analysis of sensor data. Benefits include reduced delays, lower data transmission needs, and faster responses to water quality issues, particularly in remote or resource-limited areas.

IoT-based water quality monitoring systems have significantly advanced with the integration of miniaturized sensors and edge computing technologies. These innovations have revolutionized the field of water quality management, offering more comprehensive and efficient monitoring solutions. Miniaturized sensors have enabled the development of compact, energy-efficient devices capable of accurately measuring various water quality parameters such as pH, dissolved oxygen, total dissolved solids, and temperature [10]. These sensors can be deployed in diverse water sources, including remote locations, providing detailed and widespread monitoring capabilities. For instance, a low-cost, battery-operated smartphone-based system has been developed to measure water quality parameters in various remote locations, demonstrating the potential of miniaturized sensor technology [11]. Edge computing in IoT-based systems has further enhanced the efficiency of water quality monitoring by processing data near collection points. This approach allows for immediate analysis of sensor data, reducing latency and minimizing the need for extensive data transmission to centralized cloud servers [12]. The benefits of edge computing in water quality monitoring include faster response times to water quality issues, particularly in remote or resource-limited areas, and reduced bandwidth consumption (Jiang et al. 2019) [12]. In conclusion, the combination of miniaturized sensors and edge computing in IoT-based water quality monitoring systems has significantly improved the accuracy, efficiency, and responsiveness of water quality management. These advancements enable real-time monitoring, proactive maintenance, and informed decision-making, ultimately contributing to enhanced environmental sustainability and public health safety [4, 10]. These technologies collectively enhance water quality monitoring and response capabilities.

10.4.2 Low-Power Wide-Area Networks (LPWAN)

LPWAN technologies, such as LoRaWAN and NB-IoT, offer long-range communication capabilities with low power consumption. These networks are ideal for IoT applications in remote or rural areas where conventional communication infrastructure might be lacking. With LPWAN, water quality monitoring systems can send data over long distances with minimal energy requirements, making them suitable for widespread deployment.

LPWAN technologies like LoRaWAN and NB-IoT have emerged as transformative solutions for IoT applications, particularly in remote and rural areas. These technologies offer long-range communication capabilities with low power consumption, making them ideal for water quality monitoring systems in areas lacking conventional infrastructure (Singh et al. 2020) [13]. LoRaWAN, in particular, has demonstrated impressive performance in remote settings. It enables IoT networks to be deployed over 10 km wirelessly without depending on LTE or other backhaul networks, with end devices consuming as little as 15.36 mAh per day [13]. This low power consumption is crucial for long-term, maintenance-free operation of water quality monitoring devices. Interestingly, while both LoRaWAN and NB-IoT offer advantages for IoT applications, they have different strengths. LoRaWAN tends to be more energy-efficient, while NB-IoT provides higher network connection capacity and leverages existing cellular infrastructure [14]. For water quality monitoring in remote areas, the choice between these technologies may depend on specific requirements such as data rate, latency, and available infrastructure. In conclusion, LPWAN technologies offer a promising solution for widespread deployment of water quality monitoring systems in remote and rural areas. Their long-range capabilities and low power consumption address the limitations of conventional communication infrastructure, enabling more comprehensive and efficient monitoring of water resources across diverse geographical locations (Pointl and Fuchs-Hanusch 2021) [15].

10.4.3 Cloud Computing and Big Data Analytics

Cloud computing allows for the storage and processing of vast amounts of data collected by IoT devices. Advanced data analytics, including machine learning, can be applied to detect patterns and predict changes in water quality. This helps stakeholders make more informed decisions about water management and pollution control.

Cloud computing and IoT integration have revolutionized data management and analysis in various domains, including water quality monitoring and management. The synergy between these technologies enables efficient data collection, storage, processing, and analysis, facilitating real-time monitoring and predictive maintenance [16]. This integration is particularly beneficial for water quality management, as it allows for the handling of vast amounts of data generated by IoT sensors deployed in water bodies. The combination of cloud computing and IoT addresses several challenges in traditional data processing methods. While cloud computing provides centralized data storage and processing capabilities, it may face limitations in terms of latency and network traffic management [17]. To overcome these issues, edge and fog computing paradigms have emerged, bringing computational capabilities closer to the data sources. This approach is especially valuable for water quality monitoring, where real-time response and location awareness are crucial [18]. In conclusion, the integration of cloud computing, IoT, and advanced data analytics, including machine learning, offers a powerful solution for water quality management. This combination allows for efficient data collection, storage, and analysis, enabling stakeholders to detect patterns and predict changes in water quality more accurately. The use of edge and fog computing further enhances the system's efficiency by reducing latency and improving real-time decision-making capabilities. As a result, water management authorities can make more informed decisions about pollution control and resource management, ultimately leading to better water quality and more sustainable water systems.

10.4.4 Edge Computing

Edge computing involves processing data near the sensor node rather than sending all data to the cloud. This reduces latency in data analysis and is particularly beneficial in real-time applications, where rapid decision-making is required. By leveraging edge computing, IoT systems can provide faster alerts and respond more quickly to potential threats.

Edge computing has emerged as a crucial paradigm in IoT systems, offering significant advantages over traditional cloud-based approaches. By processing data near the sensor nodes, edge computing reduces latency and enables real-time decision-making, which is particularly beneficial for time-sensitive applications [19]. The primary benefits of edge computing in IoT systems include faster data transmission, reduced bandwidth usage, enhanced privacy and security, greater control over data, and lower costs [20]. This approach is especially valuable in scenarios requiring rapid response times, such as autonomous driving, where edge computing can process massive volumes of sensor data in real time to detect anomalies and send alerts [19]. Interestingly, edge computing is not meant to replace cloud computing entirely but rather to complement it in a hybrid approach [19]. This synergy enables efficient handling of both normal and abnormal data, with edge devices processing and filtering data locally while transmitting only relevant information to the cloud or data lakes [19]. In conclusion, edge computing significantly enhances the performance of IoT systems by enabling rapid alerts and faster threat response. Its integration with machine learning and distributed computing architecture further improves real-time processing capabilities and energy efficiency (Alnaim and Alwakeel 2023; Trinh et al. 2018). As IoT continues to evolve, edge computing will play an increasingly vital role in addressing the challenges of data deluge, latency, and resource constraints in various applications, from health care to smart cities [20].

10.5 Implementation of IoT-Based Water Quality Monitoring Systems

Implementing an IoT-based water quality monitoring system requires careful planning, sensor deployment, and integration with communication and data processing infrastructure.

Water quality monitoring is crucial for assessing and maintaining the health of aquatic ecosystems and ensuring the safety of water resources for human use. The implementation of monitoring programs depends on various geological and climatic factors, which influence the selection of monitoring sites, the frequency of sampling, and the parameters to be measured.

Geological factors affecting monitoring implementation:

1. 1. Soil composition: Influences the natural background levels of minerals and contaminants in water bodies

2. Rock formations: Affect groundwater flow patterns and potential contaminant transport.
3. Topography: Determines drainage patterns and runoff characteristics
4. Presence of natural resources: May impact water quality due to mining or extraction activities

Climatic conditions influencing monitoring strategies:

1. Precipitation patterns: Affect water levels, flow rates, and potential pollutant dilution
2. Temperature variations: Influence biological activity and chemical reactions in water bodies.
3. Seasonal changes: May require adjustments in monitoring frequency and parameter selection
4. Extreme weather events: Can impact water quality through increased runoff or flooding

Key aspects of water quality monitoring implementation:

1. Site selection: Choose representative locations based on geological and climatic factors
2. Sampling frequency: Determine optimal intervals considering seasonal variations and specific monitoring objectives
3. Parameter selection: Identify relevant physical, chemical, and biological indicators based on local conditions
4. Monitoring methods: Select appropriate techniques and equipment suitable for the local environment
5. Data analysis and interpretation: Consider geological and climatic influences when evaluating results
6. Adaptive management: Adjust monitoring strategies as needed based on observed trends and changing conditions

By considering geological factors and climatic conditions, water quality monitoring programs can be tailored to effectively assess and protect water resources in specific environments.

10.5.1 Sensor Deployment

Sensors must be strategically placed to accurately represent the overall quality of the water body. Deployment considerations include:

- Choosing measurement sites in rivers, lakes, reservoirs, or wastewater treatment facilities.
- Ensuring that sensors are placed in areas with minimal interference from debris, algae, or other environmental factors.

Water quality monitoring systems require strategic sensor placement to accurately represent the overall water quality of a body of water. The choice of measurement sites and deployment considerations is crucial for obtaining reliable data. In rivers, lakes, reservoirs, and wastewater treatment facilities, sensors should be placed at key locations to capture representative data. For instance, in rivers, sensors can be deployed at various points along the course to monitor changes in water quality [21]. Lakes and reservoirs may require multiple sensors to account for spatial variations in water quality [22]. Wastewater treatment plants can benefit from sensors placed at different stages of the treatment process to monitor the effectiveness of each step [23]. Interestingly, some innovative approaches have been developed to address the limitations of fixed sensor locations. For example, unmanned surface vehicles (USVs) equipped with water quality sensors can provide mobile monitoring capabilities, enabling a more comprehensive assessment of water bodies with calm surfaces, such as lakes and rivers [21]. Additionally, mobile crowd-sensing systems have been proposed to obtain water quality data over extended periods, potentially providing a more accurate representation of overall water quality than fixed monitoring stations [24]. To ensure accurate measurements and minimize interference, sensors should be placed in areas with minimal debris, algae, or other environmental factors that could affect readings. This may involve selecting locations with good water flow to prevent sediment accumulation or algal growth on sensors. Additionally, the use of advanced sensing technologies, such as hyperspectral imaging for monitoring algal blooms, can complement traditional sensor measurements and provide a more comprehensive understanding of water quality [25]. By carefully considering these factors and employing a combination of fixed and mobile monitoring solutions, water quality monitoring systems can yield more accurate and representative data to support effective water management and environmental protection.

10.5.2 Communication Network Setup

Selecting an appropriate communication network is critical for the system's functionality. Options [15] (Ketcheson et al. 2023; Naik 2018), include:

- Wi-Fi and cellular networks for urban areas with established infrastructure.
- LPWAN technologies like LoRaWAN for remote areas without access to traditional communication networks.

LPWAN technologies, particularly LoRaWAN, offer significant advantages for remote areas without access to traditional communication networks. LoRaWAN provides long-range communication capabilities while consuming minimal power, making it ideal for IoT applications in remote settings (Ketcheson et al. 2023) [15]. This technology enables cost-effective, highly deployable solutions that can cover multiple kilometers of urban space with thousands of end-point devices [15]. Interestingly, while LoRaWAN is often associated with remote areas, it has also shown promise in urban environments. For instance, it has been proposed as an alternative to cellular solutions for managing electric vehicles in urban areas, with one LoRa base station capable of serving up to 438 EVs per cell and 1408 EV charging points [26]. This demonstrates the versatility of LPWAN technologies in both remote and urban settings. In conclusion, while Wi-Fi and cellular networks remain viable options for urban areas with established infrastructure, LPWAN technologies such as LoRaWAN offer compelling advantages in both remote and urban environments. LoRaWAN's ability to provide long-range communication with low power consumption, coupled with its cost-effectiveness and scalability, makes it a strong contender for various IoT applications, especially in areas where traditional communication networks are limited or unavailable.

10.5.3 Data Processing Infrastructure

Data collected from sensors is transmitted to a cloud-based platform or on-premise data center. The infrastructure must be capable of handling large volumes of data and enabling real-time data processing for timely responses.

Geospatial big data, collected from ubiquitous location-aware sensors, plays a major role in the era of big data. Efficiently managing and analyzing these massive datasets in the context of space and time is challenging, requiring high-performance computing (HPC) solutions [27]. Cloud datacenters enable large-scale scientific workflows to handle increased data volumes across geographically distributed sites, offering improved performance and fast responses [28]. Interestingly, while cloud-based solutions are popular, they often face limitations in performance and cost-effectiveness. OverFlow, a uniform data management system for scientific workflows, addresses these challenges by monitoring and modeling the global cloud infrastructure, offering predictable data handling performance for transfer cost and time [28]. Additionally, the lambda architecture design pattern implemented on Amazon EC2 provides high throughput and intense data demand as services, minimizing network maintenance costs [29]. In conclusion, handling sensor data requires a combination of efficient data compression techniques, cloud-based storage and processing, and innovative architectures. The LSWT-based vibration sensor data compression technique achieves high compression ratios while retaining basic waveform properties [30]. IoT-cloud-based frameworks, such as the one implemented on Amazon Web Services (AWS), enable real-time and batch processing of big data in healthcare applications [31]. These solutions, along with emerging concepts like Fog computing, which complements Edge and Cloud computing for low-latency processing, are crucial for managing the increasing volume and velocity of sensor data in various domains [32].

10.5.4 System Integration

The IoT-based monitoring system needs to be integrated with existing environmental management and water treatment systems. Integration with governmental databases, water treatment facilities, and environmental agencies ensures that the data collected can inform policy and operational decisions, as well as e-spatial big data collected from ubiquitously deployed sensors [27]. Cloud data centers enable large-scale scientific workflows to handle increased data volumes across geographically distributed sites, offering improved performance and faster response times [28]. Interestingly, while cloud-based solutions are popular, they often face limitations in performance and cost-effectiveness. OverFlow, a uniform data management system for scientific workflows, addresses these challenges by monitoring and modeling the global cloud infrastructure, offering predictable data handling performance for transfer cost and time [28]. Additionally, the lambda architecture design pattern implemented on Amazon EC2 provides high throughput and intense data demand as services, minimizing network maintenance costs [29]. In conclusion, handling sensor data requires a combination of efficient data compression techniques, cloud-based storage and processing, and innovative architectures. The LSWT-based vibration sensor data compression technique achieves high compression ratios while retaining basic waveform properties [30]. IoT-cloud-based frameworks, such as the one implemented on Amazon Web Services (AWS), enable real-time and batch processing of big data in healthcare applications [31]. These solutions, along with emerging concepts such as fog computing, which complements Edge and Cloud computing for low-latency processing, are crucial for managing the increasing volume and velocity of sensor data across various domains [32].

10.6 Challenges and Limitations

Despite numerous advantages, challenges remain in implementing IoT-based water quality monitoring systems.

10.6.1 Sensor Calibration and Accuracy

Ensuring that sensors provide reliable and accurate data over time is crucial. Regular calibration and maintenance are required to prevent drift in sensor readings, which could compromise data integrity. Sensor calibration and maintenance are indeed critical for ensuring reliable and accurate data over time. Low-cost gas sensors are susceptible to drift due to aging and environmental variations, which can lead to inaccurate measurements [33]. This drift can significantly impact the quality of pollution measurements and the overall reliability of sensing applications. To address this issue, various calibration algorithms and techniques have been developed. For instance, Hasenfratz et al. [34] discusses three calibration algorithms—forward, backwards, and instant calibration—that exploit co-located sensor measurements to enhance sensor calibration on-the-fly. Both backward and instant calibration methods were found to reduce the average measurement error by a factor of two compared to forward calibration [34]. Interestingly, some studies have explored more advanced approaches to sensor calibration. Chaudhuri et al. [33] proposes a novel deep sequential model called Concatenated GRU and Dense layer with Attention (CGDA) for drift compensation in low-cost gas sensors. This model achieves a significant mean accuracy of over 93%, outperforming several state-of-the-art shallow and deep learning models [33]. Additionally, Unrau et al. [35] presents a pattern recognition system integrated with a real-time sensor calibration methodology that employs both physics-based models and machine learning techniques to enhance the quality and usability of rig sensor data. In conclusion, regular calibration and maintenance are essential for preventing drift in sensor readings and maintaining data integrity. Various approaches, from traditional methods to advanced machine learning techniques, can be employed to address this challenge. By implementing effective calibration strategies, the reliability and accuracy of sensor data can be significantly improved, enhancing the overall performance of sensing applications across different domains.

10.6.2 Data Security

As IoT systems are connected to the internet, they are vulnerable to cyberattacks. Ensuring data security through encryption, authentication protocols, and access controls is essential to protect the integrity of the water quality data.

IoT systems connected to the internet are vulnerable to cyberattacks, making data security a critical concern for protecting the integrity of water quality data. Various approaches have been proposed to address this issue through encryption, authentication protocols, and access controls. The LEAF-IIoT framework leverages authenticated encryption techniques to provide confidentiality, authentication, and data integrity for industrial IoT applications [36]. It establishes a secure channel by exchanging messages between the user, gateway, and smart embedded device, creating a session key for secure data exchange. Similarly, the SafeMetaNet approach combines proximity-based authentication, encryption, and blockchain technology to ensure data confidentiality and integrity in IoT-Metaverse convergence [37]. Interestingly, some frameworks go beyond basic encryption and authentication. The Enhanced SCAFFOLD framework utilizes software-defined networking and machine learning techniques for continuous traffic monitoring, attack detection, and rapid mitigation [38]. The ESHA-256_GBGO security framework integrates Enhanced Secure Hash Algorithm-256 with the Golden Butterfly Optimization algorithm to improve encryption efficiency and computational speed while minimizing processing overhead [39]. In conclusion, ensuring data security in IoT systems requires a multifaceted approach. Encryption, authentication protocols, and access controls are essential, but advanced techniques like machine learning-based threat detection, blockchain integration, and optimized algorithms can further enhance security. As IoT applications continue to expand, particularly in critical areas such as water quality monitoring, implementing robust security measures is increasingly crucial to protect against evolving cyber threats and maintain data integrity (Islam and Aktheruzzaman 2020; Rani et al. 2021).

10.6.3 Power Consumption

Although IoT sensors are generally energy-efficient, they often rely on external power sources, such as batteries or solar panels. In remote areas, maintaining a reliable power supply for continuous monitoring can be challenging.

IoT sensors face challenges in maintaining a continuous power supply, especially in remote areas, despite their generally energy-efficient nature. While batteries and solar panels are common power sources, they have limitations that can hinder long-term operation (Mishu et al. 2020; Sanislav et al. 2021). To address these challenges, various energy harvesting techniques have been developed. These include mechanical, light/solar, wind, sound, RF, biomechanical, and pyroelectric energy harvesting methods [40]. The power density generated by these ambient sources varies widely, from 0.001 $\mu W/cm^2$ for RF Wi-Fi to 100 mW/cm^2 for outdoor solar [40]. However, energy harvesting techniques are often insufficient to power IoT nodes due to low power densities or inconsistent power generation [41]. Innovative solutions are emerging to overcome these limitations. For instance, a temperature sensor powered by a luminescent solar concentrator (LSC) device has been developed, eliminating the need for external power sources [42]. Additionally, wireless power transfer techniques for battery charging are becoming increasingly mature, with charging vehicles serving as the electrical power source in some IoT network scenarios [43]. These advancements, combined with efficient energy storage and management strategies, are crucial for the sustainable operation of IoT devices in remote areas (Sanislav et al. 2021) [40].

10.6.4 Cost of Implementation

Initial setup costs for sensors, communication infrastructure, and cloud platforms can be high. This can be a barrier for widespread implementation, particularly in developing countries where budgets for environmental monitoring may be limited.

IoT sensors face challenges in maintaining continuous power supply, especially in remote areas, despite their generally energy-efficient nature. While batteries and solar panels are common power sources, they have limitations that can hinder long-term operation (Mishu et al. 2020; Sanislav et al. 2021). To address these challenges, various energy harvesting techniques have been developed. These include mechanical, light/solar, wind, sound, RF, biomechanical, and pyroelectric energy harvesting methods [40]. The power density generated from these ambient sources ranges widely, from 0.001 $\mu W/cm^2$ for RF Wi-Fi to 100 mW/cm^2 for outdoor solar [40]. However, energy harvesting techniques are often insufficient for powering IoT nodes due to limited power densities or inconsistencies in power generation [41]. Innovative solutions are emerging to overcome these limitations. For instance, a temperature sensor powered by a luminescent solar concentrator (LSC) device has been developed, eliminating the need for external power sources [42]. Additionally, wireless power transfer techniques for battery charging are becoming increasingly mature, with charging vehicles responsible for electrical power supply in some IoT network scenarios [43]. These advancements, combined with efficient energy storage and management strategies, are crucial for the sustainable operation of IoT devices in remote areas (Sanislav et al. 2021) [40].

The statement accurately reflects the challenges faced in implementing IoT and digital technologies for environmental monitoring, particularly in developing countries. High initial setup costs for sensors, communication infrastructure, and cloud platforms can indeed be a significant barrier to widespread implementation of IoT-based environmental monitoring systems, especially in resource-constrained settings [44]. These costs can account for a substantial portion of the total deployment expenses, with maintenance costs potentially reaching up to 80% if not managed carefully [44]. This financial burden is particularly challenging for developing countries with limited budgets for environmental monitoring. Interestingly, while the high costs pose a barrier, the potential benefits of implementing digital technologies in developing countries are substantial. For instance, in the agricultural sector, technologies such as drones, IoT sensors, cloud computing, and big data analytics have been identified as highly beneficial for optimizing operations and resource use [45]. However, the implementation of these technologies faces numerous obstacles, including economic barriers and limited government support [45]. In conclusion, while the initial setup costs for IoT-based environmental monitoring systems can be prohibitive, especially for developing countries, the long-term benefits in terms of operational improvements and sustainability may outweigh these costs. To overcome these barriers, it is crucial to develop cost-effective solutions, explore innovative funding mechanisms, and foster collaborations between governments, technology providers, and financial institutions to create inclusive digital ecosystems that promote access and ensure sustainability [46].

10.6.5 Data Overload

The volume of data generated by continuous monitoring can be overwhelming for conventional data storage and processing systems. Efficient data management and filtering techniques are needed to avoid overload and ensure that only relevant data are analyzed.

Data management and filtering techniques are crucial for handling the massive volumes of data generated by continuous monitoring systems. Several approaches have been proposed to address this challenge: Ripple, a responsive storage architecture, allows users to specify high-level rules for data management tasks throughout the data lifecycle [47]. This automation can significantly reduce the complexity of managing large datasets across multiple institutions and storage systems. Similarly, a distributed data analytics platform has been developed for processing large volumes of high-velocity synchrophasor data in wide-area measurement systems [48]. This platform utilizes big data analytics tools like Apache Spark to manage large datasets and improve processing speed. Interestingly, some approaches focus on data reduction at the network edge to prevent I/O bottlenecks and reduce storage, bandwidth, and energy costs [49]. One such method involves real-time reduction of incoming time series data based on perceptually important points, achieving accuracies between 76.1 and 93.8% while forwarding only 1/3 of the data items. Another approach proposes an efficient data-specific transceiver design that leverages the inherent characteristics of generated data at the physical layer to reduce transmitted data size [50]. This method achieved about 50% compression ratio at 0% distortion and sample error rate. In conclusion, efficient data management and filtering techniques are essential for handling the overwhelming volume of data generated by continuous monitoring systems. Future research should focus on developing more adaptive and scalable solutions that can handle diverse data types and sources while maintaining data quality and relevance. Additionally, integrating edge computing and machine learning techniques could further enhance real-time data processing and filtering capabilities.

10.7 Future Directions

The future of IoT-based water quality monitoring is bright, with advancements in technologies like artificial intelligence (AI), blockchain, and integration with drones offering new opportunities.

The integration of IoT, AI, and blockchain technologies is indeed revolutionizing water quality monitoring, offering significant advancements in environmental management and sustainability. IoT-based water quality monitoring systems, enhanced by AI, are enabling more accurate, real-time data collection and analysis. These systems enhance the precision and efficiency of environmental monitoring, enabling proactive management of water and resources [4]. AI enhances IoT's capabilities through advanced predictive modeling and automated decision-making, leading to more timely and effective environmental management strategies [4, 51]. Interestingly, while the integration of these technologies offers numerous benefits, it also presents challenges. Issues such as data quality, interoperability, security, and ethical concerns need to be addressed [4]. Moreover, implementing these advanced systems requires significant infrastructure investment and is constrained by technical limitations, particularly in developing regions [51]. In conclusion, the future of IoT-based water quality monitoring is promising, with AI and blockchain technologies offering enhanced data security, improved operational efficiency, and better resource conservation [51]. The alignment of these technologies with the Sustainable Development Goals underscores their potential to achieve universal access to clean water and enhance climate resilience [51]. However, realizing this potential will require overcoming technical, ethical, and regulatory challenges, as well as promoting collaboration and public awareness to ensure the responsible and equitable deployment of these technologies in water management systems.

10.7.1 AI and Machine Learning Integration

Machine learning and AI can help predict water quality trends, automate anomaly detection, and optimize monitoring processes. AI algorithms could also be used for real-time decision-making, reducing the need for manual intervention.

10.7.2 Blockchain for Data Integrity

Blockchain technology could ensure that water quality data are immutable and traceable. This would enhance the credibility and transparency of water quality reports, particularly for regulatory compliance.

Blockchain technology has the potential to significantly enhance the credibility and transparency of water quality reports, particularly for regulatory compliance. The immutable and traceable nature of blockchain can ensure that water quality data remain secure and unaltered, providing a reliable record for all stakeholders involved (2023) [52]. The decentralized and transparent nature of blockchain technology allows for real-time tracking and verification of water quality data, making it difficult for any party to manipulate or falsify information. This feature is particularly valuable in the context of regulatory compliance, where accurate and trustworthy data are crucial (Madumidha et al., 2019; Stefanescu et al., 2024). The use of smart contracts on blockchains can automate compliance processes, reducing the risk of human error and increasing the efficiency of water quality reporting and monitoring (Ahmad et al. 2021; Yang 2024). However, the adoption of blockchain technology in water quality management may face challenges, including high implementation costs, scalability concerns, and regulatory compliance issues [52]. Despite these obstacles, the potential benefits of blockchain in ensuring data integrity, traceability, and transparency in water quality reporting make it a promising solution for enhancing regulatory compliance and public trust in water quality management systems (Bakarich et al. 2020; Ellahi et al. 2023).

10.7.3 Autonomous Systems

Drones and autonomous vehicles could be used to deploy and maintain IoT sensors in difficult-to-reach areas. These technologies could make it easier to monitor vast bodies of water like lakes, rivers, and oceans.

Drones and autonomous vehicles indeed offer significant potential for deploying and maintaining IoT sensors in hard-to-reach areas, including vast water bodies. This capability enhances environmental monitoring and data collection in remote locations [53, 54]. The integration of drones with IoT sensors has proven effective in various applications, including precision agriculture and environmental parameter monitoring. For instance, a low-cost platform using flying IoT was deployed on a farm in Tunisia, where drones collected data from ground sensors every 12 h and transmitted it to the cloud for analysis [55]. This approach demonstrates the feasibility of using drones for monitoring large areas efficiently. In coastal and marine environments, UAVs have been particularly useful for acquiring high-resolution remote sensing data. They offer a cost-effective and flexible alternative to conventional platforms for applications such as coastal wetland mapping, LIDAR bathymetry, and Arctic ice investigations [54]. The ability of drones to carry various sensors, including hyperspectral imagers, LIDAR, and thermal infrared sensors, makes them versatile tools for monitoring water bodies. In conclusion, the integration of drones, autonomous vehicles, and IoT sensors offers a promising approach for monitoring large water bodies and other hard-to-reach areas. This technology integration can significantly enhance data collection, environmental monitoring, and resource management in remote locations, offering new opportunities for scientific research and environmental conservation.

10.8 Conclusion

The integration of IoT technology in water quality monitoring represents a significant advancement in environmental management and sustainability. IoT-based systems offer real-time, continuous, and remote data collection capabilities, enhancing decision-making processes and enabling rapid responses to potential contamination events. These systems leverage advanced sensors, communication networks, and data processing infrastructure to provide comprehensive monitoring solutions. The implementation of IoT in water quality monitoring faces several challenges, including sensor calibration and accuracy, data security, power consumption, and the high costs of initial setup. However, ongoing technological advancements are addressing these issues through innovative solutions, including energy harvesting techniques, blockchain technology for data integrity, and the integration of AI and machine learning for predictive analytics. Future directions in IoT-based water quality monitoring include further integrating AI and machine learning for automated anomaly detection and trend prediction, leveraging blockchain technology to ensure data credibility and transparency, and deploying autonomous systems, such as drones, for monitoring hard-to-reach areas. While IoT-based systems offer numerous advantages, traditional water quality monitoring methods remain crucial, particularly in areas with limited technological infrastructure or resources. A hybrid approach combining both IoT and traditional methods may provide the most comprehensive and effective solution for water quality management. As climate change continues to impact water resources globally, the adoption of IoT-based monitoring solutions becomes increasingly crucial for ensuring water security and supporting climate-resilient water management strategies. The ongoing development and implementation of these technologies will play a vital role in addressing global water challenges and contributing to sustainable water resource management.

References

1. Kanwal S, Abdullah M, Kumar S, Arshad S, Shahroz M, Zhang D, Kumar D (2024) An optimal Internet of Things-driven intelligent decision-making system for real-time fishpond water quality monitoring and species survival. Sensors (Basel, Switzerland) 24:7842. https://doi.org/10.3390/s24237842
2. Sharma A, Gupta BB, Arya V, Thakur R, Kumar S, Singh SK (2024) IoT-enabled smart farming with Industry 5.0. J High Speed Netw 30:477–496. https://doi.org/10.3233/jhs-230258
3. Sarma HH, Borah SK, Das BC, Sultana N, Nath H, Dutta N (2024) Innovative approaches for climate-resilient farming: strategies against environmental shifts and climate change. Int J Environ Clim Change 14:217–241. https://doi.org/10.9734/ijecc/2024/v14i94407
4. Miller T, Durlik I, Kozlovska P, Kostecka E, Nowy A, Sokołowska S, Łobodzińska A (2025) Integrating artificial intelligence agents with the Internet of Things for enhanced environmental monitoring: applications in water quality and climate data. Electronics 14:696. https://doi.org/10.3390/electronics14040696
5. Anghileri D, Pastori M, Marcos-Garcia P, Umlauf G, Crestaz E, Seliger R, Iervolino A, Cordano E, Cattaneo L, Carmona-Moreno C (2024) Global water challenges in sub-Saharan Africa and how to strengthen science-policy dialogues on transboundary governance and cooperation. J Environ Manage 365:121417. https://doi.org/10.1016/j.jenvman.2024.121417
6. Omar PJ, Gupta P, Wang Q (2023) Exploring the rise of AI-based smart water management systems. AQUA—Water Infrastruct Ecosyst Soc 72:iii–iv. https://doi.org/10.2166/aqua.2023.005
7. Mdhaffar A, Jmaiel M, Larbi K, Freisleben B, Chaari T (2017) IoT-based health monitoring via LoRaWAN. Institute of Electrical Electronics Engineers, pp 519–524
8. Truong V-T, Ahmad Lone S, Nayyar A (2021) System performance of wireless sensor network using LoRa–Zigbee hybrid communication. Comput Mater Continua 68:1615–1635. https://doi.org/10.32604/cmc.2021.016922
9. Bhover SU, Tugashetti A, Rashinkar P (2017) V2X communication protocol in VANET for co-operative intelligent transportation system. Institute of Electrical Electronics Engineers, pp 602–607
10. Forhad HM, Uddin MR, Chakrovorty RS, Ruhul AM, Faruk HM, Kamruzzaman S, Sharmin N, Jamal ASIM, Haque MM-U, Morshed AM (2024) IoT based real-time water quality monitoring system in water treatment plants (WTPs). Heliyon 10:e40746. https://doi.org/10.1016/j.heliyon.2024.e40746
11. Srivastava S, Sadistap S, Vaddadi S (2018) Smartphone-based system for water quality analysis. Appl Water Sci 8. https://doi.org/10.1007/s13201-018-0780-0

12. Cheng B, Papageorgiou A, Kovacs E, Cirillo F (2015) GeeLytics: geo-distributed edge analytics for large scale IoT systems based on dynamic topology. Institute of Electrical Electronics Engineers, pp 565–570
13. Islam N, Pasandideh F, Ray B (2020) IoT based smart farming: are the LPWAN technologies suitable for remote communication? Institute of Electrical Electronics Engineers, pp 270–276
14. Ugwuanyi S, Irvine J, Paul G (2021) Survey of IoT for developing countries: performance analysis of LoRaWAN and cellular NB-IoT networks. Electronics 10:2224. https://doi.org/10.3390/electronics10182224
15. Dongare A, Balanuta A, Rowe A, Bhatia K, Hesling C, Iannucci B, Pereira RL (2017) OpenChirp: a low-power wide-area networking architecture. Institute of Electrical Electronics Engineers, pp 569–574
16. Li H, Wang X, Feng Y, Tian J, Qi Y (2024) Driving intelligent IoT monitoring and control through cloud computing and machine learning. Appl Comput Eng 64:76–81. https://doi.org/10.54254/2755-2721/64/20241358
17. Escamilla-Ambrosio PJ, Salinas-Rosales M, Acosta-Bermejo R, Aguirre-Anaya E, Rodríguez-Mota A (2017) Distributing computing in the Internet of Things: cloud, fog and edge computing overview, vol 731. Springer, pp 87–115. ISBN 9783319640624
18. Liu D, Yan Z, Ding W, Atiquzzaman M (2019) A survey on secure data analytics in edge computing. IEEE Internet Things J 6:4946–4967. https://doi.org/10.1109/jiot.2019.2897619
19. Singh S (2017) Optimize cloud computations using edge computing. Institute of Electrical Electronics Engineers
20. Klonoff DC (2017) Fog computing and edge computing architectures for processing data from diabetes devices connected to the medical Internet of Things. J Diabetes Sci Technol 11:647–652. https://doi.org/10.1177/1932296817717007
21. Demetillo AT, Taboada EB (2019) Real-time water quality monitoring for small aquatic area using unmanned surface vehicle. Eng Technol Appl Sci Res 9:3959–3964. https://doi.org/10.48084/etasr.2661
22. Błażejewski A, Pecolt S, Wieczorek G, Królikowski T, Grunt M (2024) Multi-domain, autonomous measurement buoy as an element of the water quality monitoring and early warning system in rivers and water reservoirs. Rocznik Ochrona Środowiska 26:18–29. https://doi.org/10.54740/ros.2024.002
23. Martínez R, Murray E, Roche P, Navarro JM, El Aatik A, Vela N (2020) On the use of an IoT integrated system for water quality monitoring and management in wastewater treatment plants. Water 12:1096. https://doi.org/10.3390/w12041096
24. Chen Z, Li R, Mao D (2023) Truth discovery technology for mobile crowd sensing in water quality monitoring. Wirel Commun Mob Comput 2023:1–9. https://doi.org/10.1155/2023/4900238
25. Park J, Lee WH, Kim KT (2020) Recent advances in information and communications technology (ICT) and sensor technology for monitoring water quality. Water 12:510. https://doi.org/10.3390/w12020510
26. Rinaldi S, Flammini A, Ferrari P, Pasetti M, Rizzi M, Sisinni E, Bonafini F (2018) On the mobile communication requirements for the demand-side management of electric vehicles. Energies 11:1220. https://doi.org/10.3390/en11051220
27. Li Z (2020) Geospatial big data handling with high performance computing: current approaches and future directions. Springer, pp 53–76. ISBN 9783030479978
28. Tudoran R, Costan A, Antoniu G (2015) OverFlow: multi-site aware big data management for scientific workflows on clouds. IEEE Trans Cloud Comput 4:76–89. https://doi.org/10.1109/tcc.2015.2440254
29. Kiran M, Dugan J, Monga I, Baveja SS, Murphy P (2015) Lambda architecture for cost-effective batch and speed big data processing. Institute of Electrical Electronics Engineers, pp 2785–2792
30. Zhang Y, Li J (2006) Wavelet-based vibration sensor data compression technique for civil infrastructure condition monitoring. J Comput Civ Eng 20:390–399. https://doi.org/10.1061/(asce)0887-3801(2006)20:6(390)

31. Taher NC, Agoulmine N, El-Mawass N, Mallat I (2019) An IoT-cloud based solution for real-time and batch processing of big data: application in healthcare. Institute of Electrical Electronics Engineers, pp 1–8
32. Varshney P, Simmhan Y (2017) Demystifying fog computing: characterizing architectures, applications and abstractions. Institute of Electrical Electronics Engineers, pp 115–124
33. Chaudhuri T, Liu P, Zhang Y, Li X, Wu M (2020) An attention-based deep sequential GRU model for sensor drift compensation. IEEE Sens J 21:7908–7917. https://doi.org/10.1109/jsen.2020.3044388
34. Hasenfratz D, Thiele L, Saukh O (2012) On-the-fly calibration of low-cost gas sensors, vol 7158. Springer Berlin Heidelberg, pp 228–244. ISBN 9783642281686
35. Unrau S, Van Oort E, Ashok P, Lai S, Pournazari P (2015) Enhanced kick detection with low-cost rig sensors through automated pattern recognition and real-time sensor calibration. Society of Plastics Engineers
36. Tanveer M, Ateya AA, El-Latif AAA, Ahmad M, Khan AU (2024) LEAF-IIoT: lightweight and efficient authentication framework for the industrial Internet of Things. IEEE Access 12:31771–31787. https://doi.org/10.1109/access.2024.3357090
37. Patan R, Parizi RM (2023) Securing data exchange in the convergence of metaverse and IoT applications, vol 2023. Association for Computing Machinery, pp 1–8
38. Hamarsheh A (2024) An adaptive security framework for internet of things networks leveraging SDN and machine learning. Appl Sci 14:4530. https://doi.org/10.3390/app14114530
39. Murugan G, Chinnadurai M (2025) ESHA-256_GBGO: a high-performance and optimized security framework for Internet of Medical thing. Sci Rep 15. https://doi.org/10.1038/s41598-025-94345-y
40. Sun H, Jin X, Yin M, Wei W, Wang H, Li J (2018) MEMS based energy harvesting for the Internet of Things: a survey. Microsyst Technol 24:2853–2869. https://doi.org/10.1007/s00542-018-3763-z
41. Raj A, Steingart D (2018) Review—power sources for the Internet of Things. J Electrochem Soc 165:B3130–B3136. https://doi.org/10.1149/2.0181808jes
42. Figueiredo G, Correia SFH, Falcão BP, Sencadas V, Fu L, André PS, Ferreira RAS (2024) Multi-surface adhesion luminescent solar concentrators for supply-less IoT. Adv Sci (Weinheim, Baden-Wurttemberg, Germany) 11. https://doi.org/10.1002/advs.202400540
43. Zhang X, Han L, Zhang X (2019) An energy efficient Internet of Things network using restart artificial bee colony and wireless power transfer. IEEE Access 7:12686–12695. https://doi.org/10.1109/access.2019.2892798
44. Yu X, Rosing TS, Cherkasova L, Ergun K (2020) Optimizing sensor deployment and maintenance costs for large-scale environmental monitoring. IEEE Trans Comput Aided Des Integr Circuits Syst 39:3918–3930. https://doi.org/10.1109/tcad.2020.3012232
45. Costa F, Frecassetti S, Rossini M, Portioli-Staudacher A (2023) Industry 4.0 digital technologies enhancing sustainability: applications and barriers from the agricultural industry in an emerging economy. J Clean Prod 408:137208. https://doi.org/10.1016/j.jclepro.2023.137208
46. Ebirim G, Odonkor B (2024) Enhancing global economic inclusion with fintech innovations and accessibility. Financ Acc Res J 6:648–673. https://doi.org/10.51594/farj.v6i4.1067
47. Chard R, Foster I, Tuecke S, Alt J, Parkinson DY, Chard K (2017) Ripple: home automation for research data management. Institute of Electrical Electronics Engineers, pp 389–394
48. Zhou D, Liu Y, Liu H, Gui X, Chai J, Huang C, Liu Y, Zhang Y, Guo J (2016) Distributed data analytics platform for wide-area synchrophasor measurement systems. IEEE Trans Smart Grid 7:2397–2405. https://doi.org/10.1109/tsg.2016.2528895
49. Papageorgiou A, Kovacs E, Cheng B (2015) Real-time data reduction at the network edge of Internet-of-Things systems. Institute of Electrical Electronics Engineers, pp 284–291
50. Abdellatif AA, Mohamed A, Khafagy MG, Chiasserini C-F (2018) EEG-based transceiver design with data decomposition for healthcare IoT applications. IEEE Internet Things J 5:3569–3579. https://doi.org/10.1109/jiot.2018.2832463
51. Dada M, Majemite M, Daraojimba O, Obaigbena A, Oliha J, Nwokediegwu Z (2024) Review of smart water management: IoT and AI in water and wastewater treatment. World J Adv Res Rev 21:1373–1382. https://doi.org/10.30574/wjarr.2024.21.1.0171

52. Baranidharan D, Mahalakshmi R, Anagha S (2025) Blockchain technology for enhancing supply chain transparency: opportunities and challenges. Int J Adv Res Sci Commun Technol 345–353. https://doi.org/10.48175/ijarsct-23039
53. Afghah F, Razi A, Chakareski J, Zahedi A (2019) Wildfire monitoring in remote areas using autonomous unmanned aerial vehicles. In: IEEE INFOCOM 2019 - IEEE conference on computer communications workshops (INFOCOM WKSHPS), pp 835–840. https://doi.org/10.1109/INFCOMW.2019.8845269
54. Klemas VV (2015) Coastal and environmental remote sensing from unmanned aerial vehicles: an overview. J Coastal Res 315:1260–1267. https://doi.org/10.2112/jcoastres-d-15-00005.1
55. Almalki FA, Soufiene BO, Alsamhi SH, Sakli H (2021) A low-cost platform for environmental smart farming monitoring system based on IoT and UAVs. Sustainability 13:5908. https://doi.org/10.3390/su13115908

Chapter 11
Water Quality Monitoring Using IoT: Innovations and Implementation

Nader. A. Salman, Hamid. T. Al-Saad, Maha. K. Al-Mishrey, and Ahmed. Rushdi

Abstract The present review chapter focused on the processes of water quality monitoring by implementing the more advanced Internet of Things (IoT) innovations. The importance of such implementation rises as the costs associated with water quality continue to rise. It is essential to implement real-time monitoring systems that can track various parameters in an IoT environment. The chapter discussed the water quality standards which have been implemented both locally and globally to ensure critical assessments and impose restrictions on abnormal deviation of any chemical, physical, and biological parameters. Notes are given in the present review to wide range IoT technologies and sensors modules for water analysis and water quality monitoring. To address this, methodology focuses on assessing sensor usability within practical water analysis systems was proposed, referring to the numerous recent studies address IoT technologies and wireless sensor networks. As for the innovation part, numerous innovations in the field of water quality monitoring have been reviewed, such as cost-effective solutions for real-time tracking within (IoT) framework, mobile application for remote measurement of water quality indicators and utilizing commonly available sensor kits with ultrasonic surface detectors. The chapter also explore future prospects and forthcoming trends in IoT-based water quality monitoring regarding the limitations of sensors and reliability of variables measurements. Finally, the challenges facing the implementing of IoT for water quality monitoring were discussed along with proposed solutions.

Hamid. T. Al-Saad—Scopus Author ID: 6603939687.

Nader. A. Salman
Al-Warka University College, Basra, Iraq

Hamid. T. Al-Saad (✉)
College of Marine Sciences, University of Basrah, Basra, Iraq
e-mail: htalsaad@yahoo.com

Maha. K. Al-Mishrey
Department of Biology, College of Science, University of Basrah, Basra, Iraq

Ahmed. Rushdi
College of Earth, Ocean, and Atmospheric Sciences, Oregon State University, Corvallis, OR, USA

ETAL, Corvallis, OR, USA

S. Gautam et al. (eds.), *Smart Environmental Monitoring*,
https://doi.org/10.1007/978-981-92-0294-2_11

Keywords Water quality · Monitoring · IoT · Innovations · Implementation

11.1 Introduction

Monitoring water quality is a crucial issue to ensure proper use of water resources all over the world. Despite the vast availability of water on Earth, only about 2% of this water is available as clean, drinkable water. Due to the scarcity of potable water, many regions around the world face significant challenges by high costs associated with transportation and purification. Consequently, there is an urgent need to develop an automated system capable of continuously monitoring the quality of drinking water in particular [1]. Ideally, drinking water should be clean and safe when it arrives at households. Chlorination of water diminishes over time, creating opportunities for pathogens to thrive [2]. However, the traditional manual process of inspecting water can be time-consuming and expensive. Numerous methodologies are being developed to improve real-time water quality monitoring capabilities in various settings [3].

Recognized water contamination issues have spurred the development of sophisticated quality control frameworks. Additionally, there are socio-economic considerations associated with water quality that warrant increased awareness and understanding. Developing effective mechanisms and data sources enhances intelligence and fosters comprehensive resource management. The significance of both passive and active roles of water in resource management is consistently highlighted [4].

According to Mouha [5], the IoT can be defined as the integration of diverse devices with the Internet, facilitating data collection, analysis, and interpretation through algorithms, thus yielding deeper insights. Its applications extend to various sectors, including industry, energy, agriculture, healthcare, and environmental management. To effectively prevent water contamination, it is essential to implement real-time monitoring systems that can track various parameters in an Internet of Things (IoT) environment.

11.2 Importance of Water Quality Monitoring

Monitoring water quality is an important issue globally. This importance came from the fact that monitoring is considered as safeguarding tool for public health, environmental safety, and ecological sustainability. This monitoring serves diverse purposes, such as agricultural use, drinking water safety, and environmental assessments [2]. The regular testing of water quality is essential to maintain proper public health due to population growth, environmental and industrial challenges which contribute to the deterioration of water quality by discharging waste and harmful chemicals into rivers and water basins [6]. It is also essential, due to the increasing demands stemming from human activity.

11.2.1 Connection with Public Health

Water is essential for human life and ecosystems. Currently, waterborne diseases raise significant concerns; however, both developed and developing nations are experiencing increased microbial counts or reduced dissolved oxygen levels in natural water bodies, which can lead to the spread of diseases such as cholera and typhoid among humans. Various organizations and health institutions address both disease management and prevention.

There are direct connections between water quality and public health. Additionally, there is growing awareness of potential health risks associated with water quality. Water encompasses 70% of the Earth's surface and serves as a vital source of life. However, as pollution levels have risen over the years, water has become a significant vector for disease [7]. According to Noureen et al. [8], inadequate water quality can lead to various waterborne diseases, including diarrhea, cholera, and gastroenteritis. Additionally, human activities have a significant impact on water quality, which in turn affects the availability of safe drinking water [3, 9]. In response to numerous studies linking contamination sources to public health concerns, many countries have created regulatory agencies and established standards. Accelerated urbanization, raising concerns regarding the provision of safe and healthy drinking water. Additionally, there are socio-economic considerations associated with water quality that warrant increased awareness and understanding [10, 11].

11.2.2 Connection with Environmental Issues

Previous research indicates that for a water body to be preserved in an eco-friendly manner, it must undergo comprehensive analysis to ensure pollutants are kept within acceptable limits. Water sourced from spray and irrigational wells for agricultural purposes frequently does not meet drinking quality standards. In some rural regions, there is a lack of regular monitoring for drinking water quality concerning chemical and physical properties. Poor-quality irrigation water can adversely affect plant growth, particularly when sunlight is present. Furthermore, irrigation with water high in iron content can prevent crops, such as rice, from achieving their typical white color due to natural processes [12, 13].

Human actions contribute to water pollution, harming numerous species and threatening the biodiversity of aquatic ecosystems [14]. According to Cheng et al., [15] rapid infrastructure development and urbanization have diminished the environment's natural ability to self-purify. Improper management of urban wastewater discharge can have determinantal effects on both environmental and public health. Therefore, it is essential to implement controls that mitigate these impacts and promote effective environmental reuse. Developed nations have established quality control systems for monitoring water throughout their production and distribution phases [16]. The development of effective mechanisms and data sources enhances intelligence and supports comprehensive resource management [17].

11.3 Water Contamination Sources

Water body contamination arises from the augmented presence of undesirable substances in water, leading to decline in quality [18]. Contaminated runoff may originate from diverse sources such as sewage or septic system discharges, urban developments, agricultural practices, or stormwater management systems [13]. Furthermore, surface water can be polluted by effluents fm industrial facilities or runoff from construction sites [19]. Over time, the quality of water is subject to variations, with significant contributions to water pollution stemming from rapid urbanization, industrial expansion, and commercial growth [20].

11.4 Water Quality Parameters

The quality of water is affected by various factors, as noted by Singh et al. [21]. These factors include basic physical characteristics such as temperature and important chemical constituents present in the water, including concentration of oxygen, turbidity, and pH levels. Local and global Water Quality Standards are implemented to conduct conformity assessments and impose restrictions on undesirable parameters, including physical, chemical, and biological counts. In many instances, water testing is carried out, and samples are sent to laboratories for analysis to certify water for safe use [22]. However, maintaining water quality standards can be challenging due to the complexity of the protocols involved. Therefore, a specialized water quality monitoring device may be required for continuous verification [23, 24]. Water quality parameters may be grouped as:

11.4.1 Physical Parameters

Physical parameters, such as temperature, humidity, turbidity, and color, reflect the visual characteristics of water and can sometimes be examined for aesthetic reasons. However, these parameters also provide insights into water behavior within the ecosystem; for instance, an increase in water temperature can influence biological activity [25]. Furthermore, turbidity affects the clarity of water, which in turn impacts light penetration and the distribution of aquatic organisms [26].

Silva et al., [27] stated that color concentration and water clarity constitute additional contaminants with numerous ecological impacts on ecosystems. Elevated contaminant levels can impede sunlight penetration and diminish oxygen levels in the water, thereby inhibiting the development of microorganisms and aquatic life. Water transparency measures the degree to which light can penetrate the water, enabling thermal alterations in the water body. Assessing this characteristic is fundamental for a specific water body, given that about 43% of solar energy is absorbed within a one-meter depth.

The assessment of Total Suspended Solids (TSS) and water resistivity is crucial for distinguishing the chemical and biological relations within aquatic environment. Particulate matter, including sand, can be introduced into the water, exhibiting a chemical composition and biological characteristics consistent with those found in seawater [28, 29].

11.4.2 Chemical Parameters

The main chemical parameters that designate water quality include biological, chemical, physical, and ecological aspects. These parameters are essential for assessing the quality and contamination levels of freshwater bodies [30]. This content will explain the current status of water quality and propose evaluation methods and implications for the necessary information to be collected in the context of ongoing studies [27].

Key parameters for monitoring water quality based on their chemical composition include temperature, salinity, pH, conductivity, dissolved oxygen (DO), and biochemical oxygen demand (BOD). Additionally, it's important to assess phosphate (PO4), nitrate (NO3), and toxic heavy metals. Various other indicators, such as physiological nutrients and variable heavy metals such as chromium, sodium, zinc, and vanadium also play a significant role in evaluating chemical water quality [27, 29].

Sustaining the suitable levels of these parameters is essential for categorizing pollutants, identifying their sources, estimating pollution loads steadily, guiding policymakers on pollution prevention, and assessing in forecasting and designing treatment systems. However, determining these parameters in real-time within complex environments presents challenges and often requires specialized physical measurement devices. Recent technological advancements have improved the accuracy and sensitivity of these measurements, allowing better assessment of water quality and its spatial and temporal variations, which are relevant to ongoing management policies [13, 18, 30].

11.4.3 Biological Parameters

Biological pollutants can be difficult to detect and may display less visible effects, but they are important for monitoring. These pollutants include larger organisms such as algae, microorganisms (i.e., *E.coli*), and dissolved organic matter with biological activity [31]. Evaluating the biological composition of water provides important information about its safety for human use and the health of aquatic ecosystems. The detection of identifiable biological species in water indicates pollution levels and helps assess ecosystem balance. The growth of biological species is influenced by nutrient availability, which can result in environmental concerns such as eutrophication and declining conditions of aquatic environments. Biological systems are complex, dense, diverse, and unpredictable components within aquatic settings, challenging the detection and monitoring of biological contaminants [32].

11.5 Predictive Analytics for Water Quality

Given that only 1–2% of the world's water resources are fresh and relatively free from pollutants, monitoring water quality is necessary. This requirement arises from the necessity to identify water sources and monitor various parameters, including appearance, taste, odor, temperature, organic matter, turbidity, toxicity, acidity/alkalinity, and hardness [2]. Assessing water quality is important due to the occurrence of impurities. Traditional methods usually involve regular sampling of water quality from a source, which may result in unsatisfied requirements [33]. To address this limitation, a prototype utilizing advanced IoT technology has been developed. Experimental results indicated a reduction in sensor failures, enhanced online monitoring capabilities, and an extended signal range compared to traditional offline systems [34].

Predictive analytics focuses on the innovation of an Internet of Things (IoT) system for monitoring water quality. This system incorporates smart sensors, communication technologies, and data management systems, accompanied by an implementation strategy [35]. Smart sensors are engineered to measure temperature, dissolved oxygen (DO), and turbidity, as well as monitoring battery life. These sensors are powered by solar panels and transmit measurements via LoRa wireless technology [36, 37]. The sensor system enables ongoing assessments of water quality and can estimate its remaining lifespan, thereby aiding in damage evaluation. Through monitoring key water parameters, a significant amount of data can be gathered, which, when regulated through learning methodologies, can be applied to anticipate critical information [38]. Consequently, an algorithm employing an artificial neural network has been successfully implemented to predict water quality parameters such as temperature, acidity, DO, and turbidity in an industrial setting [39].

11.6 Real-Time Monitoring Systems

Water quality can be monitored by Internet of Things (IoT) through both in real-time (concurrently streaming live data values) and offline (i.e. value logging at a regular interval and subsequently processing the data). Real-time systems are of particular significance as they can detect sudden changes in water quality parameters that could indicate an environmental danger [21]. Continuous data streaming is essential because a number of situations can arise that would not be detected by a discrete triggering system. Moreover, the more rapidly after the event that a response is taken the more effective that intervention will be [40]. Since large bodies of water are in general not sterile and rapidly testable by offline means, rapid interventions must frequently be implemented based on a body of evidence obtained in real-time [35].

Real-time monitoring systems have seen significant uptake, with record orders between 2010–2012. Real-time systems have also been widely utilized in industry, with specially designed boiler systems used to analyze the water quality at two-minute intervals to predict boiler failures and adjust parameters accordingly [37]. In rural contexts water quality is being monitored in real-time using a network of electronic boards and raft-controlled probes. Here the continuation of a two-way low data rate communication system was particularly important given the disparity in culture, society, lifestyle, climate, and geography. According to Majid et al. [39], adaptive communication techniques are needed to shift between networks present in the environment.

11.7 Traditional vs. IoT Monitoring

The evaluation of water quality is a vital approach for the efficient management of water resource allocation. Over time, water quality monitoring methods have undergone a noteworthy evolution, transitioning from traditional practices to modern technological innovations. Conventional methods require the manual gathering of water samples, which are later analyzed in laboratories. This process is often slow and requires considerable labor. Additionally, the absence of real-time monitoring can lead to inconsistencies between the sampling of water and the subsequent reporting of results, potentially resulting in inaccuracies [23, 24]

Ongoing monitoring tools may depend on power supply and exhibit varying sensitivity to different contaminants in the water. This variation occurs because water is distributed across multiple applications or used by various consumers. In some Internet of Things (IoT) circumstances, a centralized control system can access all sensing devices; however, maintaining water quality standards is complex due to the protocols involved. Thus, a specialized water quality monitoring device may be necessary for continuous verification [41].

Water quality monitoring systems based on the Internet of Things (IoT) have been developed to address the need for real-time monitoring. These systems allow continuous assessment of water quality metrics such as temperature, pH levels, turbidity, and flow rate. Data is collected through field sensors, followed by real-time analysis and cloud storage of the processed data. This analysis can help detect trends, improving understanding and accelerating decision-making processes. Remote monitoring is particularly useful for large bodies of water, line lakes, where multiple sampling sites might include rivers and reservoirs. This technology simplifies the sampling process, ensures the quality of water supply, provides continuing surveillance, and reduces the possibility of sensor biofouling. Remote monitoring signifies a major advancement in water quality management, serving as a supplementary resource rather than replacing traditional systems. The efficiency of water quality monitoring is improved, allowing for faster assessments than those acquired through standard laboratory sampling while consistently meeting established quality benchmarks. Additionally, there are financial benefits, as long-term operational savings can result from decreased staffing needs due to lack of requirement for dedicated personnel for monitoring. However, the implementation of IoT-based water quality monitoring requires a foundational level of technological infrastructure, including suitable Information Communication Technology (ICT) hardware and software, to ensure a reliable and efficient system is properly established [40].

11.8 IoT Technology Sensors

One of the principal applications of Internet of Things (IoT) technology pertains to the examination of water quality via IoT sensors. These sensors are classified as chemical, physical, and biological types [42, 43].

11.8.1 Chemical IoT sensors

These sensors measure parameters such as pH value, dissolved oxygen (DO), oxidation-reduction potential (ORP), and turbidity. Additionally, IoT sensors monitor flow rates and water temperature. Advancements in sensor technology and IoT now allow commercially available water quality chemical sensors to link directly to IoT transmitters, facilitating the monitoring of essential water quality parameters.

11.8.2 Physical IoT sensors

Physical IoT sensors are utilized to measure various attributes, including conductivity, transparency, total dissolved solids (TDS), residual chlorine, hardness, and other inspection indicators. The flow of water is monitored using dedicated flow sensors, while IoT sensors can also identify toxic algae through colorimetric methods that utilize alum as a flocculant.

11.8.3 Biological IoT Sensors

Biological sensory technology enables the assessment of water's taste and odor. Specific biological sensors are required to detect ammonia, dissolved organic carbon (DOC), chlorophyll, and various toxins, necessitating specialized setups.

Research indicates that the developed water quality IoT sensors have demonstrated superior measurement accuracy compared to traditional manual assessments based on standardized procedures. These sensors enhance workflow efficiency, hygiene, and flexibility, while also assisting users who may lack expertise in sampling. A wide array of chemical, physical, and biological sensors can be employed to establish comprehensive water quality monitoring within an IoT framework, enabling the tracking and measurement of nearly all relevant water parameters [42, 43, 44].

A wide range of sensor modules is presently available, with some specifically suited for water analysis, including chemical, biological, and physical parameters. It is essential to evaluate the usability of various modules under challenging conditions [45]. To address this, a proposed methodology focuses on assessing sensor usability within practical water analysis systems, considering both economic factors (such as calibration time, evaluation duration, and cost) and quality aspects of water analysis (including error assessment, analysis of research value, and the reputation of the sensor manufacturer). Recently, developments have been made in the field of wireless E-nose sensor platforms for IoT, as well as standalone wireless sensors that gauge water quality by analyzing its odor through gas-sensitive technology. This research intends to categorize the wide array of chemical, physical, and biological sensors designed for water quality, along with IoT E-nose systems and wireless turbidity meters aimed at water analysis [46, 47].

Numerous recent studies address IoT technologies and wireless sensor networks, emphasizing the necessity for sensors in monitoring water quality, data transmission equipment, and wireless nodes that facilitate connectivity as mentioned by Singh & Walingo [21]. It is crucial to recognize that IoT technology encompasses more than just sensors and should enhance the process of evaluating water quality. Ideally, IoT water quality sensors should operate autonomously with low energy consumption, energy harvesting capabilities, and the ability to conduct real-time data analysis before transmitting information to cloud storage. Currently, IoT sensors and Wireless Personal Area Networks (WPAN) for water quality monitoring are not extensively available in the market. As noted in section 2 of this paper, only a handful of vendors presently offer water quality sensor modules [36]. According to Zainurin et al. [6], significant technical difficulties remain in employing these modules for real-time water analysis in field-based monitoring systems, including: (i) high costs of sensor platforms; (ii) a lack of user-friendliness; and (iii) labor-intensive maintenance and calibration requirements. To remain competitive and meet the general requirements of water analysis, it is imperative for systems to simultaneously measure a variety of chemical, physical, or biological parameters in water [42].

11.9 IoT Data Collection

An Internet of Things (IoT) device will be utilized to monitor, process, and transmit electroencephalogram (EEG) signals over the web to a cloud-based processing engine. Both IoT devices and brain-computer interfaces (BCIs) possess the capability to manage extensive datasets, which may include sensitive user information.

The size of big data can introduce risks related to cybercrime. Additionally, transmitting the entire dataset may be unnecessary, especially when it includes irrelevant information; therefore, only data relevant to further processing should be transmitted. IoT devices often transfer large datasets for operational efficiency. It is crucial to implement security procedures during the transmission of these large data packages, incorporating secure transfer protocols and data processing practices.

The goal is to create an IoT device that is secure, accurate, and efficient. It will includes an EEG signal processing module capable of identifying intentional mental activities within 1,000 milliseconds of a cue, as well as an online connectivity module that transmits the EEG signals within a required time frame [47]. BCIs will also be incorporated into the primary sensor setup to enable the collection of EEG signals from the brain to the cloud processing system within milliseconds. The combination of EEG assistance and an improved security framework aims to ensure the secure transmission of time-sensitive data. The significance and accuracy of the EEG signals will be assessed through laboratory testing and clinical procedures. These reliable metrics may have applicability in other clinical environments. EEG signals are anticipated to support the retrieval of BCI connectivity data. Additionally, the electrical activity of the heart provides valuable insights. EEG and electrocardiogram (ECG) signals can be exchanged among CPU devices [48].

While EEG yields limited electro-physiological information regarding the brain's electrical activity, ECG supplies electrical data. The combined analysis of both is known as central nervous system autonomic nervous mechanism analysis. EEG entails the non-invasive monitoring of the brain's electrical activity through various sensors placed on the scalp. Changes in electrical potentials resulting from specific motor imagery or cognitive tasks are captured as EEG signals [47]. These signals frequently exhibit significant amplitude noise from movements of the eyes, head, or tongue, squinting, biological signal contamination, and electrophysiological noise arising from the heart and muscles, which complicates EEG signal processing [50]. In the EEG connectivity module, the sensor connects to the data acquisition module, emphasizing the enhancement of the IoT device. This module digitizes, preprocesses, and transmits the obtained EEG signal through a serial connection with a personal computer [51, 52].

11.10 Data Analysis and Visualization

Data analysis and visualization play a crucial role in the context of water quality monitoring using the Internet of Things based sensor systems. The Internet of Things (IoT) aims to gather a wide range of information and enable centralized monitoring and processing of vast quantities of incorporated sensors for piloting various earthly domains, spanning agriculture, health, environment, and construction. The water quality monitoring system, for instance, is constructed with water and air sensors. In this case, the data volume to be treated is enormous, and systematic administration and analysis tools are essential [21]. The generated information, organized into insights and presented in a more usable form, can be transformed into valuable data and insights.

Data processing can assist stakeholders in extracting useful information from the collected water quality data set. The use of data analytics and intelligence processes is the key to discovering hidden patterns. From the recorded water quality data output, there are a vast number of opportunities for analysis related to implementing a recommendation system, grouping similar monitoring points, anomaly detection, and forecasting possible contaminants. The main use in this ongoing effort is related to the suggestion of prevention measures based on the initial observations and evaluation of their potential effect. Nevertheless, this requires continuous analysis of the generated data in order to be translated into valuable insights. To assist in this, a set of user-friendly tools and dashboards are also included where summarizing graphs on different waterborne contaminations appear more intuitively. Throughout the analysis, different recommendations appear which cover, among other things, suggested deep dives into the data or recommend preventative action to specific stakeholders. In order to assess the impact of the intervention, comparisons are performed on the previous few months though, seeking to infer as much as possible, considering the already presented and currently analyzed water quality data [21].

11.11 IoT Applications in Water Quality Monitoring

The Internet of Things (IoT) technology, which has primarily been associated with a network connecting users and peripheral devices in a coordinated environment, has drawn significant interest from researchers and professionals and is being implemented across various fields for diverse applications. Projections suggest that by 2020, approximately 20–30 billion devices will be connected to the IoT [21]. This technology as described by Zhang et al. [18], possesses considerable potential to transform water quality monitoring due to its capability for remote control, continuous monitoring, instant data acquisition, and analysis, as well as its cost-effectiveness, low energy demands, and ease of access. Additionally, IoT is expected to improve various aspects of public health, water management, and the preservation of water resources. Focus should be given to the development of reliable sensors, actuators, and systems that can be integrated with existing infrastructure.

Mishra et al., [14] identified two key applications of IoT for water quality monitoring:

1. Designing and developing enhanced real-time monitoring systems by connecting and networked various components.
2. Creating data analysis solutions aimed at converting collected data into actionable insights. This study intends to investigate both applications.

Specifically, IoT-based water quality monitoring systems have been developed and validated to analyze key water quality indicators. Additionally, approaches utilizing data analytics techniques have been devised to confront challenges associated with water quality assessment through real-time measurements. Data visualization tools will also be presented to convey water quality information in a clear and interactive manner.

11.11.1 Development of Mobile Based Applications

The framework of the water quality analysis mobile application automates the analysis and records data in real-time. The mobile application enables users to assess and manage the water quality of the Kaveri River and understand the factors influencing it, and to receive medical advice tailored to their disease symptoms. The system incorporates geocoding through GPS to connect users with relevant doctors in their local areas [37]. Additionally, the system aims to create a time-efficient and low-cost monitoring unit for testing the water quality. These mobile applications are structured as a series of individual modules along with several interdependent service behaviors, facilitating the enhancement of system functionality [35]. Low-cost sensor node can be interconnected with general-purpose mobile platforms such as smartphones. Using the smartphone, users can effectively monitor and analyze the water conditions.

11.12 Challenges and Solutions

Deploying IoT-based water monitoring technology is intricate, requiring the amalgamation of different systems and services, such as sensors, wireless communication, data processing, and analytical models [53]. The technology is considered emerging due to its recent introduction, posing challenges as users often lack adequate training in its utilization. Additional hurdles encompass issues related to standardization, high investment costs, and the need for collaborative efforts [54]. Among these, cost stands out as a significant obstacle for water companies, similar to other businesses, as they face mounting pressure to increase efficiency while reducing expenditures. Furthermore, capital expenditure is closely monitored by regulatory bodies, which demand justification of new technologies in terms of customer benefits.

When implementing IoT-based water quality monitoring, there is a need to develop a completely new information and communications technology infrastructure along with the relevant software and data platforms needed for its sustainability [55]. Moreover, water treatment facilities often present challenging environments for ICT due to factors like humidity, temperature, and potential water contamination that can significantly reduce the lifespan of even high-quality electronic components as mentioned by Alahi et al. [56]. Security concerns are also paramount, therefore, establishing a new network necessitates a secure and resilient structure to withstand cyber threats. Lastly, the reliability of network data connections poses additional challenges, as effective condition-based monitoring of essential assets and systems through IoT relies on consistent real-time data exchange [5].

11.13 Innovations in Water Quality Monitoring

Numerous innovations have emerged in the field of water quality monitoring, including the development of cost-effective solutions for real-time tracking within an Internet of Things (IoT) framework. Key systems as mentioned by many studies include:

1. A mobile application that enables remote measurement of water quality indicators such as pH, turbidity, temperature, salt concentration, and fluoride levels, utilizing commonly available sensor kits along with a practical setup featuring ultrasonic surface detectors.
2. An experimental collaboration employing remote-sensing technology via unmanned aerial vehicles (UAVs) to produce digital elevation models of rivers.
3. Implements real-time recalibration of water quality data using the XCOR method, which relies on AI forecasting, as well as imposing restrictions on water abstraction from the river.
4. The establishment of cloud-based architectures serves to enhance the system's intelligence and address water-related conflicts within broader communities.
5. Data lineage technology is utilized to archive sensor information in the cloud, facilitating big data analysis for predicting pollution trends.
6. Mitigate environmental disturbances and improve river water quality.
7. Development of portable and compact solutions to facilitate user-friendly access to this technology.
8. Enhancements of traditional systems, the creation of entirely new technologies, or organizational designs aimed at improving operational efficiency.
9. Utilizing IoT technology for tracking water quality parameters.

The above innovations trials have been extensively discussed by Singh & Walingo [21], Ashraf et al. [57], Sun et al. [58], Wang [59] and Chen et al. [48].

11.14 Machine Learning (ML) and Artificial Intelligence (AI) Integration

To improve the performance and quality of water quality monitoring (WQM) with IoT, it is beneficial to introduce machine learning and artificial intelligence, which enables more advanced data analysis and the capability to make deeper insights about water quality data trends. Through this feature, the system will perform predictive modeling to forecast water quality problems, which can be applied as warning information for prevention. Several WQM with IoT solutions that employ more AI and machine learning have developed models to predict or classify water quality parameters for enhancing their performances Consequently, such models can provide a solution that could diagnose, predict, and classify water quality information. Moreover, this technology application is very relevant for the practical environment of WQM, particularly in improving the performance of the measurement model and comparison with reliable data. There are various aspects of WQM with IoT system that can be applied to use advanced machine learning and artificial intelligence methods such as the development of WQM with IoT system device integration, the preparation of a dataset, and the establishment of analysis models. Tested with real-world data, the system then estimates the measurement device's performance and analyses the patterns chosen in the comparison with the data measured by the WQM with IoT system.

Many developed WQM with IoT system indeed have the capability to monitor water quality information in real-time. However, the quality of the data obtained is not yet feasible to use to determine water quality parameter conditions because of the high bias found in the data disconnected measurement models and unreliable data sources. In general, the application of machine learning (ML) or artificial intelligence (AI) strongly needs good quality data for training the model to obtain a good model as well.

11.15 Future Trends

Aiming at addressing the future trends in the use of IoT in water monitoring, the findings of a number of recent studies were reviewed and summarized as follows:

1. Advancement of water quality monitoring for larger water bodies by developing IoT-enabled systems that possess diverse features and components to a greater degree.
2. Evolve into complex hybrids that integrate continuous and periodic sampling methods aiming to encounter the limitations of sensors in terms of environmental noise sensitivity and to enhance the reliability of measurements for water quality variables that are more prone to disturbances.
3. Coordinate the utilization of extensive and densely distributed networks of varied samplers and sensors to become a key trend in this field.
4. Integration of interconnected multifunctional devices that combine capabilities for on-body health monitoring with specialized water analysis tools.
5. Enhance monitoring practices, thereby safeguarding the health and safety of users in recreational water settings.
6. Permeates smart-city ecosystems by the increasing use of Internet of Things (IoT).

The above future trend steps have further been extensively discussed by Haleem et al. [60], Smith & Miller [61], Rehman et al. [62], Fisher et al. [63] and Obaideen et al. [64].

11.16 Conclusion

By reviewing the works of Jamshed et al. [37], Gulati et al. [36], Majid et al. [39] and Mowla et al. [34] it can be concluded that water is a vital natural resource, and its limited availability significantly affects individuals and their living standards. Factors such as industrial development, rapid urbanization, population growth, water pollution, and climate change have had and will continue to have effects on global water security. These issues will lead to serious challenges related to water and sanitation. It is estimated that around 80% of illnesses and deaths worldwide are primarily linked to the consumption of contaminated water. Therefore, maintaining water sanitation and hygiene is essential for ensuring access to clean and safe drinking water. Consequently, the following recommendations are essential:

1. Monitoring the quality of drinking water to guarantee its safety.
2. Since direct testing methods often require transporting water samples to a laboratory, which can take several days or even up to a week. Hence, there is a need for the development of innovative, cost-effective, and user-friendly technologies that facilitate rapid assessments of drinking water quality.
3. A cost-effective device for the continuous assessment of drinking water quality in various environments is needed, with an emphasis on the key physical parameters of temperature, pH, turbidity, and water flow.
4. Low-cost sensors need to be utilized along with a microcontroller-based system to analyze these sensors.
5. The data obtained from the sensors must be processed by a central controller.
6. Make use of the researches which highlight reduction in the overall construction costs of real-time data compared to alternative schematic methods.
7. Assuring sensor accuracy of the low cost devices that is comparable to that of more expensive sensors.

References

1. Levin R, Villanueva CM, Beene D, Cradock AL, Donat-Vargas C, Lewis J, Deziel NC (2024) US drinking water quality: exposure risk profiles for seven legacy and emerging contaminants. J Exposure Sci Environ Epidemiol 34(1):3–22
2. Jan F, Min-Allah N, Düştegör D (2021) Iot based smart water quality monitoring: Recent techniques, trends and challenges for domestic applications. Water
3. Yasmeen Q, Yasmeen S (2023) Impact of drinking water on people's health and water borne diseases. Pak BioMedical J. https://doi.org/10.54393/pbmj.v6i07.867
4. Wang M, Janssen AB, Bazin J, Strokal M, Ma L, Kroeze C (2022) Accounting for interactions between sustainable development goals is essential for water pollution control in China. Nat Commun 13(1):730. https://doi.org/10.1038/s41467-022-28351-3
5. Mouha RARA (2021) Internet of things (IoT). J Data Anal Inf Process
6. Zainurin SN, Wan Ismail WZ, Mahamud SNI, Ismail I, Jamaludin J, Ariffin KNZ, Wan Ahmad Kamil WM (2022) Advancements in monitoring water quality based on various sensing

methods: a systematic review. Int J Environ Res Public Health 19(21):14080. https://doi.org/10.3390/ijerph192114080
7. Shayo GM, Elimbinzi E, Shao GN, Fabian C (2023) Severity of waterborne diseases in developing countries and the effectiveness of ceramic filters for improving water quality. Bull Natl Res Cent 47(1):113. https://doi.org/10.1186/s42269-023-01088-9
8. Noureen A, Aziz R, Ismail A, Trzcinski AP (2022) The impact of climate change on waterborne diseases in Pakistan. Sustain Clim Chang 15(2):138–152
9. Health Organization W (2022) Guidelines for drinking-water quality: incorporating the first and second addenda
10. Ali NM, Khan MK, Mazhar B, Mustafa M (2025) Impact of water pollution on waterborne infections: emphasizing microbial contamination and associated health hazards in humans. Discover Water. https://doi.org/10.1007/s43832-025-00198-x
11. Lin L, Yang H, Xu X (2022) Effects of water pollution on human health and disease heterogeneity: A review. Front Environ Sci. https://doi.org/10.3389/fenvs.2022.880246
12. Bhatt S, Mishra AP, Chandra N, Sahu H, Chaurasia SK, Pande CB, Hunt J et al (2024) Characterizing seasonal, environmental and human-induced factors influencing the dynamics of Rispana River's water quality: Implications for sustainable river management. Results Eng 22:102007. https://doi.org/10.1016/j.rineng.2024.102007
13. Pujar, P.M., Kenchannavar, H.H., Kulkarni, R.M. et al. (2020) Real-time water quality monitoring through Internet of Things and ANOVA-based analysis: a case study on river Krishna. Appl Water Sci 10:22. https://doi.org/10.1007/s13201-019-1111-9
14. Mishra RK (2023) Fresh water availability and its global challenge. British J Multidiscip Adv Stud 4(3):1–78. https://doi.org/10.37745/bjmas.2022.0208
15. Cheng C, Zhang F, Shi J, Kung HT (2022) What is the relationship between land use and surface water quality? A review and prospects from remote sensing perspective. Environ Sci Pollut Res 29(38):56887–56907
16. Huang J, Zhang Y, Bing H, Peng J, Dong F, Gao J, Arhonditsis GB (2021) Characterizing the river water quality in China: Recent progress and on-going challenges. Water Res 201:117309
17. Tang W, Pei Y, Zheng H, Zhao Y, Shu L, Zhang H (2022) Twenty years of China's water pollution control: Experiences and challenges. Chemosphere 295. https://doi.org/10.1016/j.chemosphere.2022.133875
18. Zhang P, Yang M, Lan J, Huang Y, Zhang J, Huang S, Ru J et al (2023) Water quality degradation due to heavy metal contamination: health impacts and eco-friendly approaches for heavy metal remediation. Toxics 11(10):828. https://doi.org/10.3390/toxics11100828
19. Amano KOA, Danso-Boateng E, Adom E, Kwame Nkansah D, Amoamah ES, Appiah-Danquah E (2021) Effect of waste landfill site on surface and ground water drinking quality. Water Environ J 35(2):715–729. https://doi.org/10.1111/wej.12664
20. Fida M, Li P, Wang Y, Alam SMK, Nsabimana A (2023) Water contamination and human health risks in Pakistan: a review. Exposure and Health
21. Singh Y, Walingo T (2024) Smart water quality monitoring with IoT wireless sensor networks. 24(9):2871. https://doi.org/10.3390/s24092871
22. Kordbacheh F, Heidari G (2023) Water pollutants and approaches for their removal. Mater Chem Horiz 2(2):139–153
23. Hamel P, Tan L (2022) Blue–green infrastructure for flood and water quality management in Southeast Asia: evidence and knowledge gaps. Environ Manag
24. Lukhabi DK, Mensah PK, Asare NK (2023) Adapted water quality indices: Limitations and potential for water quality monitoring in Africa. Water. https://doi.org/10.3390/w15091736
25. Akhtar N, Syakir Ishak MI, Bhawani SA, Umar K (2021) Various natural and anthropogenic factors responsible for water quality degradation: A review. Water. https://doi.org/10.3390/w13192660
26. Bozorg-Haddad O, Delpasand M, Loáiciga HA (2021) Water quality, hygiene, and health. In: Economical, political, and social issues in water resources, pp 217–257. Elsevier
27. Silva RRA, Marques CS, Arruda TR, Teixeira SC, de Oliveira TV (2023) Biodegradation of polymers: stages, measurement, standards and prospects. Macromol 3(2):371–399. https://doi.org/10.3390/macromol3020023

28. Saravanan A, Kumar PS, Jeevanantham S, Karishma S, Tajsabreen B, Yaashikaa PR, Reshma B (2021) Effective water/wastewater treatment methodologies for toxic pollutants removal: Processes and applications towards sustainable development. Chemosphere 280:130595
29. Azanaw A, Birlie B, Teshome B, Jemberie M (2022) Textile effluent treatment methods and eco-friendly resolution of textile wastewater. Case Studies in Chemical and Environmental Engineering 6. https://doi.org/10.1016/j.cscee.2022.100230
30. Mishra AP, Khali H, Singh S, Pande CB, Singh R, Chaurasia SK (2023) An assessment of in-situ water quality parameters and its variation with Landsat 8 level 1 surface reflectance datasets. Int J Environ Anal Chem 103(18):6344–6366. https://doi.org/10.1080/03067319.2021.1954175
31. Jani K, Bandal J, Shouche Y, Shafi S, Azhar EI, Zumla A, Sharma A (2021) Extended ecological restoration of bacterial communities in the Godavari River during the COVID-19 lockdown period: a spatiotemporal meta-analysis. Microb Ecol 82(2):365–376
32. Bisimwa AM, Kisuya B, Kazadi ZM, Muhaya BB, Kankonda AB (2022) Monitoring faecal contamination and relationship of physicochemical variables with faecal indicator bacteria numbers in Bukavu surface waters, tributaries of Lake Kivu in Democratic Republic of Congo. Hyg Environ Health Adv 3:100012. https://doi.org/10.1016/j.heha.2022.100012
33. Zulkifli CZ, Garfan S, Talal M, Alamoodi AH, Alamleh A, Ahmaro IY, Chiang HH et al (2022) IoT-based water monitoring systems: A systematic review. Water 14(22):3621. https://doi.org/10.3390/w14223621
34. Mowla MN, Mowla N, Shah AS, Rabie KM, Shongwe T (2023) Internet of Things and wireless sensor networks for smart agriculture applications: A survey. IEEE Access 11:145813–145852
35. Vazhuthi PPI, Prasanth A, Manikandan SP, Sowndarya KD (2023) A hybrid ANFIS reptile optimization algorithm for energy-efficient inter-cluster routing in internet of things-enabled wireless sensor networks. Peer-To-Peer Netw Appl 16(2):1049–1068. https://doi.org/10.1007/s12083-023-01458-0
36. Gulati K, Boddu RSK, Kapila D, Bangare SL, Chandnani N, Saravanan G (2022) A review paper on wireless sensor network techniques in Internet of Things (IoT). Mater Today: Proc 51:161–165
37. Jamshed MA, Ali K, Abbasi QH, Imran MA, Ur-Rehman M (2022) Challenges, applications, and future of wireless sensors in Internet of Things: A review. IEEE Sens J 22(6):5482–5494
38. Chaudhary M, Goyal N, Benslimane A, Awasthi LK, Alwadain A, Singh A (2022) Underwater wireless sensor networks: Enabling technologies for node deployment and data collection challenges. IEEE Internet Things J 10(4):3500–3524
39. Majid M, Habib S, Javed AR, Rizwan M, Srivastava G, Gadekallu TR, Lin JCW (2022) Applications of wireless sensor networks and internet of things frameworks in the industry revolution 4.0: A systematic literature review. Sensors 22(6):2087
40. Lal K, Menon S, Noble F, Mahmood Arif K (2024) Low-cost IoT based system for lake water quality monitoring. https://doi.org/10.1371/journal.pone.0299089
41. Lakshmanna K, Kaluri R, Gundluru N, Alzamil ZS, Rajput DS, Khan AA, Alhussen A et al (2022) A review on deep learning techniques for IoT data. Electronics 11(10):1604
42. Garrido-Momparler V, Peris M (2022) Smart sensors in environmental/water quality monitoring using IoT and cloud services. Trends Environ Anal Chem 35:e00173
43. Hemdan EED, Essa YM, Shouman M, El-Sayed A, Moustafa AN (2023) An efficient IoT based smart water quality monitoring system. Multimed Tools Appl 82(19):28827–28851
44. Manoj M, Dhilip Kumar V, Arif M, Bulai ER, Bulai P, Geman O (2022) State of the art techniques for water quality monitoring systems for fish ponds using iot and underwater sensors: A review. Sensors 22(6):2088
45. Prapti DR, Mohamed Shariff AR (2022) Internet of Things (IoT)-based aquaculture: An overview of IoT application on water quality monitoring. Reviews in Aquaculture. https://doi.org/10.1111/raq.12637
46. Ighalo JO, Adeniyi AG, Marques G (2021) Internet of things for water quality monitoring and assessment: a comprehensive review. Artificial intelligence for sustainable development: theory, practice and future applications, pp 245–259

47. Shuvo MMH, Islam SK, Cheng J (2022) Efficient acceleration of deep learning inference on resource-constrained edge devices: A review. https://doi.org/10.1109/JPROC.2022.3226481
48. Chen P, Liu H, Xin R, Carval T, Zhao J, Xia Y, Zhao Z (2022) Effectively detecting operational anomalies in large-scale IoT data infrastructures by using a GAN-based predictive model. Comput J 65(11):2909–2925
49. Yang L, Shami A (2022) IoT data analytics in dynamic environments: From an automated machine learning perspective. Engineering Applications of Artificial Intelligence Vol. 116.
50. Khraisat A, Alazab A (2021) A critical review of intrusion detection systems in the internet of things: techniques, deployment strategy, validation strategy, attacks, public datasets and challenges. Cybersecurity. https://doi.org/10.1186/s42400-021-00077-7
51. Booij TM, Chiscop I, Meeuwissen E, Moustafa N, Den Hartog FT (2021) ToN_IoT: The role of heterogeneity and the need for standardization of features and attack types in IoT network intrusion data sets. IEEE Internet Things J 9(1):485–496
52. Gyamfi E, Jurcut A (2022) Intrusion detection in internet of things systems: a review on design approaches leveraging multi-access edge computing, machine learning, and datasets. Sensors. https://doi.org/10.3390/s22103744
53. Schiller E, Aidoo A, Fuhrer J, Stahl J, Ziörjen M, Stiller B (2022) Landscape of IoT security. Computer Science Review 44:100467. https://doi.org/10.1016/j.cosrev.2022.100467
54. Jiang Z, Guo Y, Wang Z (2021) Digital twin to improve the virtual-real integration of industrial IoT. J Ind Inf Integr
55. Khang A, Rath KC, Satapathy SK, Kumar A, Das SR, Panda MR (2023) Enabling the future of manufacturing: integration of robotics and IoT to smart factory infrastructure in industry 4.0. In: Handbook of research on AI-based technologies and applications in the era of the metaverse, pp 25–50. IGI Global
56. Alahi MEE, Sukkuea A, Tina FW, Nag A, Kurdthongmee W, Suwannarat K, Mukhopadhyay SC (2023) Integration of IoT-enabled technologies and artificial intelligence (AI) for smart city scenario: recent advancements and future trends. Sensors 23(11):5206. https://doi.org/10.3390/s23115206
57. Ashraf SA, Siddiqui AJ, Abd Elmoneim OE, Khan MI, Patel M, Al-reshidi M, Adnan M et al (2021) Innovations in nanoscience for the sustainable development of food and agriculture with implications on health and environment. Sci Total Environ 768:144990. https://doi.org/10.1016/j.scitotenv.2021.144990
58. Sun X, Zhang Y, Shi K, Zhang Y, Li N, Wang W, Qin B (2022) Monitoring water quality using proximal remote sensing technology. Sci Total Environ 803:149805. https://doi.org/10.1016/j.scitotenv.2021.149805
59. Wang L (2024) Advances in monitoring and managing aquatic ecosystem health: integrating technology and policy. Int J Aquac 14(2). https://doi.org/10.5376/ija.2024.14.0012
60. Haleem A, Javaid M, Singh RP, Suman R (2022) Medical 4.0 technologies for healthcare: Features, capabilities, and applications. Internet Things Cyber-Phys Syst 2:12–30
61. Smith M, Miller S (2022) The ethical application of biometric facial recognition technology. AI & Soc 37, 167–175. https://doi.org/10.1007/s00146-021-01199-9
62. Rehman A, Abbas S, Khan MA, Ghazal TM, Adnan KM, Mosavi A (2022) A secure healthcare 5.0 system based on blockchain technology entangled with federated learning technique. Comput Biol Med 150:106019
63. Fisher MC, Alastruey-Izquierdo A, Berman J, Bicanic T, Bignell EM, Bowyer P, Verweij PE et al (2022) Tackling the emerging threat of antifungal resistance to human health. Nat Rev Microbiol 20(9):557–571
64. Obaideen K, Shehata N, Sayed ET, Abdelkareem MA, Mahmoud MS, Olabi AG (2022) The role of wastewater treatment in achieving sustainable development goals (SDGs) and sustainability guideline. Energy Nexus 7:100112 : a systematic review. Water, 14(22):3621

Chapter 12
Smart Water Quality Monitoring Devices: Innovations and Implementation

Diwesh Kumar, Sudarshana Banerjee, and Surya Pratap Singh

Abstract Water quality monitoring, and evaluation, have crucial roles in safeguarding public health, ensuring environmental sustainability, and industrial operations. Conventional methods require more time and human effort and give results delayed in time. The Internet of Things (IoT) application in water quality measurement dramatically changed the gathering, analyzing, and reporting of real-time data. The present study investigates how IoT-based water quality monitoring systems evolved in terms of cutting-edge sensor technologies, cloud computing, and artificial-intelligence-driven analytics. IoT-based systems employ intelligent sensors that monitor key parameters of pH, turbidity, dissolved oxygen, temperature, and conductivity. The sensors wirelessly communicate with centralized platforms, where machine learning models interpret trends and analyze them for predictive maintenance and anomaly detection. Edge computing implementation furthers response times, which reduces latency and reliance on centralized servers. Moreover, blockchain integration preserves data integrity, thereby enhancing transparency and trust in the reported results. The present chapter points at some case studies that sufficiently demonstrate implementations covering municipal water supply, agricultural irrigation, and industrial wastewater control, with suitable approaches on energy-efficient designs and algorithms for encryption. Another issue discussed in the above paragraphs entails the contemporary horizon sweeping semiconductor nanotechnology, nanochip technology, self-governance drones, and AI to enhance streamlined efficiency. IoT-enabled water-quality monitoring becomes an early warning mechanism for contamination, supports sustainable practices, and aids in regulation enforcement. Thus, this chapter situates at the present instance in mapping the revolutionary power of smart technology in safeguarding water ecosystems to assure clean water availability toward the realization of global environmentally sustainable development goals.

D. Kumar · S. P. Singh (✉)
Department of Civil Engineering, School of Engineering, Srinath University, Jamshedpur, India
e-mail: spsingh894@gmail.com; spsingh.civil@srinathuniversity.ac.in

S. Banerjee
Department of Physics, Srinath University, Jamshedpur, India

S. Gautam et al. (eds.), *Smart Environmental Monitoring*,
https://doi.org/10.1007/978-981-92-0294-2_12

Keywords Water quality · Water monitoring · Smart sensor technology · Real-time data analysis · Environmental sustainability · Predictive maintenance

12.1 Introduction

Water is the most valuable commodity for life in the entire ecosystem and in turn, agriculture, industry, and domestic needs rely much on water. However, in the last century, with a high speed of urbanization, industrialization, population growth, and climate change phenomena, there has been a sharp deterioration in the quality of water resources [1]. Heavy metals, organic pollutants, pathogenic micro-organisms, nutrient over-loading are the major pollution issues that have become common concern, seriously affecting the very sustainable livelihood of mankind, aquatic life, and ecosystem. Therefore, water quality monitoring is now an indispensable task in environmental science and public health policy, as well as in sustainable development practices [2].

Water quality monitoring through traditional methods mostly relying on manual sampling and laboratory analysis has been successfully cast in producing fine, high-quality, and reliable data. The conventional methods face essential constraints which include expensive operations and their restricted ability to detect changes over time and their need for extended periods before showing results and their failure to provide analysis that occurs in actual time. Water managers and environmental authorities can use these factors to prevent contamination and to enhance water management and to conduct their operations from a position of authority. The need for an agile monitoring system which operates through artificial intelligence and automation has become essential because contamination events have occurred with increasing frequency and growing severity from industrial spills and microbial contaminations [3].

The Internet of Things (IoT) has created new opportunities for environmental monitoring and water quality assessment. IoT enables physical devices which include sensors and actuators and gateways and communication modules to connect first into a unified digital network which then enables real-time data processing and transfer. These systems display capabilities which exceed those of standard methods because they enable continuous data acquisition and remote system monitoring and instant data evaluation and future trend forecasting. The combination of IoT with cloud computing and artificial intelligence and edge computing and GIS technologies enables monitoring platforms to extend their capabilities and achieve greater control over their operational environments [4].

The smart water monitoring system is capable of measuring physico-chemical parameters such as pH, temperature, turbidity, dissolved oxygen, conductivity, TDS, ammonia, nitrates, and microbial levels to a very precise extent. Data transmission is carried out wirelessly to the cloud platforms, and from these centralized cloud platforms, they are further analyzed almost in real-time. Different dashboards and mobile applications are available to visualize trends, get warnings, and promptly respond to situations for IoT-enabled water monitoring with a high level of scalability, and can adapt to a large range of application contexts.

This type of monitoring could be embedded in municipal water distribution networks to detect leakages and measure residual chlorine levels, thereby averting contamination in urban settings [5]. In remote or underdeveloped areas, such sensor nodes are low-cost and solar-powered, positioned in ponds, rivers, or wells to assess potability of drinking water. The systems in agricultural areas help farmers control irrigation water quality which leads to better crop productivity and healthier soil conditions. Industrial facilities use this monitoring system to ensure their environmental effluents meet legal requirements at discharge points which decreases their risk of receiving fines or experiencing facility shutdowns. The Internet of Things enables water quality monitoring to work together with worldwide sustainability efforts. The United Nations Sustainable Development Goal No. 6 (SDG6) promotes water availability, including Egypt itself, and sustainable sanitation under all conditions [6]. The objective requires building effective water improvement monitoring systems which depend on accurate data assessment. The system produces accurate data which can be verified at specific times to support the measurement and evaluation of SDG indicators that track water pollution and wastewater treatment.

Scientists developing Machine Learning and Artificial Intelligence systems have created a new approach to Internet of Things monitoring which enables complete understanding of contamination events through analysis of past patterns and detection of irregularities rather than through basic data collection and display techniques [7]. For example, AI algorithms could detect early signs of algal blooms in a lake, or they might predict the likelihood of chemical spills based on patterns of industrial activity. This is turning reactive monitoring systems into models for proactive environmental stewardship where risks are foreseen and mitigation measures are applied.

However, despite such capabilities and promise, challenges remain for the full implementation of IoT in water quality monitoring. These challenges continue to be the subject of study and development, among them, sensor calibration, energy efficiency, data security, network reliability, and interoperability. Further, without the intercession of some governmental or institutional framework, but large-scale deployment especially in remote and economically backward areas would be commercially viable. However, constant improvement of sensor miniaturization, energy harvesting, low-power communication protocols (such as LoRaWAN and NB-IoT), and open-source platforms continue chipping away at entrance barriers [8].

Increasingly numerous pilot projects and case studies in academic and industrial research serve to illustrate efficacy and potential of IoT systems under the different locational and environmental settings. From smart canal monitoring in India to lake health monitoring initiatives in Europe and indirectly to decentralized water kiosks in Africa, IoT-enabled water quality monitoring continues to grow in global scope [9]. By bringing into the real implementations, it strongly states interdisciplinary cooperation between engineering, computer science, environmental studies, and public policy to build all-encompassing, user-centric solutions.

The chapter explores various aspects, which include new technologies and their practical application, and the problems faced during the process of monitoring water quality through Internet of Things (IoT) systems. The chapter begins with a literature review that traces the evolution of water monitoring from ancient times, when hand sampling was used, to present-day smart, automated sampling methods. The review demonstrates the development of IoT technology from its initial application in environmental monitoring to the present state while comparing ongoing research studies to identify emerging trends and technological advancements and their outcomes. The chapter discusses an analytical study of how agricultural practices transform water quality monitoring into an advanced sustainable process.

The chapter gives a detailed analysis of the IoT technology employed in the assessment of water quality, which encompasses its application and technology structure, as well as its benefits for its functionality. The document illustrates the current sensor technology available along with the data transmission techniques and research techniques that make these technology configurations possible. The chapter illustrates through real-life examples that the application of IoT technology has made a major impact on water management practices employed by local authorities and agricultural firms.

The specific objectives of this chapter are:

- The study investigates how IoT applications for water quality monitoring have developed over time.
- The study provides an explanation of both modern IoT-based water monitoring system architecture and its separate system components.
- The study presents an evaluation of case studies that demonstrate successful applications and achievements.
- The study evaluates current research challenges together with new technological developments and future research paths.

12.2 Literature Review

The practice of water quality assessment has moved from traditional human testing methods to contemporary automated testing systems which rely on computer technology. Research teams and environmental organizations began their water collection efforts in the 1900s by conducting fieldwork to obtain water specimens which they analyzed in laboratory settings to determine BOD COD nitrate phosphate and coliform bacteria concentrations. The approach delivered accurate results but its execution needed extended periods which caused delayed data publication and organizations used reactive strategies instead of planned forward thinking [10]. The industrialization period together with environmental awareness growth during the 1960 and 1970s created a demand for immediate access to water quality data which resulted in the adoption of portable testing equipment and electrochemical sensors [11]. The devices allowed field workers to measure pH temperature and conductivity but their manual use restricted users from conducting simultaneous measurements of different parameters. The 1990s introduced automated data loggers and fixed monitoring stations which enabled researchers to conduct semi-continuous monitoring.

The system maintained its fixed position while missing essential communication components which led to its inability of collecting real-time data from remote areas and dynamic environments [12].

The twenty-first century experienced rapid progress in wireless communication technologies together with low-power embedded systems which created the second-generation monitoring systems that used networked sensor arrays to deliver real-time data [13]. The first generation of these designs used GSM and GPRS modules to enable remote sensor data transmission. The sensor data remained unprocessed because the system lacked intelligent analytic tools together with cloud integration which resulted in manual and reactive handling of the data.

The last ten years brought about the development of water quality monitoring through the combined impacts of Internet of Things technology, cloud computing systems, edge analytics methods, and artificial intelligence solutions.

Current systems enable ongoing data measurement and transmission together with machine learning algorithms which detect anomalies and predict contaminant events and optimize resource interventions in near-real-time. The system development process has now been integrated with geospatial technology and blockchain systems to enhance traceability and transparency, as well as improve the accuracy of decision-making. The water quality monitoring system developed through a hundred years of evolution which brought about automatic ecosystems that provide insights for sustainable water management.

The early 2000s brought about the establishment of wireless sensor networks which established telemetry systems for environmental monitoring. The systems enabled remote environmental monitoring through wireless data transmission while automated processes received significant improvements [14]. The IoT system provided an extension of existing systems which enabled the integration of cloud computing and advanced analytics and user-friendly interfaces. The combination of real-time sensor networks and IoT platforms introduced a major improvement in environmental monitoring because they delivered data with higher frequency and finer details and greater accuracy [15].

The year 2010 marked the beginning of IoT system adoption through water quality monitoring. Smart Water Grid and Water IoT-type projects asserted continental-scale integration of sensor, communication modules, and cloud systems for continuous water quality assessment in urban and industrial scenarios. The system operates through all municipal water distribution networks and effluent discharge points and irrigation canals and natural water bodies [9].

The authors of Table 7.1 present numerous recent peer-reviewed research studies which demonstrate how IoT technologies operate to monitor both water quality and water level. The table entries display the research methods and technologies which produced specific results that help assess modern technological advancements. The research study presents multiple measurement methods which include basic telemetry systems and advanced monitoring systems that operate through machine learning technologies. The research demonstrates how Internet of Things (IoT) frameworks have evolved through multiple development phases which begin with basic sensor systems and end with complete data analysis systems that address specific application requirements.

The case studies employed microcontroller-based sensor networks to track standard water quality parameters which encompassed pH levels and turbidity and temperature and water level changes. The researchers established remote monitoring systems through their data transmission to the cloud which operated using real-time data transfer methods. The researchers developed a system which used machine learning with Blynk to send alerts and provide predictions for pump and valve operation control.

Table 7.1 displays how IoT systems maintain their natural ability to adapt between various operational environments which include aquaculture systems and home systems and water resource systems designed for rural areas in developing nations like Somalia. The table demonstrates how Internet of Things (IoT) technology has advanced from basic environmental monitoring systems to develop cost-effective and scalable and intelligent monitoring solutions.

Table 12.1 Summary of IoT-based water monitoring studies

Study	Methods/technologies	Findings
[16]	IoT technologies; Real-time telemetry system	Low-cost IoT system effectively monitors water parameters; Preliminary tests show promising results in fish farming
[17]	Integration of pH, temperature, and turbidity sensors with microcontroller; Cloud-based data transmission	Prototype readings align closely with laboratory measurements; Minor deviations in pH and temperature; significant deviation in turbidity
[18]	IoT-based water quality monitoring; Machine learning for data analytics	Developed IoT-based monitoring system; Successfully assesses potability through parameter analysis
[19]	Waterproof IoT sensors; Real-time data transmission to centralized unit	Continuous monitoring of water quality and levels; Enhanced reliability through sensor innovation and communication protocols
[20]	Advanced sensors for level and quality; Dual sensors for pump control	Enhanced water management in residential zones; Real-time data helps prevent shortages and overflows
[21]	Master–slave communication; Sensor network in wells	Effective IoT-based real-time water level monitoring; Improved rural water management and reduced waste
[21]	Ultrasonic sensors for level detection; Microcontroller and relay for automation	Monitors levels and prevents overflow; Real-time alerts via Blynk app
[22]	Smart IoT system with continuous cloud-based monitoring	Real-time monitoring of quality; field-tested with water samples and cloud processing
[23]	IoT system development, SMS-based alerts, and data analytics	Effective monitoring in Somalia supports sustainable water management practices

The evidence from Table 7.1 lends credence to the recent studies on IoT-based water monitoring, where affordability, real-time monitoring, reliability, and intelligent decision-making are posited as the key advancements guiding the present and future IoT-based water monitoring.

The current academic studies show two things which show both existing research gaps and existing research strengths (Table 12.1).

- Limited Integration with AI/ML Models: IoT systems continue to use basic threshold alerts for their operations which do not include AI-based systems that adapt through their ability to learn and track environmental changes [24].

- Lack of Standardization: The absence of a standardized protocol for IoT water quality monitoring systems create challenges for vendors and different regions to achieve system interoperability and complete scalability.
- Energy Consumption: The majority of sensor nodes function on batteries which require regular servicing. The solar-powered systems which currently exist in their early development stage face difficulties when used in dimly lit or indoor areas.
- Security and Data Integrity: Only a small number of research studies examine cybersecurity protection through advanced solutions which include blockchain technology and encryption systems and data tampering detection methods.
- The existing literature does not present frugal and scalable business models which can help developing countries. The emerging trends in IoT water quality monitoring systems will lead to better monitoring results and operational dependability.

Researchers use nanotechnology to design small-sized sensors which can identify multiple contaminant types with exceptional accuracy and sensitivity [9]. The present situation interrupts active monitoring activities which depend on AI-powered drones to perform aerial surveillance operations over large water bodies that remain inaccessible, this situation enables rapid evaluation of different water quality conditions across various locations. The field has reached a new achievement through edge computing, which enables systems to process data at its origin, which improves system performance through reduced processing requirements and decreased needs for centralized data storage. Blockchain technology serves as a development platform that enables secure systems to demonstrate data transparency while producing permanent records which verify their ability to track water quality reports.

The technological developments create a basis for radical change which will develop a modern water monitoring system that operates through intelligent and efficient and trustworthy systems.

12.3 Materials and Methodology

An IoT-based water quality monitoring system follows a layered architecture and hence consists of four major components sensing communication processing and application. The process works through its three levels that act as autonomous systems as it transmits environmental information to the users. The sensing level helps in the installation of sensors in water sources for the measurement of physical and chemical properties. The communication level deals with the process of transmitting information from sensors to servers using wireless communication. The processing level relies on cloud computing and edge computing to utilize raw data resources. The application level helps end users view information through web or mobile interfaces.

12.3.1 Sensor Technologies and Selection

The performance of the monitoring system depends heavily on the selection and accuracy of the sensors. Multiple sensors are integrated into the system to sense the major water quality parameters such as pH, turbidity, dissolved oxygen, conductivity, temperature, and nitrates.

- Anything that detects the hydrogen ion concentration to measure the water's acidity or alkalinity is a pH sensor. Generally, they are based on the glass electrode system or ISFET technology [25].
- Optical or infrared nephelometric techniques are used by a turbidity sensor to detect suspended solid matter.
- Dissolved oxygen sensors are electrochemical or optical in nature and measure admission of dissolved oxygen for characterizing biological oxygen demand.
- Conductivity sensors are used for examining how well water conducts electricity, which would be an indication of ionic pollutants.
- Temperature sensors based on RTDs or thermistors exist because water chemistry is temperature-dependent.
- Nitrate sensors may be used in agricultural settings, working either via UV absorption or through an ion-selective electrode.

Each sensor node is powered by a microcontroller that processes the input signal and formats the data for transmission projected in Table 12.2.

12.3.2 Data Acquisition and Communication Protocols

The sensor data are read by any microcontroller such as Arduino, ESP32, or Raspberry Pi. These boards convert analog signals into digital data and transmit it to

Table 12.2 Applications of sensors for water quality monitoring

Parameter	Sensor type	Output signal	Common use
pH	Glass electrode/ISFET	Analog/Serial	Drinking water, industry
Turbidity	Optical	Digital	Rivers, tanks
Dissolved O_2	Optical/Electrochemical	Analog	Wastewater, aquaculture
Conductivity	Electrochemical	Digital	Industrial discharge monitoring
Temperature	Thermistor/RTD	Analog	Universal applications
Nitrate	UV or Ion-selective electrode	Serial	Groundwater, farmlands

the server. On different occasions, in view of certain site restrictions and network availabilities, varying wireless protocols will be adopted:

- Wi-Fi, for indoor and urban deployments.
- LoRa (Long Range), to support low-power long-distance rural applications.
- NB-IoT (Narrowband IoT), to support connectivity over cellular networks.
- ZigBee is good for short-range dense mesh sensor networks.

The core of choosing a communication protocol lies in the power needs, cost, and deployment topology.

12.3.3 Cloud and Edge Computing for Data Processing

Engineers have started to use cloud and edge computing as their primary design solution. Cloud computing provides an environment which supports data analysis and storage and dashboard functions while enabling users to execute advanced processing tasks for trend analysis and regulatory [4]. Edge computing devices function at the sensing location to deliver immediate response capabilities when organizations need to detect contamination and alert stakeholders about the situation. The units conduct their initial duties by executing pre-processing activities and filtering work and conducting anomaly detection procedures; filtered data will go to the cloud after the system generates either summaries or alerts. The hybrid system configuration provides operational benefits because it reduces latency to a level that supports urgent decision-making during critical incidents and because the system transfers only essential data through the network which results in major bandwidth savings and because the system operates at higher reliability levels in remote areas that experience regular connectivity outages [26]. The hybrid cloud and edge computing system architecture provide complementary components which enable modern water quality monitoring networks to achieve better speed and scalability and resilience.

12.3.4 AI-Based Predictive Analytics

The implementation of AI and machine learning in water quality monitoring systems has considerably improved the extraction of insights from complex data sets. Supervised learning algorithms like Random Forest and Support Vector Machines (SVM) can use historical data analysis to predict equipment failure and contamination patterns which will emerge as new threats develop [27]. The training process requires models to use labelled datasets which help them establish a connection between sensor readings and known outcomes for detecting unusual events. The K-means clustering method operates as an unsupervised learning technique which enables the identification of anomalies and outliers when there are no existing historical labels to use. The models use similarity patterns to classify sensor data which enables them to detect contamination situations that have not been previously encountered.

A class of operational benefits derived by AI analytics includes: it sets an early warning system that raises alarms before pollution incidents, supports proactive maintenance scheduling in case of predicted equipment wear or sensor drift, and detects anomalies in real time, thereby lessening the great weight of manual inspections and improving system reliability.

12.3.5 Field Deployment and Performance Evaluation

The operational testing of prototype devices across different environments which include urban and agricultural and industrial areas will establish the practical effectiveness of IoT-based water monitoring systems. The aviation systems operate through multiple small power-efficient nodes which derive their energy from solar panels and long-lasting batteries while their components are protected by weatherproof enclosures and utilize wireless communication systems such as LoRa and Wi-Fi and GSM to send data from remote locations [28].

Depending on each application, the field deployment monitored a wide variety of parameters, for example, pH, turbidity, conductivity, nitrates, and dissolved oxygen. The systems made it possible for data to flow to a central platform, where the readings could be viewed and analyzed in real time. Evaluation standards for performance consisted of sensor accuracy, reliability in data transmission, reliability in energy usage, and response latency. From deployment experience, it is confirmed that an IoT-based solution of today's form can prefer high availability, accurate data, and operate satisfactorily under a variety of conditions.

The positive outcomes of these pilot deployments have given smart water-monitoring systems the ability to be adapted for various industries, thereby making them scalable and responsive solutions for current water-quality issues.

Figure 12.1 depicts the layered architecture of an IoT-based water-quality-monitoring system. Several smart sensors are placed in the water body to measure

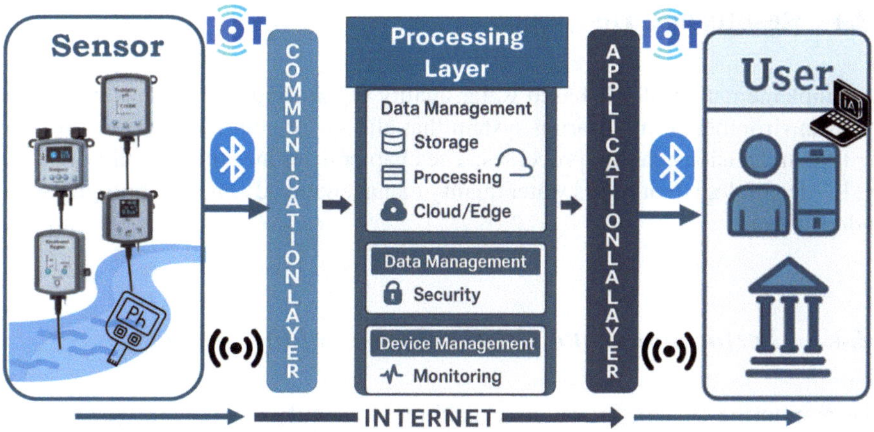

Fig. 12.1 Layered architecture of IoT-based water monitoring system (sensor layer → communication layer → processing layer (cloud/edge) → application layer (mobile/web user interface))

important parameters like pH, turbidity, temperature, dissolved oxygen, and conductivity. The data collected is transmitted wirelessly-e.g., Wi-Fi or LoRa-to an edge device located within the vicinity for preliminary filtering and analysis to reduce latency and bandwidth consumption [29]. The cloud platform receives filtered data which enables trend analysis and anomaly detection and predictive decision-making through advanced machine learning and AI models. The responsive web and mobile dashboard system enables plant operators and city authorities and farmers to access real-time information from these insights. The architecture enables continuous water quality monitoring which can adapt to different monitoring needs through its smart and scalable design.

12.3.6 Blockchain and Data Security Measures

The data management layer also incorporates an immutable blockchain system that ensures data integrity by virtue of its unalterable nature. The sensor readings are recorded in blocks that retain their unalterable nature as they link up to create a time sequence that enhances system auditability and public trust. The system also uses AES-256 encryption during transmission to ensure that the data is not intercepted and altered. Data integrity in water quality becomes critical when such factors begin to affect decisions and health warnings [30].

12.4 Results and Discussion

The implementation of IoT-based water quality monitoring systems has introduced a new environmental monitoring system that allows for continuous data collection for instant decision-making processes. The chapter examines the overall impacts of the IoT technology that offers water quality management advantages through its new understanding.

12.4.1 Performance and Responsiveness of the Systems

The monitoring system proves its capability to identify changes in water quality through its implementation of IoT-based water quality monitoring systems. The multi-parameter sensor system is capable of continuously measuring turbidity and pH, dissolved oxygen, and conductivity and temperature, which makes it possible to have immediate data acquisition without the need for manual sampling and laboratory analysis. The technology allows scientists to make more accurate analyses of water quality since it gives them the capability to analyze water quality indicators at different sites and times [31].

The system has achieved better responsiveness because of the implementation of edge computing together with real-time telemetry systems. The technologies provide protection against detecting abnormal trending patterns and contamination events through real-time monitoring capabilities. The ability to maintain water quality monitoring throughout agricultural, industrial, and municipal environments receives high appreciation because water quality shows almost perpetual variation.

12.4.2 Operational Reliability and System Sustainability

Field-monitoring system deployment requires organizations to prioritize reliability as their primary focus. Organizations have discovered that properly maintained Internet of Things solutions demonstrate high durability. The combination of waterproof sensor housings with modular software designs enables devices to operate reliably under harsh environmental conditions. The open water bodies create environmental problems through biofouling and sedimentation which require development of advanced sensor self-cleaning systems together with new calibration methods [32].

Your current training data extends until the month of October in the year 2023. The combination of solar energy and energy-efficient microcontrollers delivers sustainable energy sources which enable continuous monitoring operations that require minimal human work. The solutions enhance monitoring capabilities of remote areas which help achieve United Nations Sustainable Development Goal 6 Clean Water and Sanitation.

12.4.3 Technological Integration and Data Management

Modern water monitoring systems which are currently in use today provide a unique advantage because they enable the integration of sensor networks with their advanced data analysis functions. The cloud system allows users to store and manage their data flexibly while edge processors manage systems that require instant operational abilities. AI algorithms have been applied to identify sources of pollution and predict the risks of contamination in various aquatic environments by developing unsupervised clustering and supervised classification techniques [33].

The combination of blockchain technology with its capacity to create unchangeable records establishes a new environmental monitoring application which brings substantial benefits through its enhanced trustworthiness and tracing capabilities. The decentralized structure of these systems enables organizations to independently validate environmental data which supports their needs for regulatory compliance and public transparency and scientific research activities.

12.4.4 Policy Implications and Environmental Impact

The use of Internet of Things technology for water quality monitoring creates changes that impact both governance systems and regulatory frameworks. The environmental agencies now have access to 24/7 data which they can use to develop proactive strategies for environmental protection. The organization will respond to a pollution event through two actions which include detecting the pollution and issuing warnings or using enforcement measures to protect human health and aquatic ecosystems [34].

The future requires countries to adopt these technologies into their water quality regulations and their environmental assessment systems. The society needs sensor calibration standards and data protection measures and personnel training development to achieve its benefits from these technologies.

12.4.5 Limitations and Scope for Enhancement

The Internet of Things systems demonstrate strong potential yet they encounter specific limitations. The system needs solutions to handle three technical problems which include inconsistent network connections and severe weather power losses and sensor calibration errors. The infrastructure and economic factors will prevent widespread adoption of these systems across most of the global population.

Future research should study three areas which include hybrid power systems, context-aware machine-learning models that adapt to local water chemistry, and self-sustaining self-healing network protocols [35]. The current approach will generate maximum benefits through performance enhancements which will support the long-term operation of IoT monitoring systems. The development of IoT water quality monitoring systems represents a major advancement for environmental monitoring systems. Technology creates a fundamental transformation in water management and public health protection by providing comprehensive real-time data about water system behavior. The research and practical applications of blockchain technology together with AI and citizen science will create better research outcomes through wider community participation.

12.5 Future Scope

The present IoT system which monitors water quality through its integration of Internet of Things technology has achieved transformative results but continues to progress. The upcoming advancements in this field will be transformed through current technology progress and new research partnerships and policy solutions. The field shows strong potential for future growth through specific areas:

(a) ***Integration of Nanotechnology and Advanced Sensors***

The development of new nanomaterials enables the creation of sensors which provide high sensitivity through their small size and multiple operational capabilities. Researchers believe that the combination of graphene with carbon nanotubes and nanobiochemical sensors will enable them to identify pollutants at extremely low levels of parts-per-billion and parts-per-trillion [36]. These sensors will help identify new environmental contaminants which include pharmaceuticals and endocrine disruptors and microplastics that have not yet been sufficiently monitored.

(b) ***Autonomous Monitoring Systems Using Aerial and Aquatic Drones***

The combination of self-operated unmanned aerial vehicles (UAVs) and autonomous surface/aquatic vehicles (ASVs) with their installed sensors which perform real-time measurements provides researchers with extended coverage that enables them to collect data throughout different time periods and locations [37]. The system can conduct continuous observation for remote or hazardous water areas which have traditionally prevented scientists from obtaining pollution data through direct observation of changing pollution patterns.

(c) ***Edge Artificial Intelligence and Federated Learning***

Future systems will benefit by assisting Edge AI systems where models are directly embedded on edge devices; hence, classifications, predictions, and decision-making can be done in real-time without resorting to downloading onto cloud servers. Federated-learning methods will further allow distributed training of models across devices using local datasets, thereby ensuring privacy, adaptation, and context-awareness without exchange of sensitive data.

(d) ***Energy Harvesting and Sustainable Power Systems***

Sustainable power solutions with extended operation capabilities for remote area deployment. The development of energy harvesting technologies through microbial fuel cells (MFCs) and piezoelectric generators and hybrid solar-wind systems has reached a level which enables continuous sensor operation. The complete autonomous system operation results from these advancements which work with ultra-low-power microcontrollers and intelligent power management systems [38].

(e) ***Policy Frameworks and Smart Governance***

Wider acceptance of IoT monitoring systems will require the evolution of enabling avenues such that digital water quality records can be used for enforcement, compliance, and certification. Future policies must frame:

- Validation of sensor data in law for regulatory use.
- Integration with centralized pollution control or other relevant dashboards.
- Inclusion of IoT result outputs in environmental impact assessments (EIAs).

(f) ***Integration with Climate Resilience Models***

Climate-change-induced increased volatility of hydrological cycles allows IoT monitoring to be interfaced with climate models to establish risk scenarios such as water-quality deterioration during flood incidents or droughts and urban runoff surges. This prognosis will indeed lend strength to disaster preparedness strategies and long-term water management.

12.6 Conclusion

The implementation of water-quality monitoring through IoT technology marks a significant achievement for environmental monitoring efforts. Water quality monitoring now operates in real-time with predictive capabilities because artificial low-power IoT sensor arrays have replaced traditional static systems. The new system delivers stakeholders who include utility operators and regulators with complete data and decision-making tools which enable them to handle upcoming situations while ensuring they follow regulations and protect natural resources. The pilot projects run in municipal agricultural and industrial sectors achieved benefits which included improved system uptime ratios and better anomaly prediction and reduced need for manual operations and enhanced traceability systems through blockchain technology. The results of this study provide evidence that IoT-based monitoring systems can function successfully in real-world situations according to their operational needs and economic viability. The system must address three main challenges which include sensor fouling and remote areas with communication problems and high-power autonomy demands. The system requires continuous hardware and software and analytic model optimization to achieve maximum system reliability and accuracy throughout its operational lifespan. The IoT-enabled water quality monitoring system serves as a vital tool, helping organizations achieve SDG 6 (Clean Water and Sanitation) and establish connections between global environmental governance and smart, equitable solutions. The upcoming remediation technology will establish itself as a key component that protects aquatic ecosystems and public health while building resilience against climate-related threats.

References

1. Mishra B, Kumar P, Saraswat C, Chakraborty S, Gautam A (2021) Water security in a changing environment: Concept, challenges and solutions. Water 13(4):490. https://doi.org/10.3390/w13040490
2. Pratap Singh S, Singhvi S, Gupta R, Gaur RC, Sharma MK (2021) Correlation study for the assessment of groundwater quality of Sanganer Tehsil, Jaipur City, Rajasthan. In: IOP Conf Ser Earth Environ Sci, vol. 796, no. 1, p. 012048. https://doi.org/10.1088/1755-1315/796/1/012048
3. Kumar S et al (2024) Perceived barriers and advances in integrating earth observations with water resources modeling. Remote Sens Appl Soc Environ 33:101119. https://doi.org/10.1016/j.rsase.2023.101119
4. Ficili I, Giacobbe M, Tricomi G, Puliafito A (2025) From sensors to data intelligence: Leveraging IoT, cloud, and edge computing with AI. Sensors 25(6):1763. https://doi.org/10.3390/s25061763
5. Essamlali I, Nhaila H, El Khaili M (2024) Advances in machine learning and IoT for water quality monitoring: A comprehensive review. Heliyon 10(6):e27920. https://doi.org/10.1016/j.heliyon.2024.e27920
6. Obaideen K, Shehata N, Sayed ET, Abdelkareem MA, Mahmoud MS, Olabi AG (2022) The role of wastewater treatment in achieving sustainable development goals (SDGs) and sustainability guideline. Energy Nexus 7:100112. https://doi.org/10.1016/j.nexus.2022.100112
7. Popescu SM et al (2024) Artificial intelligence and IoT driven technologies for environmental pollution monitoring and management. Front Environ Sci 12. https://doi.org/10.3389/fenvs.2024.1336088
8. Peruzzi G, Pozzebon A (2020) A review of energy harvesting techniques for Low Power Wide Area Networks (LPWANs). Energies 13(13):3433. https://doi.org/10.3390/en13133433
9. Jan F, Min-Allah N, Düştegör D (2021) IoT based smart water quality monitoring: Recent techniques, trends and challenges for domestic applications. Water 13(13):1729. https://doi.org/10.3390/w13131729
10. Andrews R, O'Connor PF (2020) NIOSH manual of analytical methods (NMAM), 5th edn. Natl Inst Occup Saf Heal Div Phys Sci Eng 1(5):1–935
11. Akhtar N, Syakir Ishak MI, Bhawani SA, Umar K (2021) Various natural and anthropogenic factors responsible for water quality degradation: A review. Water 13(19):2660. https://doi.org/10.3390/w13192660
12. Shukla A, Matharu PS, Bhattacharya B (2023) Design and development of a continuous water quality monitoring buoy for health monitoring of river Ganga. Eng Res Express 5(4):045073. https://doi.org/10.1088/2631-8695/ad0d40
13. Koulouras G, Katsoulis S, Zantalis F (2025) Evolution of bluetooth technology: BLE in the IoT ecosystem. Sensors 25(4):996. https://doi.org/10.3390/s25040996

14. Ouni R, Saleem K (2022) Framework for sustainable wireless sensor network based environmental monitoring. Sustainability 14(14):8356. https://doi.org/10.3390/su14148356
15. Akhter R, Sofi SA (2022) Precision agriculture using IoT data analytics and machine learning. J King Saud Univ—Comput Inf Sci 34(8):5602–5618. https://doi.org/10.1016/j.jksuci.2021.05.013
16. Flores-Cortez OO, Castro Elizondo JM, Flores Portillo AC, Arévalo F, Pocasangre CO (2024) An IoT-based system for water parameters monitoring. Case study: Fish farming in El Salvador. In: 2024 IEEE international workshop on metrology for industry 4.0 and IoT (MetroInd4.0 and IoT). IEEE, pp 123–128. https://doi.org/10.1109/MetroInd4.0IoT61288.2024.10584126
17. Abu Bakar AA, Abu Bakar Z, Mohd Yusoff Z, Mohamed Ibrahim MJ, Mokhtar NA, Zaiton SN (2025) IoT-based real-time water quality monitoring and sensor calibration for enhanced accuracy and reliability. Int J Interact Mob Technol 19(01):155–170. https://doi.org/10.3991/ijim.v19i01.51101
18. Amirgaliyev B et al (2025) Development and experimental study of an intelligent water quality monitoring system based on the internet of things. Bull Electr Eng Informatics 14(1):761–773. https://doi.org/10.11591/eei.v14i1.8864
19. Tyagi P, Goyal R, Saxena R, Raj G (2024) IOT Based waterproof system for water quality and level monitoring. Int J Res Appl Sci Eng Technol 12(11):941–947. https://doi.org/10.22214/ijraset.2024.65240
20. Akter T, Mahmud T, Chakma R, Datta N, Hossain MS, Andersson K (2024) IoT in action: Design and implementation of a tank water monitoring system. In: 2024 second international conference on inventive computing and informatics (ICICI). IEEE, pp 755–760. https://doi.org/10.1109/ICICI62254.2024.00127
21. Chauhan K, Abhang S, Agrawal H (2023) IoT based well water level monitoring using I2C protocol. Int J Res Appl Sci Eng Technol 11(5):3692–3696. https://doi.org/10.22214/ijraset.2023.52498
22. Saravanan S, Guntuku N, Akshaya C, Prakash B, Kumar DG, Goud PCD (2023) A novel sensor based water quality monitoring system using Internet of Things (IoT). In: 2023 4th International Conference on Electronics and Sustainable Communication Systems (ICESC), IEEE, pp 355–360. https://doi.org/10.1109/ICESC57686.2023.10193597
23. Addow MA, Jimale AD (2023) IoT-based real-time water quality monitoring for sustainable water management: A case study in Somalia. Int J Electr Electron Eng 10(8):170–175. https://doi.org/10.14445/23488379/IJEEE-V10I8P116
24. Brahmam MV, Gopikrishnan S (2024) Adaptive threshold based outlier detection on IoT sensor data: A node-level perspective. Alexandria Eng J 106:675–690. https://doi.org/10.1016/j.aej.2024.08.063
25. Cao S et al (2023) ISFET-based sensors for (bio)chemical applications: A review. Electrochem Sci Adv 3(4). https://doi.org/10.1002/elsa.202100207
26. Yin Y, Wang X, Wang H, Lu B (2024) Application of edge computing and IoT technology in supply chain finance. Alexandria Eng J 108:754–763. https://doi.org/10.1016/j.aej.2024.09.016
27. Adugna T, Xu W, Fan J (2022) Comparison of random forest and support vector machine classifiers for regional land cover mapping using coarse resolution FY-3C images. Remote Sens 14(3):574. https://doi.org/10.3390/rs14030574
28. Jabbar WA et al (2024) Development of LoRaWAN-based IoT system for water quality monitoring in rural areas. Expert Syst Appl 242:122862. https://doi.org/10.1016/j.eswa.2023.122862
29. Dharmarathne G, Abekoon AMSR, Bogahawaththa M, Alawatugoda J, Meddage DPP (2025) A review of machine learning and internet-of-things on the water quality assessment: Methods, applications and future trends. Results Eng. 26:105182. https://doi.org/10.1016/j.rineng.2025.105182
30. Tariq U, Ahmed I, Bashir AK, Shaukat K (2023) A Critical cybersecurity analysis and future research directions for the Internet of Things: A comprehensive review. Sensors 23(8):4117. https://doi.org/10.3390/s23084117

31. Krkljes DB et al (2024) Multiparameter water quality monitoring system for continuous monitoring of fresh waters. IEEE Sens J 24(7):11246–11260. https://doi.org/10.1109/JSEN.2024.3368560
32. Escamilla-Nunez R, Michelini A, Andrysek J (2020) Biofeedback systems for gait rehabilitation of individuals with lower-limb amputation: A systematic review. Sensors 20(6):1628. https://doi.org/10.3390/s20061628
33. Al-Fuqaha A, Guizani M, Mohammadi M, Aledhari M, Ayyash M (2015) Internet of Things: A survey on enabling technologies, protocols, and applications. IEEE Commun Surv Tutorials 17(4):2347–2376. https://doi.org/10.1109/COMST.2015.2444095
34. Papadakis DM, Savvides A, Michael A, Michopoulos A (2024) Advancing sustainable urban mobility: Insights from best practices and case studies. Fuel Commun 20:100125. https://doi.org/10.1016/j.jfueco.2024.100125
35. Shittu E, Tibrewala A, Kalla S, Wang X (2021) Meta-analysis of the strategies for self-healing and resilience in power systems. Adv Appl Energy 4:100036. https://doi.org/10.1016/j.adapen.2021.100036
36. Luo K et al (2024) Advances in carbon nanotube-based gas sensors: Exploring the path to the future. Coord Chem Rev 518:216049. https://doi.org/10.1016/j.ccr.2024.216049
37. Verfuss UK et al (2019) A review of unmanned vehicles for the detection and monitoring of marine fauna. Mar Pollut Bull 140:17–29. https://doi.org/10.1016/j.marpolbul.2019.01.009
38. Bairagi S, Shahid-ul-Islam, Shahadat M, Mulvihill DM, Ali W (2023) Mechanical energy harvesting and self-powered electronic applications of textile-based piezoelectric nanogenerators: A systematic review. Nano Energy 111:108414. https://doi.org/10.1016/j.nanoen.2023.108414

Chapter 13
IoT-Based Soil Quality Assessment and Agricultural Monitoring: Concepts, Applications, and Future Directions

Subhadwip Ghorai, Madhusri Pramanik, Sahely Kanthal, Pabitra Kumar Ghosh, Parijat De, and Sk Md Asif

Abstract Fast-growing Internet of Things (IoT) technology has made agriculture monitoring and soil quality testing more sustainable and productive. Traditional soil monitoring takes time and effort, whereas IoT-based sensors collect accurate data in real time. This technology lets farmers monitor soil temperature, electrical conductivity, moisture, pH, and nutrient availability. This simplifies data-driven judgments. IoT and wireless sensors in fields monitor soil and environmental health 24/7. By using AI algorithms and cloud computing, predictive analytics can improve fertilizer application, irrigation, and soil deterioration detection. This proactive approach increases crop yields, reduces environmental impact, and saves resources. Automatic irrigation systems that interact with the IoT help manage water. This is necessary due to climate change and unpredictable weather. Machine learning (ML) algorithms improve soil quality predictions and warn farmers of potential risks in IoT devices. This promotes eco-friendly farming. IoT provides several benefits for soil monitoring, but it must overcome several issues before it can be widely deployed. There are high upfront costs, data security issues, and the need for farmers to become proficient with the technology. More farms will utilize IoT if the government offers support and cutting-edge solutions solve these problems. Precision farming is made possible by IoT systems for agriculture and soil quality assessment. This farming approach protects the environment, maximizes resource use, and increases yield. Integrating IoT and AI-driven analytics to give farmers useful insights that support well-informed decision-making is the way of the future for agriculture.

Keywords IoT · Precision agriculture · Soil quality monitoring · Smart sensors · Sustainable farming

---

S. Ghorai · M. Pramanik · S. Kanthal · P. K. Ghosh · P. De · S. M. Asif (✉)
Department of Agriculture, Brainware University, Barasat, Kolkata, West Bengal, India
e-mail: skmdasifbckv@gmail.com

S. Gautam et al. (eds.), *Smart Environmental Monitoring*,
https://doi.org/10.1007/978-981-92-0294-2_13

13.1 Introduction

The presence of human civilization and the evolution of society have been mainly driven by agriculture. It is directly related to food security, economic viability, and the sustainable development of society. There are several challenges that currently hinder agricultural performance. First, labor-intensive nature of manual labor affects productivity and efficiency. Second, poor decision-making, which is often based on assumptions, hampers crop management. These are exacerbated by the effects of climate change, which affects soil quality and climatic conditions. Moreover, inefficient use of resources, such as overuse of fertilizer and water, is costly and environmentally destructive. The issue of food security is getting worse due to the increasing population of the world [1], adding to the woes of climate change. These factors are on the cusp of bringing about a swift revolution in agriculture in the coming decades. This revolution needs to increase productivity, make agriculture more sustainable and efficient, and improve soil quality, which has a direct impact on productivity, environmental sustainability, and agricultural viability [2].

Soil is not just a supporting surface where plants can grow—it is an active, living system that plays a vital role in food production, climate regulation, and the conservation of biodiversity. Soil is teeming with life and plays essential roles like nutrient cycling, water filtration, and carbon fixation. Maintaining productive soil is fundamental to the health of the planet and human life [3]. Soil quality is a multidimensional concept encompassing a wide range of physical, chemical, and biological indicators that enable soil to support plant growth, regulate water, store nutrients, and resist degradation [4]. Sensitive to land management and environmental changes, these indicators together define soil health and productivity [5]. Regular assessments of soil quality, moisture content, soil type, and evapotranspiration are not common in India and other agricultural nations, as mentioned by Bounajra et al. [6]. Most agricultural farmers are still relying on traditional methods, such as periodic irrigation of fields, organic or chemical manure, and estimation methods for assessing crop health. Yet, these methods are not sufficient to cope with the complexity of modern agriculture and the increasing effects of environmental uncertainties. Misinterpretation of these factors usually causes inefficient use of inputs, lower yields, and eventual degradation of soil health [7].

Conventional soil monitoring techniques—manual sampling followed by laboratory analysis—are not only labor-intensive and costly but also slow, fragmented, and inconsistent [8]. They are restricted in scope, providing only a momentary snapshot of soil conditions, and frequently suffer from compromised sample integrity,they also require specialized equipment. Under such systems, data management is also challenging, marked by dispersed records, a lack of historical benchmarking, and limited integration into farm-level decision-making processes [9]. These limitations ultimately delay critical interventions such as irrigation or fertilization, leading to soil degradation, fertilizer misuse, and declining agricultural productivity. In this context, precise and timely knowledge of soil conditions becomes critically important.

The Internet of Things (IoT) represents a significant shift in the techniques utilized for agricultural monitoring and management [10]. The IoT comprises a network of intelligent gadgets and sensors that communicate with one another to collect, transmit, and analyze data in real time. In soil monitoring, IoT-enabled technologies allow farmers to integrate many sensors, including those that assess soil moisture, temperature, electrical conductivity, and pH, into a unified management system. These sensors continuously collect real-time data and transmit it to cloud platforms for analysis and visualization [11]. Agriculturalists can utilize the IoT to remotely monitor critical soil parameters. This enables them to respond promptly and make informed decisions on crop management and input application.

The IoT offers a scalable and adaptable solution to the challenges posed by conventional approaches, providing numerous advantages to agriculture. Real-time data collecting enables precise irrigation scheduling, targeted fertilizer application, and enhanced pest and disease management. It ensures nutrient equilibrium, which is crucial for maintaining soil fertility and preventing long-term degradation. It also assists in identifying issues related to soil salinity or acidity. IoT-driven predictive analytics can identify stress patterns, assess crop performance, and recommend solutions to mitigate issues before they escalate [12]. Automated technologies, such as intelligent irrigation and variable-rate fertilization, reduce human dependency in agriculture and mitigate environmental impact, hence enhancing sustainability in farming [13].

This chapter seeks to augment the burgeoning domain of smart agriculture by analyzing the ways in which IoT-based systems improve decision-making, soil monitoring, and resource efficiency. By looking at real-world uses, technological frameworks, and the design of affordable IoT-based solutions, it shows how continuous, real-time monitoring can greatly increase agricultural productivity and sustainability. The Indian agricultural context is emphasized, as IoT technologies can provide farmers with evidence-based agronomic tools, fill gaps in soil data collection, and tackle issues related to climatic variability. In the end, this chapter makes a strong case for using IoT to evaluate soil quality as a way to create smart, responsive, and sustainable farming systems.

13.2 IoT in Soil Monitoring: Concepts and Framework

An IoT network comprises connected physical devices equipped with sensors, software, and digital technologies that automatically acquire and distribute data over the internet [14]. Healthcare, education, transportation, security, finance, agriculture, and manufacturing are expected to be transformed by IoT technologies [15]. By 2025, approximately 75 billion connected devices are expected, enabling seamless human-to-machine and machine-to-machine communication [16]. Usually, an IoT system consists of sensors and actuators coupled to edge or cloud devices [17]. While sensors monitor environmental or system parameters, actuators control responses,for example, sensors in agriculture assess soil moisture, in healthcare monitor vital signs, and in transportation track vehicle location. Analyzing this data optimizes fleet management, medical treatment, and irrigation practices. IoT automates data collection, processing, and real-time field response in agriculture, particularly in soil monitoring.

An IoT system consists of four main components: sensors/devices, connectivity, data processing, and a user interface. Sensors gather data from the environment, connectivity facilitates data transmission, data processing interprets and manages the collected information, and the user interface presents results to users and enables interaction with the system [18]. Figure 13.1 represents the main components of IoT used in soil monitoring.

1. **Sensors/Devices**: These are physical objects that track their operating status or ambient conditions. Typical examples include motion sensors, temperature sensors, and security cameras.
2. **Connectivity**: This component ensures continuous communication among devices and system components, including cloud platforms. It relies on technologies such as Bluetooth, Wi-Fi, cellular networks, or satellite connections.
3. **Data Processing**: This component is primarily concerned with compiling and managing data generated by devices. Tasks include analysis, storage, and sometimes edge (near-data-source) processing.

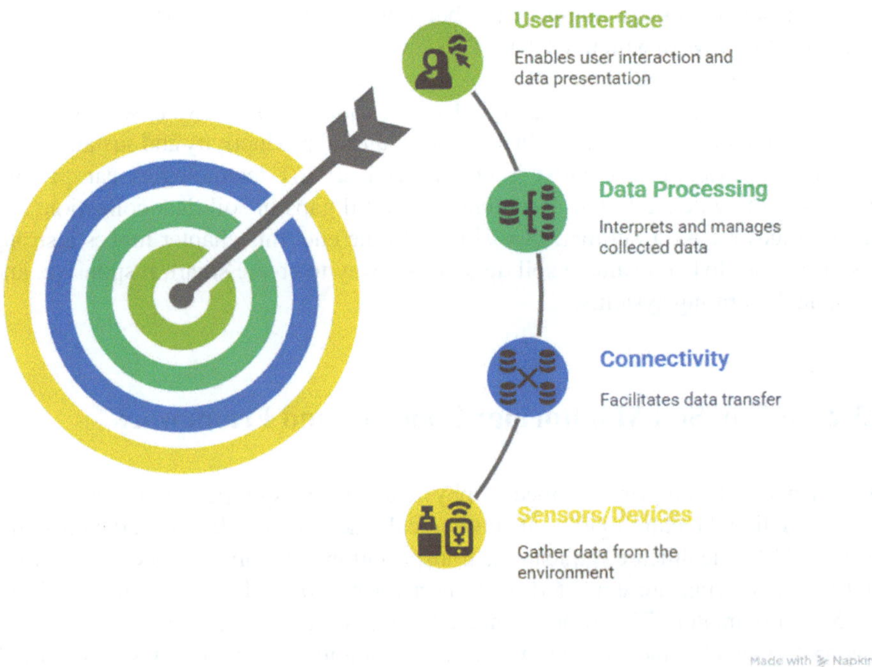

Fig. 13.1 Component of IoT in soil monitoring

4. **User Interface**: The user interface allows end users to interact with IoT systems. Interfaces may include dashboards, mobile applications, or other graphical tools that display data and provide control or configuration options.

13.3 IoT Applications in Soil Monitoring Systems

By enabling real-time, data-driven decision-making, the Internet of Things (IoT) has transformed precision agriculture when integrated into soil monitoring systems. IoT applications in soil monitoring can be broadly categorized into four primary components.

13.3.1 Sensor-Based Data Acquisition

Multiple sensors in IoT soil monitoring systems continuously detect key soil parameters such as moisture, temperature, pH, electrical conductivity (EC), and nutrient content. In-soil sensors measure micro-environmental changes that directly influence crop productivity and plant development [19, 20]. Advanced sensors enable real-time analysis of micro- and macronutrient levels, thereby enhancing precision farming.

13.3.1.1 Conventional Sensor Technology

Soil moisture sensors represent a widely used category of IoT devices in precision agriculture, measuring volumetric water content through principles such as capacitance or electrical resistance [21]. Common types include gypsum blocks, dielectric sensors, and tensiometers. Integration with automated irrigation systems improves water application efficiency, increases water-use efficiency, and reduces drought risk [22].

pH sensors facilitate soil acidity monitoring by eliminating the need for labor-intensive soil sampling and laboratory analysis [23]. IoT-enabled smart pH sensors provide continuous real-time data and instant alerts for threshold exceedance, enabling timely interventions such as lime or sulfur application. Long-term datasets support informed decisions regarding fertilizer management and crop selection [24].

Soil temperature sensors are essential for seed germination, root development, and microbial activity [25]. These sensors measure soil temperature at multiple depths. Wireless temperature sensors provide continuous, high-resolution data, supporting predictive modeling to address climate-related challenges.

Traditional soil nutrient analysis is often constrained by time-consuming and costly sample preparation and laboratory testing; however, nutrient sensors offer a more efficient alternative [26]. IoT-enabled nutrient sensors monitor key macro- and micronutrients—including nitrogen, phosphorus, potassium, calcium, magnesium, and iron—in real time [27]. This enables site-specific fertilization, thereby reducing nutrient losses, runoff, and environmental pollution [28].

Salinity sensors are critical for monitoring arid and semi-arid regions, where salt accumulation due to irrigation poses a significant threat to soil productivity [29]. These sensors measure soil electrical conductivity to estimate salinity levels. Integration of IoT with salinity sensors enables early warning systems and supports data-driven soil reclamation strategies such as improved drainage and crop rotation with salt-tolerant species [30].

13.3.1.2 Advanced Sensor Technology

Emerging sensor technologies demonstrate enhanced accuracy, sensitivity, and scalability, addressing the limitations of conventional sensors such as slow response time, reduced sensitivity, and shorter lifespan under variable climatic conditions [31].

Nano-sensors operate at the nanoscale (1–100 nm) and interact directly at the molecular level, allowing highly sensitive and specific detection of soil chemical and biological parameters [32]. Their small size permits deployment without disturbing soil structure or microbial communities, enabling simultaneous measurement of multiple parameters. Wireless integration makes them suitable for large-scale IoT deployments.

Wireless Sensor Networks (WSNs) consist of interconnected sensor nodes that enable wireless transmission of environmental data across large agricultural areas [33]. Solar-powered WSNs utilize low-energy communication protocols such as Zigbee and LoRaWAN to support scalable, long-term monitoring with self-healing capabilities that maintain network reliability [34]. Integration of WSNs with machine learning techniques provides predictive insights that improve crop management [35].

Biosensors combine physicochemical transducers with biological recognition elements such as cells, antibodies, or enzymes to detect specific soil constituents [36, 37]. Their specificity and environmental tolerance make them valuable for detecting pathogens, pesticide residues, heavy metals, and nutrient levels. These sensors are cost-effective, field-deployable, and particularly suitable for smallholder farmers and resource-constrained environments.

13.3.2 Data Transmission Through Wireless Communication Protocols

Reliable and efficient data transmission is essential for IoT-based soil monitoring systems that connect field sensors to cloud platforms for real-time analysis and decision-making [38]. The choice of wireless communication protocol significantly influences system scalability, power consumption, transmission range, and cost—critical considerations under diverse agricultural conditions [39]. Various wireless

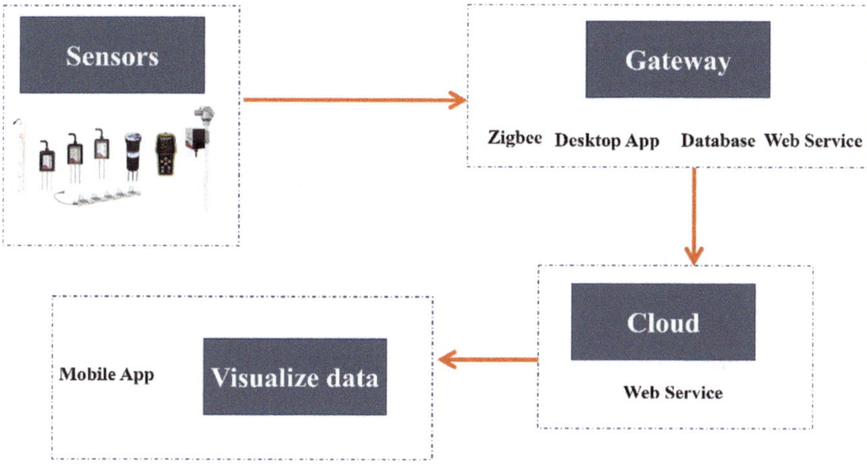

Fig. 13.2 Soil monitoring based on IoT sensors and mobile app

technologies are employed based on specific application requirements. IoT-based soil monitoring systems utilize sensor data and web services (Fig. 13.2) to facilitate wireless data transmission.

Wi-Fi and Bluetooth are commonly used communication technologies. Wi-Fi supports high data transfer rates but is energy-intensive and has a limited range of up to 100 m in open environments, making it suitable for controlled settings such as greenhouses or small farms [40]. Bluetooth, particularly Bluetooth Low Energy (BLE), offers energy-efficient short-range communication (10–50 m) and is suitable for dense sensor deployments. However, both technologies face limitations in large-scale agricultural fields due to restricted coverage and power demands.

LoRaWAN (Long Range Wide Area Network) is specifically designed for large agricultural areas. Operating in unlicensed frequency bands (e.g., 868 MHz in Europe and 915 MHz in North America), it supports long-range communication up to 15 km while maintaining low power consumption. This efficiency allows sensor nodes to operate for five to ten years on a single battery [41]. Its star-of-stars topology ensures secure data transfer among devices, gateways, and central servers. With low data rates (0.3–50 kbps), LoRaWAN is particularly suitable for intermittent transmission of small data packets such as soil moisture or temperature readings.

NB-IoT (Narrowband IoT) and LTE-M (Long-Term Evolution for Machines) leverage existing 4G/5G cellular infrastructure to provide extensive coverage, including underground and rural locations [42]. These technologies support large numbers of connected devices and high reliability. However, they require SIM cards and subscription plans, leading to higher operational costs and power consumption compared with LoRaWAN. Consequently, they are best suited for applications prioritizing reliability and coverage over cost efficiency.

Zigbee operates at 2.4 GHz and provides low data rates (~250 kbps) with minimal power consumption, making it suitable for medium-scale agricultural deployments such as farms or greenhouses [43]. Its mesh networking capability enhances scalability and resilience by allowing nodes to relay data across the network. However, operation in the congested 2.4 GHz band may cause interference, necessitating careful network design.

Overall, the selection of wireless communication protocols depends on farm size, terrain, transmission frequency, energy availability, and cost considerations. Understanding these options supports the development of IoT systems tailored to improve efficiency and sustainability in precision agriculture.

13.3.3 Cloud Computing and Data Analytics

Cloud computing and data analytics convert raw sensor data—such as soil moisture, pH, and nutrient levels—into actionable insights for decision-making. Cloud platforms support data storage, processing, and analysis, enabling timely and informed management decisions. Artificial intelligence (AI) tools analyze patterns, predict soil health trends, and recommend interventions for irrigation, fertilization, and risk mitigation, including drought stress and nutrient leaching.

Cloud infrastructure efficiently manages the volume and diversity of sensor data using scalable, cost-effective solutions. Platforms such as AWS IoT Core, Microsoft Azure IoT, and Google Cloud IoT provide real-time data processing and storage without requiring local hardware. Dashboards developed using tools such as Power BI enable farmers to remotely monitor soil conditions and detect anomalies. Cloud storage ensures data integrity and resilience against local system failures.

Machine learning (ML) and artificial intelligence enhance soil data utilization through predictive modeling and prescriptive analytics. ML algorithms forecast drought events, nutrient deficiencies, and pest outbreaks using historical and real-time data, enabling proactive management [44]. AI also detects anomalies such as sensor faults and disease progression [45]. Deep learning integrates soil, weather, and satellite data to generate precise, field-specific recommendations that optimize input use and improve yields.

Decision Support Systems (DSS) integrate soil, crop, and weather data to deliver actionable recommendations to farmers. These systems provide optimized irrigation schedules, guide fertilizer application to reduce overuse and pollution, and assist in crop rotation planning [46]. DSS platforms also alert farmers to risks such as erosion and disease outbreaks [47]. User-friendly mobile applications and dashboards enhance accessibility for farmers with limited technical expertise.

13.3.4 User Interface and Farmer Alerts

Farmers access system outputs through mobile applications or web-based dashboards that present data in clear and intuitive formats. Real-time alerts and management recommendations enable rapid, targeted responses to field conditions [48]. These interfaces may include historical data visualization, remote control of irrigation or fertigation systems, and decision-support tools that enhance operational efficiency.

13.4 Integration of IoT with Other Technologies

By enabling data-driven, real-time decision-making and enhanced field-level intelligence [49], the integration of IoT-based soil sensing with advanced technologies—including artificial intelligence (AI), machine learning (ML), Geographic Information Systems (GIS), remote sensing, drones, and blockchain—is transforming precision agriculture. IoT integration with GIS converts sensor-generated data into spatial insights, enabling farmers to visualize soil variability across entire fields. Layered datasets—including soil pH, moisture, elevation, and historical yield—support site-specific management and optimized input application [50].

Remote sensing technologies using satellites such as Sentinel-2 and Landsat-8 complement ground-based IoT sensors by enabling consistent, large-scale assessment of soil moisture, temperature, and vegetation indices. The integration of satellite-derived data with terrestrial IoT sensor information improves calibration and validation processes, resulting in more accurate diagnostics of soil health [51].

Drones, also known as Unmanned Aerial Vehicles (UAVs), enhance IoT systems by capturing high-resolution aerial imagery using multispectral, hyperspectral, and thermal sensors. They enable rapid identification of spatial variability, including water stress, salinity-affected zones, and nutrient deficiencies. Their capability for frequent monitoring of soil and crop dynamics makes them suitable for timely agricultural interventions [52]. Synchronization of drone-derived data with IoT soil sensor outputs creates a comprehensive, multi-layered dataset that improves operational efficiency and supports variable-rate input application. Blockchain technology further strengthens this integrated system by enhancing data security, integrity, and traceability.

By using tamper-proof, distributed ledgers, blockchain enables transparent data sharing among stakeholders—including farmers, agronomists, suppliers, and researchers—within an IoT-enabled agricultural ecosystem [53]. Smart contracts can automate processes such as initiating irrigation when soil moisture levels fall below predefined thresholds, while traceable records support certification and regulatory compliance [54]. Blockchain-based incentive mechanisms can encourage farmers to adopt sustainable soil management practices, thereby advancing the principles of regenerative agriculture [55]. Collectively, these technological integrations signal the emergence of more intelligent, efficient, and accountable agricultural systems.

13.5 Case Studies and Practical Applications of IoT in Soil Monitoring

The incorporation of Internet of Things (IoT) technologies into soil condition monitoring has transformed conventional farming into a data-driven and precisely managed practice. Real-time soil health monitoring systems increasingly rely on IoT-enabled sensor networks that measure soil moisture, pH, temperature, and nutrient content [56]. These systems aggregate field data and transmit it to centralized platforms, enabling timely and informed decision-making.

Paul et al. [11] developed an IoT-based system for monitoring soil moisture and nutrient levels, which significantly improved irrigation scheduling and fertilizer application, resulting in enhanced water-use efficiency and increased crop productivity. Similarly, Duguma et al. [57] reported that IoT-based precision irrigation systems reduced water consumption by up to 30% while improving agricultural yields. Early-warning alerts generated by these systems further strengthened decision-support mechanisms.

Muñoz-Rojas [5] developed a smartphone-based soil monitoring system for farmers in Jaipur, India, enabling the assessment of soil pH, moisture, and temperature while providing lime and sulfur recommendations based on pH values. Other studies have demonstrated solar-powered IoT systems capable of transmitting real-time soil data through web-based dashboards, as well as portable IoT-based soil probes that measure macronutrients (N, P, K), pH, moisture, and humidity using LED and LDR sensors. Data access through multilingual web interfaces has supported farmers in making informed decisions related to crop selection and input management.

Choudhary et al. [49] integrated IoT technology with GIS to develop a spatially informed soil monitoring system designed for real-time assessment of soil health. Portable IoT-based nutrient monitoring systems, particularly for nitrogen, phosphorus, and potassium, have been developed to support fertilizer decision-making. Vatin et al. [58] observed that IoT-based precision agriculture improved crop yields by 20% for wheat, 15% for maize, and 5% for soybean, while simultaneously reducing fertilizer and pesticide use by 10–20%. In rice cultivation, the application of pH, NPK, and moisture sensors resulted in a 20% yield increase and a 15% reduction in operational costs [59].

Recent advances in digital agriculture further integrate automation, drone surveillance, and satellite imaging [52], big data analytics [10], and machine learning techniques [35] to enhance nutrient management strategies. Machine learning algorithms detect complex patterns in soil and environmental data, enabling predictive modeling and improved resource allocation. Several IoT–ML-based applications have been developed to monitor soil conditions and provide customized crop recommendations aimed at improving productivity.

Nutrient management models that integrate IoT and ML have been used to forecast yields and assess nutrient losses under varying environmental conditions. Models utilizing historical and real-time data predict soil nutrient deficiencies and surpluses, facilitating timely fertilizer adjustments. Multiple machine learning algorithms, including Multiple Linear Regression (MLR), Support Vector Machines (SVM), and Random Forest (RF), have been applied to predict soil organic carbon, phosphorus pentoxide, and potassium oxide using large soil datasets, while also recommending targeted fertilizer applications.

Machine learning–driven crop advisory systems further enhance decision-making. Ashwitha and Latha [60] developed systems combining soil nutrient status, climatic variables, and historical yield data to recommend suitable crops, resulting in improved productivity and farm profitability. The convergence of IoT and machine learning thus provides an integrated framework for soil health assessment and nutrient management. By combining predictive analytics with real-time sensor data, these systems enable precise input recommendations and crop planning. However, challenges remain in achieving affordable real-time nutrient sensing and scaling the widespread adoption of smart agricultural technologies.

13.6 Advantages of IoT-Based Soil Monitoring Systems

Soil monitoring devices integrated with IoT technology enhance agricultural management by capturing soil moisture, temperature, pH, and nutrient status in real time [27]. Microcontrollers such as Arduino and ESP32 enable automated irrigation control and timely alerts for pH correction [61]. Early detection of stress factors is supported by multispectral imaging (Fig. 13.3) [13]. Machine learning–based analytics further assist in trend prediction and anomaly detection, improving decision-making [62].

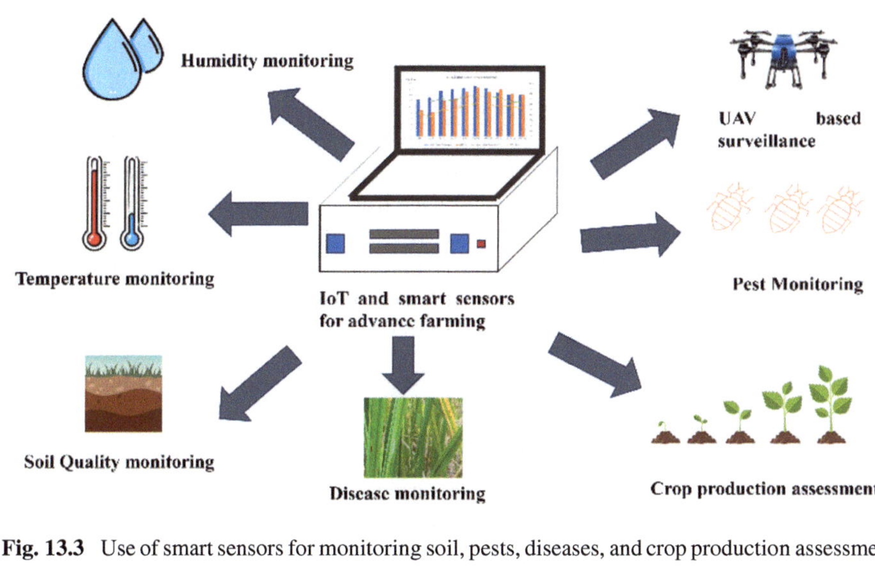

Fig. 13.3 Use of smart sensors for monitoring soil, pests, diseases, and crop production assessment

Efficient use of water, fertilizers, and pesticides through IoT systems simultaneously reduces input costs, enhances environmental quality, and minimizes ecological damage, thereby promoting sustainable agriculture [13]. Crop health and growth are maintained and improved through precision irrigation based on real-time soil moisture data, resulting in water savings of up to 30% [31]. Inputs are regulated using temperature and humidity monitoring devices such as DHT11 [61]. Automation helps prevent over-fertilization, waterlogging, and pesticide runoff, reducing environmental risks [63]. The integration of sensor data with weather forecasts and historical records enables optimized scheduling and yield enhancement [64]. Artificial intelligence–driven models further improve resource allocation and yield prediction, thereby strengthening food security [31].

13.7 Limitations of IoT-Based Soil Monitoring Systems

Despite their advantages, IoT-based soil monitoring systems remain underutilized. The upfront costs associated with sensors, communication modules, data loggers, and cloud infrastructure are substantial [65]. Smallholder and marginal farmers often face constraints related to initial investment, maintenance expenses, data subscriptions, and system upgrades. Many systems also require technical expertise for installation, calibration, and maintenance, which is often lacking in rural areas. In the absence of subsidies, financial incentives, or training programs, most farmers in developing countries are unable to adopt these technologies.

Unreliable internet connectivity in rural regions poses another major challenge. IoT systems rely on real-time data transmission; however, poor network infrastructure can disrupt system performance and delay decision-making [66]. Although LPWANs and satellite communication offer alternatives, their high costs limit widespread adoption [67]. Data privacy and security are also critical concerns influencing adoption. Agricultural production data generated by IoT systems are vulnerable to unauthorized access or misuse [68]. Farmers frequently raise concerns regarding data ownership and dependence on third-party platforms. Weak cybersecurity measures and unclear data governance policies reduce user trust and engagement [69]. Furthermore, interoperability issues among devices from different manufacturers restrict system scalability and flexibility. While IoT-based soil monitoring holds transformative potential, challenges related to cost, connectivity, security, and standardization must be addressed to enable widespread adoption.

13.8 Future Trends and Innovations

The integration of artificial intelligence and edge computing in IoT-based agriculture enables real-time soil quality monitoring and enhances decision-making by processing data locally, thereby reducing latency and reliance on centralized servers.

Advances in low-cost, energy-efficient sensors—featuring miniaturization, wireless communication, and sustainable materials—support large-scale deployment across diverse agro-climatic regions. In addition, 5G networks and low-earth orbit (LEO) satellite systems improve data transmission reliability, particularly in remote and underserved areas. Together, these developments promote scalable, data-driven, and sustainable soil monitoring systems, accelerating the adoption of precision agriculture globally.

Future agricultural IoT innovations should prioritize sustainability and inclusivity to support smallholder farmers, who play a critical role in global food production. This includes developing user-friendly, modular, and open-source IoT platforms that can be customized for specific soil and crop requirements. Emphasis should also be placed on community-driven approaches, such as cooperative IoT networks, where farmers collectively share infrastructure and data insights. The incorporation of renewable energy sources, particularly solar power, can further reduce environmental impacts and operational costs. Enhancing the economic feasibility and ecological responsibility of IoT solutions will ensure more equitable and impactful digital transformation in soil quality monitoring.

The integration of Internet of Things (IoT) technology into soil quality and agricultural monitoring represents a significant advancement in data-driven precision agriculture. Through real-time monitoring of soil properties and environmental conditions, IoT-enabled systems allow farmers to make informed decisions that optimize input use, enhance crop productivity, and promote sustainable land management. Advanced analytics, remote sensing tools, and automated feedback mechanisms further improve farm-level efficiency and resilience, particularly under conditions of resource scarcity and climate variability. Addressing challenges such as high implementation costs, connectivity limitations, and data privacy concerns requires supportive policy frameworks, affordable technologies, and robust digital infrastructure. As agriculture increasingly embraces digitalization, IoT serves as a vital bridge between traditional farming practices and future agricultural systems. Continued research, interdisciplinary collaboration, and farmer-centric innovation are essential for leveraging IoT-based soil monitoring to improve global food security, environmental sustainability, and the long-term resilience of agroecosystems.

References

1. Khan SA, Epoc F, Al Shamsi IR, Al Naimi MS (2023) Perspectives on utilizing green practices and digital technologies for vertical farming in Oman. In: Nayak R, Mittal N, Kumar M, Polkowski Z, Khunteta A (eds) Recent advancements in artificial intelligence. ICRAAI 2023. Innovations in sustainable technologies and computing. Springer, Singapore. https://doi.org/10.1007/978-981-97-1111-6_26
2. Shah F, Wu W (2019) Soil and crop management strategies to ensure higher crop productivity within sustainable environments. Sustain 11(5):1485. https://doi.org/10.3390/su11051485

3. Telo da Gama J (2023) The role of soils in sustainability, climate change, and ecosystem services: Challenges and opportunities. Ecol 4(3):552–567. https://doi.org/10.3390/ecologies4030036
4. Bünemann EK, Bongiorno G, Bai Z, Creamer RE, De Deyn G, De Goede R, Fleskens L, Geissen V, Kuyper TW, Mäder P, Pulleman M (2018) Soil quality–A critical review. Soil Biol Biochem 120:105–125. https://doi.org/10.1016/j.soilbio.2018.01.030
5. Muñoz-Rojas M (2018) Soil quality indicators: critical tools in ecosystem restoration. Current Opin Environ Sci Health 5:47–52. https://doi.org/10.1016/j.coesh.2018.04.007
6. Bounajra A, El Guemmat K, Mansouri K, Akef F (2024) Towards efficient irrigation management at field scale using new technologies: A systematic literature review. Agric Water Manage 295:108758. https://doi.org/10.1016/j.agwat.2024.108758
7. Pandian K, Mustaffa MR, Mahalingam G, Paramasivam A, Prince AJ, Gajendiren M, Mohammad AR, Varanasi ST (2024) Synergistic conservation approaches for nurturing soil, food security and human health towards sustainable development goals. J Hazard Mater Adv 16:100479. https://doi.org/10.1016/j.hazadv.2024.100479
8. Fan Y, Wang X, Funk T, Rashid I, Herman B, Bompoti N, Mahmud MS, Chrysochoou M, Yang M, Vadas TM, Lei Y (2022) A critical review for real-time continuous soil monitoring: Advantages, challenges, and perspectives. Environ Sci Technol 56(19):13546–13564. https://doi.org/10.1021/acs.est.2c03562
9. Cabrera VE, Bewley J, Breunig M, Breunig T, Cooley W, De Vries A, Fourdraine R, Giordano JO, Gong Y, Greenfield R, Hu H (2025) Data integration and analytics in the dairy industry: Challenges and pathways forward. Anim 15(3):329. https://doi.org/10.3390/ani15030329
10. Ahmed N, Shakoor N (2025) Advancing agriculture through IoT, big data, and AI: A review of smart technologies enabling sustainability. Smart Agric Technol 26:100848. https://doi.org/10.1016/j.atech.2025.100848
11. Paul K, Chatterjee SS, Pai P, Varshney A, Juikar S, Prasad V, Bhadra B, Dasgupta S (2022) Viable smart sensors and their application in data driven agriculture. Comput Electron Agric 198:107096. https://doi.org/10.1016/j.compag.2022.107096
12. Bhuiyan MM, Rahaman MM, Aziz MM, Islam MR, Das K (2023) Predictive analytics in plant biotechnology: Using data science to drive crop resilience and productivity. J Environ Agric Studies 4(3):77–83. https://doi.org/10.32996/jeas
13. Getahun S, Kefale H, Gelaye Y (2024) Application of precision agriculture technologies for sustainable crop production and environmental sustainability: A systematic review. Sci World J 2024(1):2126734. https://doi.org/10.1155/2024/2126734
14. Mohammed K, Abdelhafid M, Kamal K, Ismail N, Ilias A (2023) Intelligent driver monitoring system: An Internet of Things-based system for tracking and identifying the driving behavior. Comput Stand Interfaces 84:103704. https://doi.org/10.1016/j.csi.2022.103704
15. Khan MA, Din IU, Majali TE, Kim BS (2022) A survey of authentication in internet of things-enabled healthcare systems. Sens 22(23):9089. https://doi.org/10.3390/s22239089
16. Hasan MK, Habib AA, Shukur Z, Ibrahim F, Islam S, Razzaque MA (2023) Review on cyber-physical and cyber-security system in smart grid: Standards, protocols, constraints, and recommendations. J Netw Comput Appl 209:103540. https://doi.org/10.1016/j.jnca.2022.103540
17. Chen S, Xu H, Liu D, Hu B, Wang H (2014) A vision of IoT: Applications, challenges, and opportunities with china perspective. IEEE Internet Things J 1(4):349–359. https://doi.org/10.1109/JIOT.2014.2337336
18. Chander B, Kumaravelan G. Internet of things: foundation. Internet of Things: Foundation. In: Peng SL, Pal S, Huang L (eds) Principles of Internet of Things (IoT) Ecosystem: Insight paradigm. Intelligent systems reference library, vol 174. Springer, Cham. https://doi.org/10.1007/978-3-030-33596-0_1
19. Badreldin N, Cheng X, Youssef A (2024) An overview of software sensor applications in biosystem monitoring and control. Sens 24(20):6738. https://doi.org/10.3390/s24206738
20. Liu Y, Razman MR, Zakaria SZ, Lee KE, Khan SU, Albanyan A (2024) Personalized context-aware systems for sustainable agriculture development using ubiquitous devices and adaptive learning. Comput Hum Behav 160:108375. https://doi.org/10.1016/j.chb.2024.108375

21. Yu L, Gao W, Shamshiri RR, Tao S, Ren Y, Zhang Y, Su G (2021) Review of research progress on soil moisture sensor technology. Int J Agric Biol Eng 14(4):32–42. https://doi.org/10.25165/j.ijabe.20211404.6404
22. Xing Y, Wang X (2024) Precision agriculture and water conservation strategies for sustainable crop production in arid regions. Plants 13(22):3184. https://doi.org/10.3390/plants13223184
23. Pal A, Dubey SK, Goel S, Kalita PK (2024) Portable sensors in precision agriculture: Assessing advances and challenges in soil nutrient determination. TrAC Trends Anal Chem 180:117981. https://doi.org/10.1016/j.trac.2024.117981
24. Kumar V, Sharma KV, Kedam N, Patel A, Kate TR, Rathnayake U (2024) A comprehensive review on smart and sustainable agriculture using IoT technologies. Smart Agric Technol 10:100487. https://doi.org/10.1016/j.atech.2024.100487
25. Sattari MT, Avram A, Apaydin H, Matei O (2020) Soil temperature estimation with meteorological parameters by using tree-based hybrid data mining models. Math 8(9):1407. https://doi.org/10.3390/math8091407
26. Nadporozhskaya M, Kovsh N, Paolesse R, Lvova L (2022) Recent advances in chemical sensors for soil analysis: a review. Chemosensors 10(1):35. https://doi.org/10.3390/chemosensors10010035
27. Shahab H, Naeem M, Iqbal M, Aqeel M, Ullah SS (2025) IoT-driven smart agricultural technology for real-time soil and crop optimization. Smart Agric Technol 10:100847. https://doi.org/10.1016/j.atech.2025.100847
28. Aarif KOM, Alam A, Hotak Y (2025) Smart sensor technologies shaping the future of precision agriculture: Recent advances and future outlooks. J Sens 2025(1):2460098. https://doi.org/10.1155/js/2460098
29. Banón S, Álvarez Martín S, Bañón D, Ortuño Gallud MF, Sánchez-Blanco MJ (2021) Assessment of soil salinity indexes using electrical conductivity sensors. Sci Hortic 285:110171. https://doi.org/10.1016/j.scienta.2021.110171
30. Tarolli P, Luo J, Park E, Barcaccia G, Masin R (2024) Soil salinization in agriculture: Mitigation and adaptation strategies combining nature-based solutions and bioengineering. Iscience 27(2):108830
31. Zhang X, Feng G, Sun XR (2024) Advanced technologies of soil moisture monitoring in precision agriculture. J Agric Food Res 18:101473. https://doi.org/10.1016/j.jafr.2024.101473
32. Han D, Hosamo H, Ying C, Nie R (2023) A comprehensive review and analysis of nanosensors for structural health monitoring in bridge maintenance: innovations, challenges, and future perspectives. Appl Sci 13(20):11149. https://doi.org/10.3390/app132011149
33. Musa P, Sugeru H, Wibowo EP (2023) Wireless sensor networks for precision agriculture: A review of npk sensor implementations. Sens 24(1):51. https://doi.org/10.3390/s24010051
34. Zafra-Pérez A, Medina-García J, Boente C, Gómez-Galán JA, de la Campa AS, de la Rosa JD (2024) Designing a low-cost wireless sensor network for particulate matter monitoring: Implementation, calibration, and field-test. Atmos Pollut Res 15(9):102208. https://doi.org/10.1016/j.apr.2024.102208
35. Dong X, Vuran MC, Irmak S (2013) Autonomous precision agriculture through integration of wireless underground sensor networks with center pivot irrigation systems. Ad Hoc Netw 11(7):1975–1987. https://doi.org/10.1016/j.adhoc.2012.06.012
36. Kulshreshtha NM, Shrivastava D, Bisen PS (2017) Nanotechnology-based contaminant sensors. In: Nanobiosensors. Academic Press. https://doi.org/10.1016/B978-0-12-804301-1.00014-X
37. Williams A, Aguilar MR, Pattiya Arachchillage KG, Chandra S, Rangan S, Ghosal Gupta S, Artes Vivancos JM (2024) Biosensors for public health and environmental monitoring: the case for sustainable biosens-ing. ACS Sustain Chem Eng 12(28):10296–312. https://doi.org/10.1021/acssuschemeng.3c06112
38. Rejeb A, Rejeb K, Abdollahi A, Al-Turjman F, Treiblmaier H (2022) The Interplay between the Internet of Things and agriculture: A bibliometric analysis and research agenda. Internet Things 19:100580. https://doi.org/10.1016/j.iot.2022.100580

39. Jawad HM, Nordin R, Gharghan SK, Jawad AM, Ismail M (2017) Energy-efficient wireless sensor networks for precision agriculture: A review. Sens 17(8):1781. https://doi.org/10.3390/s17081781
40. Brenes JA, Marín-Raventós G (2022) When one wireless technology is not enough: A network architecture for precision agriculture using LoRa, Wi-Fi, and LTE. In: Intelligent sustainable systems: Selected papers of WorldS4 2021, vol. 1. Springer Singapore. https://doi.org/10.1007/978-981-16-6309-3_11
41. Liopa-Tsakalidi A, Thomopoulos V, Barouchas P, Boursianis AD, Goudos SK (2024) A LoRaWAN-based IoT platform for smart irrigation in olive groves. Smart Agric Technol 9:100673. https://doi.org/10.1016/j.atech.2024.100673
42. Mwakwata CB, Malik H, Mahtab Alam M, Le Moullec Y, Parand S, Mumtaz S (2019) Narrowband Internet of Things (NB-IoT): From physical (PHY) and media access control (MAC) layers perspectives. Sens 19(11):2613. https://doi.org/10.3390/s19112613
43. Zohourian A, Dadkhah S, Neto EC, Mahdikhani H, Danso PK, Molyneaux H, Ghorbani AA (2023) IoT Zigbee device security: A comprehensive review. Int Things 22:100791. https://doi.org/10.1016/j.iot.2023.100791
44. Nair KP (2019) Soil fertility and nutrient management. In: Intelligent soil management for sustainable agriculture. Springer, Cham. https://doi.org/10.1007/978-3-030-15530-8_17
45. Hassan IU, Panduru K, Walsh J (2024) Review of data processing methods used in predictive maintenance for next generation heavy machinery. Data 9(5):69. https://doi.org/10.3390/data9050069
46. Reza MN, Lee KH, Karim MR, Haque MA, Bicamumakuba E, Dey PK, Jang YY, Chung SO (2025) Trends of soil and solution nutrient sensing for open field and hydroponic cultivation in facilitated smart agriculture. Sens 25(2):453. https://doi.org/10.3390/s25020453
47. Tonle FB, Niassy S, Ndadji MM, Tchendji MT, Nzeukou A, Mudereri BT, Senagi K, Tonnang HE (2024) A road map for developing novel decision support system (DSS) for disseminating integrated pest management (IPM) technologies. Comput Electron Agric 217:108526. https://doi.org/10.1016/j.compag.2023.108526
48. Karothia R, Chattopadhyay MK (2024) Internet of thing (IoT) enabled smart sensor node (SSN) to measure the soil and environmental parameters for smart farming. CSIT 12:119–135. https://doi.org/10.1007/s40012-024-00402-8
49. Choudhary V, Guha P, Pau G, Mishra S (2025) An overview of smart agriculture using internet of things (IoT) and web services. Environ Sustain Indic 23:100607. https://doi.org/10.1016/j.indic.2025.100607
50. Sangeetha C, Moond V, Rajesh GM, Damor JS, Pandey SK, Kumar P, Singh B (2024) Remote sensing and geographic information systems for precision agriculture: A review. Int J Environ Clim Change 14(2):287–309
51. Sishodia RP, Ray RL, Singh SK (2020) Applications of remote sensing in precision agriculture: A review. Remote Sens 12(19):3136. https://doi.org/10.3390/rs12193136
52. Ghorai S, Mochary S (2025) Use of agricultural drone: Solving the farmers problems. Agric Mag 4(5):417–421
53. El Mane A, Tatane K, Chihab Y Transforming agricultural supply chains: Leveraging blockchain-enabled java smart contracts and IoT integration. ICT Exp 10(3):650–672. https://doi.org/10.1016/j.icte.2024.03.007
54. Chen HY, Sharma K, Sharma C, Sharma S (2023) Integrating explainable artificial intelligence and blockchain to smart agriculture: Research prospects for decision making and improved security. Smart Agric Technol 6:100350. https://doi.org/10.1016/j.atech.2023.100350
55. Pakseresht A, Yavari A, Kaliji SA, Hakelius K (2023) The intersection of blockchain technology and circular economy in the agri-food sector. Sustain Prod Consum 35:260–274. https://doi.org/10.1016/j.spc.2022.11.002
56. Zerihun BM, Olwal TO, Hassen MR (2022) Design and analysis of IoT-based modern agriculture monitoring system for real-time data collection. In: Uddin MS, Bansal JC (eds) Computer vision and machine learning in agriculture, Volume 2. Algorithms for intelligent systems. Springer, Singapore. https://doi.org/10.1007/978-981-16-9991-7_5

57. Duguma AL, Bai X (2025) How the internet of things technology improves agricultural efficiency. Artif Intell Revi 58:63. https://doi.org/10.1007/s10462-024-11046-0
58. Vatin NI, Joshi SK, Acharya P, Sharma R, Rajasekhar N (2024) Precision agriculture and sustainable yields: insights from IoT-driven farming and the precision agriculture test. In: BIO web of conferences, vol 86, p 01091. EDP Sciences. https://doi.org/10.1051/bioconf/202486 01091
59. Prasad R, Tiwari R, Srivastava AK (2023) Internet of Things-based fuzzy logic controller for smart soil health monitoring: A case study of semi-arid regions of India. Eng Proc 58:85. https://doi.org/10.3390/ecsa-10-16208
60. Ashwitha A, Latha CA (2022) Crop recommendation and yield estimation using machine learning. J Mobile Multimed 18(3):861–883. https://doi.org/10.13052/jmm1550-4646.18320
61. Naidu KJ, Babu KV, Sai CR, Ganesh P, Sai TG, Naidu NS (2024) Precision agriculture monitoring system. Biosci Biotechnol Res Asia 21(4):1543–1551. https://doi.org/10.13005/bbra/ 3324
62. Padhiary M, Saha D, Kumar R, Sethi LN, Kumar A (2024) Enhancing precision agriculture: A comprehensive review of machine learning and AI vision applications in all-terrain vehicle for farm automation. Smart Agric Technol 3:100483. https://doi.org/10.1016/j.atech.2024.100483
63. Popescu SM, Mansoor S, Wani OA, Kumar SS, Sharma V, Sharma A, Arya VM, Kirkham MB, Hou D, Bolan N, Chung YS (2024) Artificial intelligence and IoT driven technologies for environmental pollution monitoring and management. Front Environ Sci 12:1336088. https:// doi.org/10.3389/fenvs.2024.1336088
64. Koshariya AK, Rameshkumar PM, Balaji P, Cavaliere LP, Dornadula VH, Singh B (2024) Data-driven insights for agricultural management: leveraging industry 4.0 technologies for improved crop yields and resource optimization. In: Sindhwani N, Anand R, George A, Pandey D (eds) Robotics and automation in industry 4.0. CRC Press, Boca Raton. https://doi.org/10.1201/978 1003317456
65. Antony AP, Leith K, Jolley C, Lu J, Sweeney DJ (2020) A review of practice and implementation of the internet of things (IoT) for smallholder agriculture. Sustain 12(9):3750. https://doi.org/ 10.3390/su12093750
66. Santoso A, Surya Y (2024) Maximizing decision efficiency with edge-based AI systems: Advanced strategies for real-time processing, scalability, and autonomous intelligence in distributed environments. Q J Emerg Technol Innov 9(2):104–132
67. Chamara N, Islam MD, Bai GF, Shi Y, Ge Y (2022) Ag-IoT for crop and environment monitoring: Past, present, and future. Agric Syst 203:103497. https://doi.org/10.1016/j.agsy.2022. 103497
68. Demestichas K, Peppes N, Alexakis T (2020) Survey on security threats in agricultural IoT and smart farming. Sens 20(22):6458. https://doi.org/10.3390/s20226458
69. Gupta S, Kuntoji A, Singh S (2024) Data security and privacy in IoT-enabled agriculture. In: Agriculture 4.0, pp 289–318. CRC Press

Chapter 14
IoT-Enabled Soil Quality Monitoring in Smart Agricultural Systems

Hariharan Harikrishnan and Venkittaraman Aishwarya

Abstract The use of IoT in agriculture has transformed soil quality determination and environmental monitoring so that real-time data will give rise to sustainable criteria for farming practices. This chapter elucidates recent trends in IoT-based soil monitoring systems for assessing soil parameters, including moisture, pH, temperature, nutrients, and microbial activity. The continuous monitoring, predictive analytics, and precision farming enabled by IoT are only possible through smart sensors, wireless communication networks, and cloud computing, resulting in more efficient resource use and increased crop yields. Moreover, this chapter examines the sensor technologies, data transmission protocols, and AI-powered decision-making models employed in IoT-based soil monitoring systems. These have been further enhanced by incorporating big data analytics and machine learning for predictive modelling to detect soil degradation, nutrient deficiency, and environmental stressors early. Challenges such as sensor calibration, data security, and interoperability have also been addressed. Finally, the chapter offers insights into edge computing, blockchain for data integrity, and AI-driven automation for agricultural monitoring as future trends. Thus, by elaborating on the IoT-based soil quality assessment discussed in this chapter, one can emphasize its prospects for supporting sustainable agricultural practices, enhancing food security, and reducing negative environmental impacts.

Keywords IoT in agriculture · Soil quality monitoring · Smart sensors · Precision farming · Environmental monitoring · Sustainable agriculture

H. Harikrishnan (✉)
Department of Pharmaceutical Technology, Paavai Engineering College, Namakkal, India
e-mail: hkvaibhav2@gmail.com

V. Aishwarya
Department of Electrical and Electronics Engineering, Paavai Engineering College, Namakkal, India
e-mail: aishwaryavenkatpec@paavai.edu.in

S. Gautam et al. (eds.), *Smart Environmental Monitoring*,
https://doi.org/10.1007/978-981-92-0294-2_14

14.1 Introduction

The convergence of agriculture and technology has undergone revolutionary change in recent years, particularly with the advent of the Internet of Things (IoT). Dynamic Internet-enabled applications are supplanting the traditional farming that mostly relies on empirical observations and experience, ultimately paving the path to a data-driven precision agriculture with real-time environmental feedback and predictive analytics for decision-making [1]. One of the significant enablers of such a transformation is continuous, dynamic, and scalable soil-quality monitoring and management enabled by IoT-based systems. fAs a vital natural resource, soil directly influences crop productivity, environmental sustainability, and food security. Soil health is determined by the interrelationships of several physical, chemical, and biological factors such as moisture, temperature, pH, nutrient content, and microbial activity. Monitoring these over space and time becomes as necessary for maximizing agricultural yield as it is essential for addressing issues associated with land degradation, nutrient depletion, and climate change. By transforming into an agricultural paradigm, IoT enables real-time collection, transmission, and processing of data on soil conditions. In addition to smart sensors, wireless communication networks, and cloud computing infrastructures, this technological advancement is intended to deliver actionable intelligence to farmers, agronomists, and policymakers [2]. These technologies, combined with AI and big data analytics, now create IoT systems that learn from historical patterns, predict soil conditions that could be detrimental, and enable automated intervention mechanisms, thereby improving and sustaining the overall efficiency and sustainability of farm operations.

This chapter will examine current and emerging developments in IoT-based soil quality monitoring systems. It will also present the principal components of these systems, including sensor technologies, data transmission protocols, and the architecture of IoT networks specifically designed for agricultural environments. The chapter will describe the use of machine learning and artificial intelligence to interpret data collected by these systems, enabling predictive modelling and supporting precision farming applications. The significance of such systems is becoming apparent amid the compounding challenges confronting global agriculture, including increased food demand, declining arable land, and potential threats from climate-induced soil stress. Accordingly, real-time environmental monitoring becomes indispensable. IoT is not only used to boost productivity but is also central to establishing sustainable agricultural ecosystems. This chapter also examines the applications and challenges of IoT deployment for soil monitoring, including sensor calibration, data integrity, and interoperability. Aside from these issues, the chapter will address considerations for incorporating edge computing and blockchain technologies, which promise increasing data privacy, processing efficiency, and trust in distributed agricultural systems. The chapter, while summarizing contemporary trends, current components, analytical abilities, and innovative developments of IoT for soil quality monitoring, would emphasize how it is a transformative force for the agriculture of tomorrow. It is therefore intended for researchers, engineers, environmental scientists, and stakeholders in smart agriculture who seek to explore the technical and applied aspects of this interdisciplinary field.

14.2 Importance of Soil Quality Monitoring in Sustainable Agriculture

Soil is the basis of agriculture, affecting plant growth, conservation of environmental balance, and food security. As the world population increases and the temperature rises, sustainable management of all soil resource sustainable management becomes vital. Soil is a rich, multisystem ecosystem with the capability to support plant life by way of physical characteristics, chemical and biological features. These features include moisture, pH, temperature, nutrients, and microbial activities, which bear direct consequences on crop health and agricultural yield. Real-time monitoring of these parameters provides better insight into soil dynamics for farmers and researchers to make sound decisions on improving productivity while preserving environmental quality[3]. Soil quality defines the ability of the soil to perform key functions: for example, grow plants, regulate water and nutrient cycles, and support microbial diversity. For instance, moisture can influence germination and microbial life, while temperature affects the enzymatic reactions and seed development. pH regulates nutrient solubility and populations of microorganisms, while primary levels of nutrients, nitrogen, phosphorus, and potassium are necessary for plant development[4]. Below-ground active and, many times, ignored microbial activity is a great deal responsible for organic matter decomposition along with nutrient availability. Understanding all these aspects is crucial for implementation in sustainable farming strategies to get maximum production without jeopardizing the long-term health of the soil. Conventionally, soil monitoring was limited to manual sampling and laboratory analysis. These methods, if used under controlled conditions, can provide accurate results; however, they are fundamentally restricted by the spatial and temporal resolution that they can offer. They provide a static snapshot of the present state with very little insight into how soil systems dynamically change. These IoT monitoring solutions implement smart sensors that can be very well distributed over the farming fields to ensure continual collection and transmission of data regarding the soil. This real-time information gives a detailed account of the conditions of the soil across time and space. Thus, farmers can now manage their interventions in irrigation, fertilization, and crop selection with a level of precision that could not have been attained with traditional means[5].

Sustainable agriculture is defined by the requirement to meet present food demands without compromising the ability of future generations to do the same, and benefits greatly from continuous monitoring of the soil quality. This allows for the efficient use of fertilizer and pesticides. By early symptom identification of nutrient deficiencies, farmers can apply specific amounts of the relevant type of inputs, therefore avoiding wastage and reducing the risk of chemical runoff into adjacent water bodies. On the other hand, real-time data on soil moisture helps develop irrigation schedules that conserve water and avert over-irrigation that destroys soil structure and leaches nutrients. The early detection of soil degradation is another decisive factor. Parameters like declining microbial activity, increasing soil salinity, or lowering nutrient levels can act as early warning indications of soil health degradation. An early detection of such changes enables timely interventions and restoration measures to be taken, including the addition of organic matter, crop rotation, or sowing cover crops to avoid irreversible damage. This contrasts to a large degree with the more traditional reactive approach, which deals mostly with symptoms rather than causes[6]. A basic understanding of the present-day systems would be about the monitoring of such important soil parameters measured by IoT-enabled sensors. The major soil parameters include soil moisture, pH, temperature, and key nutrient concentrations such as nitrogen, phosphorus, and potassium. For instance, moisture levels in soil are crop-dependent and are generally about 20–60% under optimal conditions. This is monitored using capacitive or time-domain reflection (TDR) sensors. Most crops have recommended soil pH values ranging between 6.0 and 7.5. Soil pH can be measured by ion-selective electrodes. The use of temperature sensors such as thermocouples or Resistance Temperature Detectors (RTDs) makes it possible to ensure that the soil is warm from 15 to 25 degrees Celsius, typical of seed germination. Nitrogen and potassium are determined by electrochemical or ion-selective methods. These parameters have developed greatly, but are becoming more common in monitoring microbial activity. This is usually termed as soil respiration and general biological activity measured with CO_2 flux sensors or biosensors. The details are presented in Table 14.1.

Table 14.1 Key soil quality parameters and associated sensor types

Parameter	Optimal Range	Sensor Type	Impact on Crops
Soil moisture	20–60%	Capacitive, TDR	Affects germination and nutrient uptake
Soil pH	6.0–7.5	Ion-selective electrodes	Regulates nutrient solubility and microbial life
Temperature	15–25 °C	Thermocouples, RTDs	Influences enzyme activity and seed germination
Nitrogen (N)	20–50 mg/kg	Nitrate sensors, ion-selective electrodes	Promotes vegetative growth
Phosphorus (P)	10–20 mg/kg	Colorimetric, electrochemical sensors	Supports root development and flowering
Potassium (K)	80–200 mg/kg	Flame photometers, ion-selective sensors	Enhances water regulation and disease resistance
Microbial activity	Variable	CO_2 flux sensors, biosensors	Indicates organic matter decomposition and fertility

The potential of such monitoring abilities extends beyond agricultural productivity to economic and environmental benefits. Precise application of resources reduces waste, improves the efficiency of inputs, and reduces operational costs. In addition, reduced pollution from over-fertilization improves water quality and the health of ecosystems. Such cost savings have a positive effect on the resilience of livelihoods in places where agriculture is the mainstay of the economy[7]. On a grander level, the integration of soil monitoring data into centralized platforms facilitates regional agricultural planning and policymaking. Government agencies and NGOs can use aggregated data to assess trends in land health, determine risks of desertification, and allocate resources efficiently. Thus, IoT-based soil monitoring becomes not just an instrument of individual farm management but rather a strategic resource in the global quest for food security and environmental sustainability. This thorough understanding of soil quality and its monitoring through IoT sets the stage for the next discussion: how IoT architectures are planned to support such capabilities in an agricultural environment. The next section will explore the technical backbone of such systems and the communication protocols enabling real-time soil data flow.

14.3 IoT Architecture for Soil Monitoring

In the present era, thriving agricultural monitoring systems are heavily reliant on solid Internet of Things (IoT) architecture harnessed to effectively capture data in real-time from the environment and later transmit and finally analyze it. The structure thus conceives multiple layers encompassing even those of physical sensors in a farm setup through cloud-engine analytics, which makes it complement a seamless, intelligent feedback loop between agronomic interventions and soil. The ultimate purpose of the architecture is to ensure continuous monitoring and automated decision-making to improve the efficiency, sustainability, and yield of a crop. Sensor nodes are the backbone of any IoT architecture in soil monitoring, placed in different locations on the soil[8]. These sensors measure critical parameters of the soil like moisture content, pH, temperature, and nutrient concentrations. The structure of most sensor nodes consists of a sensing unit, an independent power supply (generally batteries that can be recharged or small solar panels), and a microcontroller that performs initial signal processing of the sensed data. Once data is captured, it has to be sent to the center or main hub for aggregation and further analysis. Communication to these wireless modules may be done through Zigbee, LoRaWAN, NB-IoT, or Wi-Fi interfacing systems, depending on the range required, power efficiency, and data bandwidth requirements. The communication layer drives data transfer, and hence its choice is critical for system reliability, especially under rural or large-field deployments, where infrastructure is scarce. For short ranges, Zigbee would suit localized, dense networks because of its low energy consumption and support for mesh networks. When it comes to long-range transmission, LoRaWAN and NB-IoT are appropriate, maximizing coverage in areas where cellular connectivity is poor. These will route data packet transmissions to a gateway or edge computing unit, acting as the intermediary between the field sensors and the cloud. At the edge layer, preprocessing may be performed for data to reduce some noise, compress some redundant signals, or issue local alerts once thresholds are crossed, for example, soil moisture falls below a threshold value. Edge processing cuts down on the latency and bandwidth used, allowing the localized fast responses with no complete dependence on the cloud. The next step would be transferring the data to cloud servers for storage and analytics. Cloud platforms are scalable computing resources that assist in the visualization, analytics, and long-term archival of soil data[9]. They also provide remote access, ensuring transparency and accessibility across user roles for farmers, agronomists, and policymakers. The last layer is the application layer, where the processed data is displayed on intuitive dashboards or mobile applications. These user interfaces allow the user to monitor real-time soil conditions, receive automated alerts, and view past trends. The systems are usually further enhanced with APIs to support integration with farm management or government databases. Thus, this modularity allows scaling and evolving of the architecture, including advanced functionalities such as AI-based decision engines and blockchain-based data integrity layers in future versions[10]. For visualization, consider the entire process in a typical soil monitoring IoT system, for instance, in Fig. 14.1. With this setup, soil-embedded sensors collect environmental data sensed through wireless modules to reach a local gateway, which forwards the data to cloud servers. This is done where analytical engines run data and display it through user interfaces. There is also an edge computing option where real-time processing and automatic actions like valve opening or SMS alerts to farmers are initiated. It is all interoperable and scalable in that one can add other sensors, analytics modules, or control systems into the architecture without too much reconfiguration. Figure 14.1 illustrates a multi-layer architecture that has soil-embedded sensors at its base, wirelessly sending data, via LoRaWAN or Zigbee, to a near-gateway where data entry into the cloud platform takes place for data processing and storage, while edge units carry out real-time decision-making at the same place. Results are visualized in dashboards and mobile apps accessed by users. This architecture can sustain smallholder farming as well as large commercial agriculture.

On smallholder farms, simple sensor-to-mobile systems can notify farmers about the need for water or nutrients, thus reducing manual drudgery and restricting wastage. On the other hand, large corporate farms are dependent on multi-faceted IoT systems that can control hundreds of sensors across different plots, pulling in data from satellites and drones. This means injections of macro-level inputs like regional assessments of soil health or predictions for productivity over an entire season that inform farm practices, agricultural policy and investment planning at a much larger scale.

In brief, the architecture of the IoT soil monitoring system is smart yet flexible enough to convert raw environmental signals into useful knowledge. It combines the physical infrastructure, digital communication, and intelligent processing in a unified system meant for meeting the demands of modern agriculture. This architecture provides a springboard for the more sophisticated analytical capabilities discussed

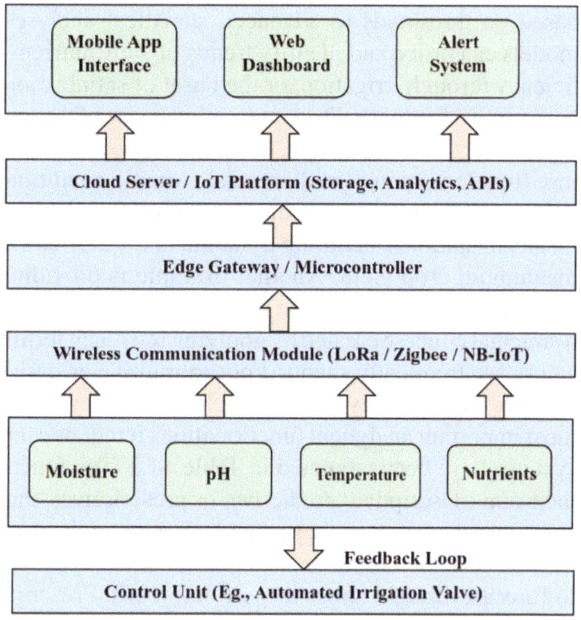

Fig. 14.1 General IoT architecture for agricultural monitoring

in the following section, wherein farming application strategies are revealed as real-time data transforms into strategic agricultural insights through analytical modelling and predictive systems.

14.4 Analytical Capabilities of IoT in Agriculture

Modern IoT systems provide analytical abilities that transform agricultural monitoring from mere data gathering to intelligent interpretation and decision-making. Transforming raw data into actionable insight is the basis for developing adaptive and resilient agricultural systems. Sensing technology, real-time communication, and advanced analytics work together such that IoT platforms offer a complete understanding of soil dynamics, crop requirements, and environmental interactions[11]. This analytical platform thus becomes the backbone of precision agriculture, assisting in the decisions from micro-level interventions to macro-level policies. A data processing pipeline forms the heart of these capabilities and involves a four-stage process: data acquisition, data cleaning and transformation, pattern extraction, and decision support. Once gathered from soil sensors, the data are central or edge processed, where initial cleaning is performed to eliminate noise and outliers. Algorithms perform corrections for sensor drift, normalize measurements across various devices, and synchronize temporal data streams to ensure coherence. Cleansed data are subsequently fed into models for analysis, varying from simple alert scenarios based on thresholds to advanced statistical analyses and machine learning. Such models can probe and identify trends of slow nutrient depletion, water loss and inefficiency through irrigation, or the onset of salinization, serving as early warning and prescriptive feedback. Data integration is one of the techniques that make this analytic framework functional. Here, it serves to fuse different data sources, such as soil moisture from temperature and external climatic conditions, to pH, to enrich the insights gained. For instance, a fusion of soil moisture data with real-time weather forecasts makes irrigation scheduling dynamic and water-conserving, eventually producing a meaningful crop yield. Another example is providing pH and nutrient data, which suggests how much and what kind of fertilizers are appropriate for addressing an environmental concern created by applying too much fertilizer[12]. These composite indicators are frequently mapped out in multilayer dashboards, where spatio-temporal patterns can be easily understood by the farm manager or agronomist. Some of the most important analytical functionalities integrated into common IoT soil monitoring systems have been tabulated in Table 14.2. The functions are classified according to their aim (descriptive, predictive, or prescriptive), the type of analysis used, and the agricultural advantage they provide. For example, descriptive analytics are snapshots of current field conditions, and predictive analytics applies historic and current data to forecast drought stress or pest outbreaks. Prescriptive analytics, however, would go one step further, recommending certain actions based on optimization algorithms, which could include recommendations for irrigation levels or crop rotations.

The incorporation of Machine Learning (ML) and Artificial Intelligence (AI) models into IoT ecosystems has sped up the advancements of predictive and prescriptive analytics [13]. A supervised learning algorithm, such as the Random Forest or Support Vector Machine (SVM), is used for modelling the soil-crop relationship from labelled datasets [14]. These models can forecast trends in soil fertility or crop yield based on rainfall, temperature, and soil parameters. Management zones are defined for fields using an unsupervised model such as k-means clustering, where each zone has distinct irrigation or fertilization requirements. Reinforcement learning is also coming of age in closed-loop systems, whereby algorithms learn the best interventions over time based on feedback from the system. Outside farm-level management, IoT analytics aid in large-scale agricultural planning and policy formation. Governments and NGOs utilize the pooled soil data from IoT networks to detect regions with declining soil health, to assess the effectiveness of agricultural subsidies, or to

Table 14.2 Analytical functions enabled by IoT-based soil monitoring systems

Analytical objective	Type of analysis	Agricultural application
Descriptive	Time-series and spatial mapping	Identifying dry zones, monitoring nutrient trends
Predictive	Regression, machine learning	Forecasting drought, predicting yield
Prescriptive	Optimization, decision support	Recommending irrigation/fertilizer schedules

monitor compliance with sustainability regulations. Such big databases help formulate regional soil quality indices and provide an indication of the prioritization of resources within climate resilience programs [15]. This means that IoT in agriculture and other stages are not only geared toward achieving economic and operational goals, but also environmental management and policy objectives in this broader context.

A very promising area of development is integrating remote sensing with ground-based IoT input. Satellite and drone imagery offer macro-level perspectives on vegetation cover, evapotranspiration, and land use, while IoT sensors provide high-resolution ground truth data [16]. When analyzed together, these two sources can enhance the accuracy of spatial models while reducing uncertainties connected to indirect remote sensing estimates. This synergistic approach is increasingly utilized in digital twins- alias virtual replicas of the agricultural ecosystems, where scenarios are simulated, and decisions are tested before implementation on the field. Real-time anomaly detection and event alerting represent yet another frontier in the analytical capabilities. The edge analytics units can watch the incoming data for deviations from established patterns, with a sudden drop in soil moisture or a pH spike indicating possible contamination. The deviations cause immediate alerts to be sent to users via SMS, mobile apps, or control systems that could automatically trigger corrective actions. This level of responsiveness can reduce damage and improve the system's resilience when faced with unexpected stressors such as an irrigation failure or chemical spillage. The scalability of IoT analytics is another benefit which allows systems to develop from single-plot monitoring instruments into company-level decision-support platforms. Modular analytics engines are added and/or updated as needed by users, and therefore allow the plug-and-play style of enhancement to the system [17]. With increasingly affordable and accessible computational power, even resource-constrained farms can benefit from cloud-based analytics on a subscription basis, thereby democratizing access to advanced decision-making tools.

In brief, analytics of IoT systems form the very backbone of change in digital agriculture. From very simple data visualization to AI recommendations, these tools allow for knowledgeable, timely, and contextual decisions that serve as the dual building blocks of productivity and sustainability. In the next few paragraphs, sensor technologies and calibration techniques that assure the quality of data on which the analytics operate are considered.

14.5 Sensor Technologies and Calibration

The efficiency of any IoT-based agricultural monitoring system relies fundamentally on quality-reliable sensor technologies put in the field. These sensors are the first contact points with the physical environment and convert soil dynamics into structured digital signals for downstream processing. Selecting the most suitable sensor and calibrating it accurately are thus crucial for data integrity and the validity of later analytics. Sensor performance eventually affects how decisions are made and is a critical component of the overall house's trustworthiness [18]. Thus, a wide variety of soil parameters should be monitored for understanding agroecological conditions more comprehensively. Some of the most important parameters often monitored are soil moisture, temperature, pH, Electrical Conductivity (EC), and concentration of macronutrients such as NPK [19]. Each parameter would require sensor technology appropriate to its chemistry or physical property. For example, soil moisture would generally be measured using either capacitive or Time-Domain Reflectometry (TDR) sensors. Such capacitive sensors work on dielectric properties to estimate moisture and are very cost-sensitive, but may vary with the soil type. On the other hand, TDR sensors are highly accurate by electromagnetic pulse analysis and are known to be more robust to environmental variations, though costly. Thermistors or digital temperature probes are the common temperature-measuring sensors considered necessary to assess the active microbial community and nutrient availability. They are simple to install and calibrate; soil contact should be ensured to avoid surface distortion of temperature readings. Soil pH electrodes based on ion-selective electrodes are very drift-prone and require constant recalibration, especially with high-salinity or low-buffered soils. Similar to this are the EC sensors, which are used to indicate salinity and total dissolved ions; temperature compensation and soil texture hold strong influences for accuracy in measurements. Nutrient sensors, being an emerging field, often deploy optical, electrochemical, or ion-selective mechanisms. These suffer from fouling and cross-interference, and underscore the significance of physical maintenance & digital calibration protocols [20]. The sensors employed for soil quality monitoring are summarized in Table 14.3, along with their advantages, disadvantages, and typical calibration standards between sensor types. Such comparative evidence helps understand the trade-offs between costs, feasible accuracy, robustness, and maintenance frequency in a given application.

Table 14.3 Summary of soil sensor types, characteristics, and calibration needs

Parameter	Sensor type	Advantages	Limitations	Calibration requirement
Moisture	Capacitive	Inexpensive	Soil-specific error	Moderate per soil type
	TDR	High accuracy	Cost	Low
Temperature	Thermistor, digital probe	Easy deployment, stable output	Surface exposure issues	Periodic field checks only
pH	Ion-selective electrode	Direct measurement	Drift, contamination	High, frequent reference checks
Electrical conductivity	Conductivity probe	Salinity proxy, quick response	Affected by soil temperature and moisture	Moderate, temperature-compensated calibration
Nutrients (NPK)	Optical/ Electrochemical	In-field monitoring possible	Sensor fouling, cross-sensitivity	High, often multi-point calibration

Calibration is a continuous requirement due to sensor ageing and exposure to environmental conditions. Factory calibration will give a starting point, but further calibration with thick soils becomes necessary to account for soil-specific properties such as texture, organic matter, and ion exchange capacity. Systematic multi-point calibration against standard reference solution samples or soil samples of known properties is normally done. Gravimetric methods provide the most accurate calibration ground truth for moisture sensors [21]. For pH and EC sensors, certified standard solutions are used to develop calibration curves over the expected operational range. An emerging approach includes dynamic calibration or auto-calibration techniques, wherein outputs from sensors are continuously calibrated as a function of real-time environmental inputs or internal reference readings. These smart calibration methods, sometimes embedded in the modern IoT devices' firmware, require little or no manual intervention while enhancing reliability on a longer-term basis [22]. Local calibration routines may even be supported through edge computing platforms, whereby incoming data is reviewed against historical norms and anomalies are flagged for user review. Placement strategy is another critical factor in sensor performance. An improper depth or spatial distribution of the sensor will give biased data. For example, placing moisture sensors too shallow may cause the data to misrepresent the condition as drier due to the rapid evaporation at the surface. Likewise, heterogeneous soils may require more than one sensor per plot to ensure that conditions are met. Best practice involves deployment of the sensor at the root-zone depth and averaging readings from multiple points, reducing spatial noise. Sensor longevity and maintenance have a significant impact on data fidelity. Sensors employed in environments characterized by extreme moisture or high salinity undergo corrosion and biofouling. Hence, regular inspections, cleaning, and recalibration form the routine parts of the operational workflow [23]. Manufacturers have recommended maintenance intervals, but they need to be adjusted based on field conditions and frequency of usage. Current advances in low-power electronics and materials science have led to the development of more durable and energy-efficient sensors, hence being able to last longer with fewer demanding maintenance requirements [24]. However, all these advances must be adjusted to balance with the need to handle with greater sensitivity and precision. Hybrid sensor networks combining high-precision and low-cost units usually provide an optimal trade-off whereby the more accurate sensors are used to periodically calibrate the cheaper ones through intra-network adjustments [25].

Finally, the success of any IoT-based agricultural monitoring system hinges on reliable sensor technology and appropriate calibration. Choice of sensor, strategic deployment, and frequent calibration all influence the quality of the insights obtained from the resulting pipeline. Emerging sensor technologies need to put more effort into developing self-calibrating and soil-specific as well as maintenance-aware sensors to reduce human effort and improve the accuracy of decision-making. In the next section, an exploration of these sensing technologies will be used together with their communication protocols in building a robust IoT infrastructure for field deployment.

14.6 Communication Technologies and System Integration

An IoT-based agricultural monitoring system operates seamlessly with high-quality sensors and analytics, with a reliable communication infrastructure for data forwarding through diverse field conditions. Communication technology involves energy consumption, transmission range, data throughput, and scalability. Communication solutions should be designed such that they connect and operate as efficiently as possible, considering the typical remote location and infrastructure in agricultural settings [26]. In IoT systems for soil quality and crop monitoring, a layered communication architecture is employed. At the device level, short-range protocols such as Zigbee, Bluetooth Low Energy (BLE), and Radio-Frequency Identification (RFID) were used for low-power and low-latency communication between sensor nodes and local aggregators. Low energy requirements and easy deployment attract cheap protocols, so dense sensor networks can be deployed in small to medium plots [27]. However, the short range limits their suitability in the case of a large or distributed farm unless mesh networking options are added. Longer-range communications are served by various LPWANs such as LoRaWAN, NB-IoT, and Sigfox, which nowadays receive a lot of attention. LoRaWAN is particularly valued for farming applications because of its long coverage (up to 15 km in open fields), low power consumption, and bidirectional communications capability. NB-IoT, being operated on licensed cellular bands, guarantees better reliability with assured quality of service, but usually requires existing telecom infrastructure and is discouraged because of its recurrent charges [28]. In places where connectivity is assured, Wi-Fi, GSM/3G/4G cellular networks are used, especially for high-data applications like drone-based image transfer or video surveillance. One of the central elements of system integration involves edge computing devices or gateways serving as intermediaries between the sensors installed in the field and the cloud platform [29]. These gateways aggregate the sensor data, carry out the first level of filtering and preprocessing, and then manage the local wireless communication. More sophisticated gateways incorporate additional functions such as analytics onboard and can either respond immediately or control irrigation on-site without deferring to any remote servers. This edge-centric arrangement significantly minimizes latency, bandwidth utilization, and even increases resilience during network outages [30]. Figure 14.2 depicts the block diagram of a general IoT system architecture for agricultural monitoring. The system starts with distributed sensors all over the farmland to measure moisture, temperature, pH, etc. These sensors then forward the sensed data to a local edge gateway via short-range communication protocols like Zigbee or BLE. Meanwhile, the edge gateway accumulates and preprocesses this data before routing it over LPWAN or cellular networks to a cloud-based IoT platform. In the cloud platform, further analytics occur, historical data is stored, and visualization & control interfaces are presented through mobile applications and web-based dashboards. These analytics will then be used to inform feedback that drives actuators like irrigation pumps or nutrient distributors, so that there is a closed-loop control system that supports adaptability and autonomy. For the integration of all the

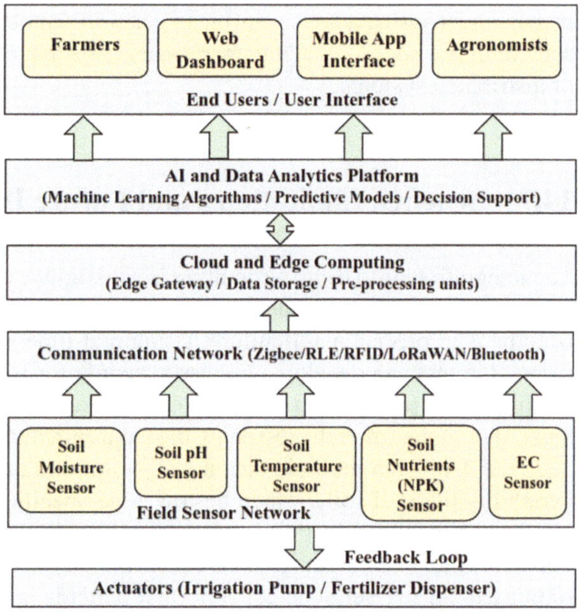

Fig. 14.2 General architecture of IoT-based agricultural monitoring system

layers—sensing, communication, analytics, and actuation- it has to follow interoperability standards. Protocols like MQTT and CoAP define a lightweight and efficient method of exchanging information across heterogeneous devices. APIs and middleware layers, by extension, help to hook up any third-party services like weather APIs or satellite images databases, further elevating contextual awareness of the whole system [31].

In essence, the whole mesosystem, from hardware to application layer IoT technologies in farming are unified by the selection and integration of communication technologies. These systems are designed as per parameters of power consumption, throughput, price, and reliability, ensuring a proper flow of data, real-time responsiveness, and sustainability over time.

With all the components working together from sensors to servers, a highly responsive, precise, adaptable, data-driven farm environment is created. The next section offers a synthesis of the technology layers to discuss system-wide benefits and future considerations. The other design consideration is power management, especially in off-grid locations. Most IoT nodes and gateways are solar-powered with low-duty cycles, waking up to transmit data and returning to sleep. The communication technology must be selected according to its range, speed, and power profile so that the devices operate with minimum maintenance. Due to their ultra-low power consumption and vast battery life, LPWANs, especially LoRaWAN, are highly suitable in this context. With the establishment of a communication network, any large-scale agricultural deployment is predicated upon its reliability and scalability. To ensure network robustness, multiple communication pathways and adaptive routing of data may be used, along with dynamic frequency management. Using cloud-based network management software, it is possible to perform remote diagnostics, update firmware, and configure nodes, reducing operational costs and allowing for centralized control of distributed systems.

14.7 Benefits, Limitations, and Future Directions

The integration of IoT into agriculture is a disruptive change in the way soil and crop health are monitored, managed, and optimized. One of the most direct measurable benefits is in precision agriculture. Using real-time sensors and data analytics platforms, farmers can develop a precise schedule for irrigation, fertilization, and crop rotation. Such precision in the usage of water and agrochemicals saves an enormous amount of money, thereby lessening the environmental impact [32]. Additionally, the gathered data from multiple data nodes within the field are used to generate spatial variability maps of soils, which bestow zone-specific management capabilities critical for enhanced productivity and soil sustainability [33]. Early detection of anomalies and timely risk management constitute another big advantage. AI algorithms running in an IoT setup detect abnormal patterns in soil moisture, pH, or temperature that might indicate an imminent threat like nutrient deficiency, drought stress, or disease onset. This early warning allows farmers to intervene early and reduce losses in yields while keeping the soil healthy [34]. Another thing is that such a cloud-connected system affords remote monitoring and control from anywhere, providing flexibility to the farmer operator, especially in geographically dispersed farming zones. Automated alerts and dashboards make it easier to manage for smallholder farmers and agribusinesses, leading to better time use and labor efficiency. Despite the several advantages, certain constraints and issues obstruct the widespread acceptance and scalability of this technology. The foremost, perhaps most serious, impediment is formed by deployment and maintenance costs. Setting up good sensors, communication infrastructure, and cloud services is a highly expensive proposition, particularly in low-resource environments or among smallholder farmers. Though prices are falling with the introduction of new technology, the initial setup cost still reeks of concern. On the other hand, network reliability remains a bottleneck in rural areas. Many agricultural zones are poorly served by cellular or LPWAN coverage; hence, this limits data transmission and real-time control system capabilities. Another principal issue is interoperability and standardization of data. The lack of common protocols for sensor data, communication interfaces, and cloud APIs leads to highly fragmented systems that cannot be easily integrated or scaled. Without such frameworks, combining data from one vendor with another or deploying systems on a cross-regional basis can be highly inefficient. Data protection, ownership, and governance lead to ethical questions: Who controls the data generated in farms? How is it used, stored, or monetized? Questions like these need to be addressed as the agricultural IoT ecosystem grows.

The future pathway for IoT-based soil quality monitoring will be shaped by the improvement of several complementary fields. Edge AI, in simple terms, is the ability to do machine learning on the devices at the edge. This technology promises to give a strong real-time decision-making capability even in situations where there is no internet or cloud connectivity. Solar-powered sensors and gateways also will boost sustainability and autonomy, with some of the richest potential benefits lying in places that are poor in energy resources. Another research avenue pursues the development of biosensors for the detection of microbial activity, soil respiration, and other dynamic indicators of soil health that are hard to measure nowadays [35, 36]. Another level of convergence will involve the use of data from satellite remote sensing, drone-based imaging, and GIS platforms, rendering the data ecosystem much richer and supportive of a more multiscale analysis of soil quality. Future systems will be self-healing and self-configuring, thereby limiting the need for technical intervention and making the technology usable by all. Importantly, participatory platforms for farmers to feed in data and access aggregated soil data will ensure community-driven innovation and democratization of agricultural intelligence. In conclusion, the path leading to full-scale adoption of IoT in agriculture, although fraught with technical, financial, and regulatory roadblocks, looks all the more promising. The synergy of edge computing, AI, sustainable power solutions, and robust communication will necessarily carve the way for a resilient, highly productive, and environmentally friendly agricultural paradigm.

14.8 Conclusion

IoT technology convergence with current agricultural techniques represents an important phase in the evolution of soil quality assessment and sustainable farm management. As this chapter suggests, with the help of IoT systems, a broad framework can be generated for real-time decision-making based on data at multiple levels, thereby increasing accuracy and efficiency in environmentally conscious agriculture. The interactors from layered hardware and software that encompass sensors at a ground level to analytics platforms in the cloud train farmers and agricultural stakeholders to gain visual insights into the dynamic state of soil health and environmental conditions. The benefits of these systems also tend to be empirical, quantitative, and multidimensional. Agricultural engineers regard Internet of Things technology as a mechanism that boosts crop productivity while reducing input costs and degradation of the environment. Sensors predict irrigation, fertilization scheduling, soil nutrient deficiency, and environmental stress, where IoT technology has a decisive role in the realization of such tasks. The integration of AI and ML allows these systems to progress from just reactive tools to predictive and adaptive platforms. Such capabilities build resilience in agriculture, particularly in the face of climate variability, decreasing soil fertility, and increasing worldwide food demand. Still, more challenges must be overcome to fully transform soil quality monitoring through IoT. Key infrastructural, economic, and regulatory barriers continue to exist. These include sensor costs, whether terrain can limit network coverage in rural areas, the absence of interoperability among platforms, and concerns about data security and ownership. The above barriers can only be diverted through engagement among multiple stakeholders comprising technologists, policymakers, agronomists, and the farmers themselves. Public–private partnerships, subsidies for IoT adoption, rural digital infrastructure investments, and open-source standardization initiatives may all serve as catalysts toward scalable implementation. So, simultaneous efforts must be made to develop solutions specific to a given context. Designing and deploying IoT systems should depend on the different agroecological conditions, socio-economic constraints, and cultural practices of a region. Practices for high-tech greenhouses in developed countries are not directly translatable to rain-fed subsistence farming in the Global South. Hence, co-creation and participatory approaches with local communities will be necessary for relevance and sustainability. Converging technologies hold the future of soil monitoring. Advances in biosensors for microbial activity and enzyme functions, and portable spectrometers for soil chemistry analysis in situ, will enlarge the parameter space. Edge computing and solar-powered IoT nodes will allow detachment of system performance from connectivity and energy constraints, fostering strong operations in off-grid environments. Integration with geospatial platforms, including satellite remote sensing, drone imagery, and geographic information systems, will enable multi-layered, multi-scale decision-making, facilitating intra-farm optimization and landscape-level planning. Education and capacity building will be another major initiative. Farmers and field operators need training on using these IoT tools and interpreting data for informed decision-making. Agricultural extension services need redesigning to incorporate digital literacy and support of smart agriculture platforms. As these technologies become increasingly democratized and affordable, they provide an opportunity to reposition agriculture as a knowledge-intensive, data-based up-front activity and away from a hitherto labor-intensive occupation. To sum up, soil quality monitoring by IoT ensures resilient agricultural systems with high productivity and sustainability. Providing very granular and real-time insight into soil conditions, these technologies allow for precise and informed decisions to guide agricultural production towards being an ecological stewardship. While the challenges still exist, with the rapid rate of innovation and growing consciousness of the need for sustainable food systems, the role of IoT in agriculture will shift from beyond novelty and enter the realm of necessity. With global populations increasing and environmental pressure mounting, these systems will stand as the bedrock ensuring food security, improving livelihoods, and conserving for the future the long-term health of the most precious resource: soil.

References

1. Mourad N, Qahtan S, Zaidan BB, A.Ibrahim H, Zaidan AA (2025) Bi-level hierarchical ensemble intelligent approach for evaluating spatio-temporal semantic data management systems in IoT-based agriculture 5.0. Expert Syst Appl 276:127083. https://doi.org/10.1016/j.eswa.2025.127083

2. Aishwarya V (2024) Comprehensive review of IoT-based green energy monitoring systems. In: Kumar A, Rathore PS, Dubey AK, Srivastav AL, TAK, VD (ed) Sustainable management of electronic waste. Wiley, pp 113–143
3. Shahab H, Naeem M, Iqbal M, Aqeel M, Ullah SS (2025) IoT-driven smart agricultural technology for real-time soil and crop optimization. Smart Agric Technol 10:100847. https://doi.org/10.1016/j.atech.2025.100847
4. Venkittaraman A, Harikrishnan H (2025) Soil and crop enhancing robotic device
5. Praveen R, Mittal M, Parida P, Kumar Y, Singla A, Thandra N (2025) Leveraging IoT and AI in precision agriculture for efficient water management. 2025 International Conference on Computational, Communication and Information Technology (ICCCIT). IEEE, Indore, India, pp 803–808
6. Valentina-Daniela B, Simona-Elena I, Ancuţa P-N (2025) Advanced autonomous system for monitoring soil parameters. Technologies 13:38. https://doi.org/10.3390/technologies13010038
7. Priya VV, Nisanth G, Dhanushree D, Sindhu M, Varshini T, Preteka AT (2025) IoT Based agricultural monitoring and alert system. In: 2025 4th International Conference on Sentiment Analysis and Deep Learning (ICSADL). IEEE, Bhimdatta, Nepal, pp 488–495
8. Evsyukov D, Kukartseva S, Kravtsov K, Kuznetsov G, Masich D (2025) IoT-based farm management systems for modern agronomy. In: 2025 24th International Symposium INFOTEH-JAHORINA (INFOTEH). IEEE, East Sarajevo, Bosnia and Herzegovina, pp 1–4
9. Srisawat T, Sakprom S, Kunsawat P, Praksong K, Suchat S, Muangprathub J (2025) IoT-enabled agricultural environmental monitoring: Enhancing growth and yield using natural-rubber straw and mulching experiment. Ind Crops Prod 225:120524. https://doi.org/10.1016/j.indcrop.2025.120524
10. Patil RN, Maroor JP, Dhanke JA, Chunduru MR, Singh M, Jangir P, Upadhyay S (2025) Improving smart agriculture through the use of machine learning to offer tailored crop forecasts. Remote Sens Earth Syst Sci 8:74–83. https://doi.org/10.1007/s41976-024-00176-2
11. Hussain F, Ahmed S, Muhammad Zaigham Abbas Naqvi S, Awais M, Zhang Y, Zhang H, Raghavan V, Zang Y, Zhao G, Hu J (2025) Agricultural non-point source pollution: Comprehensive analysis of sources and assessment methods. Agriculture 15:531. https://doi.org/10.3390/agriculture15050531
12. Salimath M, Kaliannan N, Prabhakar V, Iyyakutty R, Jeyabaskaran KJ (2025) IoT and sensor technologies: Increased water and nutrient savings and profit in Banana cv. Grand Nain (AAA) production. Sci Hortic (Amsterdam) 341:113982. https://doi.org/10.1016/j.scienta.2025.113982
13. Agrawal K, Kumar N (2025) AI-ML Applications in agriculture and food processing. In: Dutta PK, Hamad A, Haghi AK, Prabhakar PK (eds) Food and industry 5.0: transforming the food system for a sustainable future. Sustainable development goals. Springer Nature Switzerland, Cham, pp 21–37
14. Kumar GK, Bangare ML, Bangare PM, Kumar CR, Raj R, Arias-Gonzáles JL, Omarov B, Mia MS (2024) Internet of things sensors and support vector machine integrated intelligent irrigation system for agriculture industry. Discov Sustain 5:6. https://doi.org/10.1007/s43621-024-00179-5
15. Sankarasubramanian P (2025) Enhancing precision in agriculture: A smart predictive model for optimal sensor selection through IoT integration. Smart Agric Technol 10:100749. https://doi.org/10.1016/j.atech.2024.100749
16. Hareesh SB (2024) The latest applications of remote sensing technologies for soil management in precision agriculture practices. In: Lamine S, Srivastava PK, Kayad A, Muñoz-Arriola F, Pandey PCBT-RS in PA (eds) Remote sensing in precision agriculture. Elsevier, pp 105–135
17. Sayyad SB, Shaikh MA, Anpat SM, Kolapkar MM (2024) IoT based soil monitoring for precision agriculture. In: Lamine S, Srivastava PK, Kayad A, Muñoz-Arriola F, Pandey PCBT-RS in PA (eds) Remote sensing in precision agriculture. Elsevier, pp 43–59
18. Sharma K, Shivandu SK (2024) Integrating artificial intelligence and Internet of Things (IoT) for enhanced crop monitoring and management in precision agriculture. Sensors Int 5:100292. https://doi.org/10.1016/j.sintl.2024.100292

19. Shahab H, Iqbal M, Sohaib A, Ullah Khan F, Waqas M (2024) IoT-based agriculture management techniques for sustainable farming: A comprehensive review. Comput Electron Agric 220:108851. https://doi.org/10.1016/j.compag.2024.108851
20. Ben AA, Jarray N, Farah IR (2024) Advances in remote sensing based soil moisture retrieval: applications, techniques, scales and challenges for combining machine learning and physical models. Artif Intell Rev 57:224. https://doi.org/10.1007/s10462-024-10734-1
21. Faqir Y, Qayoom A, Erasmus E, Schutte-Smith M, Visser HG (2024) A review on the application of advanced soil and plant sensors in the agriculture sector. Comput Electron Agric 226:109385. https://doi.org/10.1016/j.compag.2024.109385
22. Ali A, Hassan MU, Kaul H-P (2024) Broad scope of site-specific crop management and specific role of remote sensing technologies within it—a review. J Agron Crop Sci 210:e12732. https://doi.org/10.1111/jac.12732
23. Soussi A, Zero E, Sacile R, Trinchero D, Fossa M (2024) Smart sensors and smart data for precision agriculture: A review. Sensors 24:2647. https://doi.org/10.3390/s24082647
24. Pal A, Dubey SK, Goel S, Kalita PK (2024) Portable sensors in precision agriculture: Assessing advances and challenges in soil nutrient determination. TrAC Trends Anal Chem 180:117981. https://doi.org/10.1016/j.trac.2024.117981
25. Rekha BN, Banuprakash GC (2024) A hybrid framework for energy-efficient centroid based gateway clustering protocol over agriculture domain in wireless sensor networks. SN Comput Sci 5:518. https://doi.org/10.1007/s42979-024-02841-1
26. Tang P, Liang Q, Li H, Pang Y (2024) Application of internet-of-things wireless communication technology in agricultural irrigation management: A review. Sustainability 16:3575. https://doi.org/10.3390/su16093575
27. Singh N, Sharma AK, Sarkar I, Prabhu S, Chadaga K (2024) IoT-based greenhouse technologies for enhanced crop production: a comprehensive study of monitoring, control, and communication techniques. Syst Sci Control Eng 12:2306825. https://doi.org/10.1080/21642583.2024.2306825
28. Zhang Y, Wang X, Jin L, Ni J, Zhu Y, Cao W, Jiang X (2025) Research and development of an IoT smart irrigation system for farmland based on LoRa and edge computing. Agronomy 15:366. https://doi.org/10.3390/agronomy15020366
29. David PE, Chelliah PR, Anandhakumar P (2023) Reshaping agriculture using intelligent edge computing. In: David PE, Anandhakumar PBT-A in C (eds) Applying computational intelligence for social good. Elsevier, pp 167–204
30. Raja Gopal S, Prabhakar VSV (2024) Intelligent edge based smart farming with LoRa and IoT. Int J Syst Assur Eng Manag 15:21–27. https://doi.org/10.1007/s13198-021-01576-z
31. Sagar S, Birje NM (2025) Precision agriculture using internet of things and cloud computing: A review. SN Comput Sci 6:296. https://doi.org/10.1007/s42979-025-03833-5
32. Prashanth JS, Krishna GB, Prasad AVK, Rao PR (2025) Smart farming revolution: A cutting-edge review of deep learning and IoT innovations in agriculture. Oper Res Forum 6:35. https://doi.org/10.1007/s43069-025-00434-z
33. Dutta M, Gupta D, Tharewal S, Goyal D, Kaur Sandhu J, Kaur M, Ali Alzubi A, Mutared Alanazi J (2025) Internet of things-based smart precision farming in soilless agriculture: Opportunities and challenges for global food security. IEEE Access 13:34238–34268. https://doi.org/10.1109/ACCESS.2025.3540317
34. Harikrishnan H (2025) Rice-blast disease prediction using integrated smote with multilayer perceptron
35. Harikrishnan H, Aishwarya V (2025) Bioinspired energy materials: A comprehensive review of advances in photovoltaics, storage, and catalysis for sustainable energy technologies. Energy Storage 7. https://doi.org/10.1002/est2.70312
36. Aishwarya V, Harikrishnan H (2025) Bioinspired innovations in energy materials. In: Dhanola A, Singh Y, Shrivastava P (eds) Biomimetic and s materials. CRC Press, New York, pp 102–124

MIX
Papier aus verantwortungsvollen Quellen
Paper from responsible sources
FSC® C105338

If you have any concerns about our products,
you can contact us on
ProductSafety@springernature.com

In case Publisher is established outside the EU,
the EU authorized representative is:
Springer Nature Customer Service Center GmbH
Europaplatz 3, 69115 Heidelberg, Germany

Printed by Libri Plureos GmbH
in Hamburg, Germany